Hydrogen Energy

Hydrogen Energy

Principles and Applications

Vincent J. DelGatto
Prince Consulting
East Williston, NY, USA

Louis Theodore
Theodore Tutorials
East Williston, NY, USA

R. Ryan Dupont
Utah Water Research Laboratory
Utah State University
Logan, UT, USA

Matthew C. Ogwu
Appalachian State University
Boone, NC, USA

WILEY

Published by John Wiley & Sons, Inc., Hoboken, New Jersey.
Published simultaneously in Canada.

For general information on our other products and services or for technical support, please contact our Customer Care Department within the United States at (800) 762-2974, outside the United States at (317) 572-3993 or fax (317) 572-4002.

Wiley also publishes its books in a variety of electronic formats. Some content that appears in print may not be available in electronic formats. For more information about Wiley products, visit our web site at www.wiley.com.

Library of Congress Cataloging-in-Publication Data applied for:

Hardbook ISBN: 9781394172269

Cover Design: Wiley
Cover Image: © aryos/Getty Images

Set in 9.5/12.5pt STIXTwoText by Straive, Pondicherry, India

"The energies of our system will decay, the glory of the sun will be dimmed, and the earth, tideless and inert, will no longer tolerate the race which has for a moment disturbed its solitude. Man will go down into the pit, and all his thoughts will perish."

ARTHUR BALFOUR, The Foundations of Belief (1895)

"The future is green energy, sustainability, and renewable energy."

ARNOLD SCHWARZENEGGER (2018)

"Energy, like the Biblical grain of mustard seed, will move mountains (because) a man doesn't need brilliance or genius, all he needs is energy."

HOSEA BALLOU and ALBERT GREENFIELD (2022)

Contents

Preface

Since the late 1960s, there has been an increased awareness of a wide range of energy issues covering all sources: fossil fuels, nuclear, solar, geothermal, etc. More and more people are becoming aware of energy concerns. It is now increasingly important that all individuals develop capabilities in this area, including those who do not possess an understanding of many energy problems or have the proper information available when involved with energy issues. In addition, all professionals should have a basic understanding of the technical and scientific terms related to energy issues. One of the key issues that has emerged in this century is how hydrogen can effectively evolve as a green energy alternative. Hopefully, this text will serve the needs of those interested in increasing their knowledge of this form of energy and its relation to the energy problems facing society.

The current human population on Earth is approximately seven billion, and it will increase in the future. The influence and effects of human activities on energy use have become increasingly evident at the local, state, national, and international levels. Particular interest in hydrogen energy has recently surfaced and the net result is that there has been a more significant increase in both awareness and analysis of hydrogen energy options since the turn of this decade.

Hydrogen may well emerge as a very important fuel by the middle of this century. Since hydrogen is not a basic energy resource except in the sun, it must be obtained by using some other basic energy resource to generate hydrogen from water or other hydrogen-containing chemical compounds. It is an ecologically friendly fuel that is presently used in the propulsion of spacecraft and can potentially be mass-produced and commercialized for electric generation, energy storage and all forms of transportation.

It should be noted that free hydrogen is found only in very small traces in the atmosphere, but solar and stellar spectra show that it is abundant in the sun and other stars. In combination with other elements, it is widely distributed on Earth, where the most important and abundant compound of hydrogen is water, H_2O. It is a component of all the constituents of living matter as well as of many minerals. It forms an essential part of all hydrocarbons and a vast variety of other organic substances. All acids contain hydrogen; the distinguishing characteristic of an acid is its dissociation, upon going into solution, to yield hydrogen ions.

In addition, hydrogen is known to exist in three isotopic forms. The nucleus of each atom of ordinary hydrogen is composed of one proton. Deuterium, present in ordinary hydrogen to the extent of 0.02%, contains one proton and one neutron in the nucleus of each atom and has an atomic mass of 2. Tritium, an unstable, radioactive isotope, contains one proton and two neutrons in the nucleus of each atom and has an atomic mass of 3.

Given the above, hydrogen (H_2) fuel is certain to play a major role in the U.S.'s transition towards a "carbon-neutral" economy. No doubt more of the secondary energy sources for the future will be hydrogen produced from both renewable and nonrenewable sources. It is anticipated that in the near future, hydrogen energy will support numerous jobs with revenues well over $100 billion and could become a major player in satisfying up to 15–20% of the US energy demands.

The authors considered and then decided to write a text on hydrogen energy that highlighted pragmatic rather than theoretical issues. It contains three parts and provides material that can also be used in teaching a course sequence intended to serve as a training tool for those individuals involved with and those interested in entering the hydrogen energy field. Although the literature is inundated with texts emphasizing theory and theoretical derivations, the goal of this book is to present this subject of hydrogen energy from a strictly applied perspective.

The book is the result of over 4 years of effort by the authors. The first rough draft was prepared during the years 2020 through 2022. The manuscript underwent significant revisions during the past year, some of it based on the experiences gained from earlier written material. It should also be noted that the authors cannot claim sole authorship of all of the written material in this text. Although the bulk of the material is original or taken from sources that the authors have been directly involved with, every effort has been made to acknowledge material drawn from other sources.

As noted in the Table of Contents, this hydrogen energy book is divided into three parts:

Part I: Energy Overview
Part II: Select Hydrogen Energy Topics
Part III: Technical Engineering Issues

Part I provides an introduction and broad overview of energy-related topics. Following Chapter 1 that contains a glossary of terms, the reader is introduced to energy issues, energy resources, and environmental policy and regulations. Part I concludes with chapters concerned with thermodynamics and fuel cells. Part II – the heart of the book – provides select hydrogen energy topics. Chapter headings include a hydrogen energy overview, government programs, hydrogen physical/chemical properties, hydrogen-bearing compounds, processing, storage, transportation/transmission, conversion, uses, and the hydrogen byproduct water. Part II concludes with a chapter on safety consideration for hydrogen technologies. Part III is concerned with engineering applications. It features chapters on such topics as environmental health and hazard risk assessment, energy–environmental interactions, ethical considerations, and economics. This is followed by a chapter concerned with system optimization. The final chapter of the book provides nearly 50 illustrative examples beginning with elementary thermodynamic calculations and culminating with three optimization applications employing linear programming in their solution. To summarize, the proposed text provides information on the:

Properties
Fundamentals
Engineering Principles
Regulations
Thermodynamics
Sources
Processes
Ethical Considerations
Economics
Optimization Calculations Associated with Hydrogen Fuel

Thus, this book can serve as either a guide or reference for those engineers and applied scientists working in this field. It can also serve as a text for those currently enrolled in academic curricula concerned with energy and energy-related topics.

The authors hope that this book will place in the hands of academic, industrial, and government personnel – including policymakers and energy resource/management planners – a book that covers the fundamental principles and applications of hydrogen energy thoroughly and clearly. The authors further hope that, on completion of the text, readers will have acquired not only a working knowledge of hydrogen energy, but also experience in their application; and, that they will find themselves approaching advanced texts, engineering and science literature, and industrial applications (even unique ones) with more confidence.

Last, but not least, the authors believe that this modest work will help the majority of individuals working and/or studying in the field of hydrogen energy to obtain a more complete understanding of this subject matter. If you have come this far and read through this Preface, you have more than just a passing interest in this subject. We strongly recommend that you take advantage of the material available in this book. We think it will be a worthwhile experience.

Smithfield, Utah, October 2024

Vincent J. DelGatto
Louis Theodore
R. Ryan Dupont
Matthew C. Ogwu

Part I

Energy Overview

The first part of the book provides a broad overview of energy issues, resources, policies, and regulatory considerations, and includes a glossary of some important terms. The chapters are organized as presented below.

Chapters

1) Glossary of Key Energy Terms
2) Introduction to Energy and Energy Issues
3) Energy Resources
4) Environmental Policy and Regulatory Considerations for Hydrogen Energy
5) Thermodynamic Considerations
6) Fuel Cells

Those readers solely interested in hydrogen energy and hydrogen energy issues may choose to bypass Part I and proceed directly to Part II.

Hydrogen Energy: Principles and Applications, First Edition. Vincent J. DelGatto, Louis Theodore, R. Ryan Dupont, and Matthew C. Ogwu.

1

Glossary of Key Energy Terms

1.1 Introduction

This first chapter of the book provides the definition of hundreds of terms in use throughout the energy field, particularly as they apply to hydrogen. As one might suppose, the reader is primarily introduced to energy and energy-related terms. As noted in the Preface, this book is concerned with hydrogen energy and as such, this chapter primarily addresses terms related to this topic. Thus, this chapter may be used whenever and wherever information is needed about words and/or terms in the hydrogen energy field.

Some additional points regarding this glossary that deserve mention are presented below:

1) Each definition is presented in nontechnical jargon.
2) Where necessary, common scientific and engineering units have been included.
3) A conscious attempt was made to only include terms that are often used in practice.
4) Only one spelling is used for words with multiple accepted spellings but with shared meaning.
5) Different terms with the same meaning are cross-referenced by using the wording "see also".
6) Important acronyms are also included in the chapter.

This chapter defines many but not all of the terms that the reader will encounter in this book. The following list is therefore not exhaustive as a complete index of the book or glossary of all the terms that appear in the energy field. Finally, the purpose of this chapter is to explain the meanings of technical terms that may be unfamiliar to the reader. Definitions given here are for the purposes of this text and are not necessarily complete or exhaustive. Words or phrases included here are those for which there may be some confusion as to the meaning intended.

Section titles for this energy glossary chapter are presented below:

Hydrogen Energy: Principles and Applications, First Edition. Vincent J. DelGatto, Louis Theodore, R. Ryan Dupont, and Matthew C. Ogwu.

1.2 Importance of Energy Literacy

Why a glossary chapter? A good question. The answer? For some readers, this chapter might be the most important one in the book. Understanding energy and energy-related words and terms can motivate some individuals and prepare others to understand the roles (sustainable) energy plays in their lives. Hence, some people will refer to this chapter as "energy literature" of sorts. Informed individuals on energy topics:

1) Can trace energy flow and flow patterns and think in terms of energy systems and sustainability.
2) Know about their energy usage in terms of the amounts, for what, and where that energy originates from.
3) Can purposefully assess the credibility of energy information, regulations, and policies and decisions about sustainable energy futures.
4) Can communicate generally about energy sources and usage in meaningful ways.
5) Can make informed energy-related decisions based on an understanding of impacts and consequences.
6) Continues to learn about energy.

Thus, energy-related understanding can also help:

1) Lead to more informed decisions
2) Improve the security of a nation
3) Promote economic development
4) Lead to sustainable energy use
5) Reduce environmental risks and negative impacts
6) Help individuals and organizations save/make money

Without this information, individuals and communities will have difficulty making informed decisions on topics including sustainable or smart energy use at home, consumer choices, and national and international energy regulations and policies. Energy education should begin with basic energy literacy connected with the most frequently used terminologies.

Finally, the bulk of the material in this chapter was drawn from the earlier work of Theodore, Reynolds, and Morris (1997), as well as others (Holmes, Singh, and Theodore 1993; Theodore, Ricci and VanVlet 2007; Skipka and Theodore 2014; Dupont, Ganesan, and Theodore 2016; Theodore and Theodore 2021).

1.3 Glossary

- **Above ground tank** – any tank or other container, the volume of which is completely above the plane of the adjacent surrounding surface, and whose surface can be visually inspected.
- **Absolute humidity** – the amount of water vapor present in a unit mass of air, usually expressed in kilograms of water vapor per kilogram of dry air or pounds of water vapor per pound of dry air.
- **Absolute pressure** – the actual pressure exerted on a surface that is measured relative to zero pressure; it equals the gauge pressure plus the atmospheric pressure.
- **Absolute temperature** – the temperature expressed in Kelvin (K) or degrees Rankine (°R).
- **Absolute temperature scale** – a scale (e.g., K, °R) in which temperatures are measured relative to absolute zero.

- **Absolute vacuum** – a void that is completely empty of matter.
- **Absolute zero** – the temperature of zero on either the Kelvin or Rankine scale at which molecular motion is thought to cease.
- **Absorbate** – a substance that is taken up and retained by an absorbent.
- **Absorbent** –any substance that takes in or absorbs other substances.
- **Absorber** – a device in which a gas is absorbed by contact with a liquid.
- **Absorption** – the process in which one material (the absorbent) takes up and retains another (the absorbate) to form a homogenous solution; it often involves the use of a liquid to remove certain gas components from a gaseous mixture.
- **Absorption tower** –a vertical tube in which a rising gas is partially absorbed by a liquid in the form of falling droplets.
- **Acceleration** – the rate of change of velocity with time, ft/s^2.
- **Acceleration due to gravity** – the acceleration of a free-falling body in a vacuum under the influence of gravity; on Earth, this value is taken as $32.2\,ft/s^2$ or $9.8\,m/s^2$.
- **Accident** – an unexpected, undesirable event that is caused by the presence of a hazard and adversely affects humans and/or the environment.
- **Accuracy** – the measure of agreement between a measured value and an accepted or true value; it is expressed as the percentage difference between the true and measured values relative to the true value.
- **Acetogen** – a microorganism involved in the anaerobic transformation of organic matter to methane during biogas production.
- **ACFM** – see actual cubic feet per minute.
- **Acid** – a material containing hydrogen that produces at least one hydrogen ion when dissolved in a water solution; it can react with and neutralized by a base to form a sault.
- **Acid dew point** – the dew point of flue gases that contain any significant quantity of sulfur trioxide, SO_3; this temperature is generally 300°F.
- **Activated alumina** – a highly porous, granular form of aluminum oxide with absorptivity for moisture and odor contained in gases and some liquids.
- **Activated carbon** – a highly adsorbent, amorphous form of carbon employed to remove odors, gaseous pollutants, and toxic substances from gaseous emissions or to remove dissolved organic material from wastewater.
- **Activated sludge** – a suspension of aerobic microorganisms with the ability to remove polluting material from a stream of wastewater.
- **Activation energy** – the minimum amount of energy that a molecule must require before it can be regarded as being activated (i.e., able to undergo a reaction).
- **Actual cubic feet per minute (acfm)** – a unit of flow rate measured under actual pressure and temperature conditions.
- **Additives** – specialty chemicals incorporated into fuels and lubricants that enhance the performance of the finished product.
- **Adiabatic** – a term used to describe a system in which no gain or loss of heat is allowed to occur.
- **Adiabatic flame temperature** – the maximum temperature that a combustion system can reach.
- **Adiabatic process** – a process in which there is no transfer of heat across the system boundary.
- **Adiabatic temperature** – the temperature attained by a reaction undergoing a volume or pressure change in which no heat enters or leaves the system.
- **Adsorbent** – a substance (e.g., activated carbon, activated alumina and silica gel) that can condense or hold molecules of other substances on its surface.
- **Adsorber** – an apparatus in which molecules of gas or liquid are captured by a solid.

- **Adsorption** – the physical or chemical bonding of molecules of gas, liquid, or dissolved solid to the external or internal (if porous) surface of a solid; it is an advanced method of treating waste that is employed to remove odor, color, or organic matter from a system.
- **Advanced wastewater treatment** – any process that is employed for the treatment of wastewater that follows other physical, chemical, or biological treatment and serves to improve the quality of effluent prior to reuse or discharge; this includes charged helium nucleus
- nutrients, suspended solids, nonbiodegradable organics, etc.
- **Aerobic digestion** – the breakdown of suspended and dissolved organic matter in the presence of oxygen.
- **Afterburner** – a secondary burner, located so that combustion gases from the primary incinerator are further burned to remove smoke, odors, and other pollutants.
- **Agent** – a biological, physical, or chemical entity capable of causing disease.
- **Agglomeration** – the process by which smaller particles grow larger by collision or contact with other particles.
- **Air heater** – a heater exchanger through which air passes and is heated or cooled by a medium at a different temperature.
- **Air preheater** – an air heater that increases the temperature of an air stream entering a process.
- **Air quality** – the degree of cleanness of the air with regard to lack of pollution.
- **Aliphatic** – one of the major groups of organic compounds characterized by straight or branched chain arrangement of constituent carbon atoms; they are divided into alkanes, alkenes, or branched compounds.
- **Alkali** – any basic substance that is bitter in water solution and irritating or caustic to the skin and mucous membranes; it turns litmus blue, has a pH value greater than 7.0, and is typically employed in wastewater treatment processes.
- **Alkaline electrolyzer** – the oldest technology for creating hydrogen from water and electricity.
- **Alkaline fuel cell** – one of the oldest and cheapest fuel cell technologies that is highly conductive of electrolyte and highly reactive to electrodes.
- **Alkalinity** – a measure of the degree of reaction by a compound with hydrogen ions (e.g., the capacity of water to neutralize acids).
- **Alkane** – a member of a series of saturated, aliphatic hydrocarbons.
- **Alkene** – a member of a series of unsaturated, aliphatic hydrocarbons that are characterized by double bonds between carbon atoms.
- **Alkyl** – a paraffinic hydrocarbon group that may be derived from an alkane by removing one hydrogen from the molecule.
- **Alkylation** – a process in which a high-octane blending component for gasoline is derived from catalytic combination of an isoparaffin and an olefin.
- **Alpha (α) particle** – a positivelycharged helium nucleus (i.e., two protons and two neutrons) that is emitted spontaneously from radioactive elements.
- **Ambient temperature** – the temperature of the surrounding air.
- **Amine** – a class of organic compounds of nitrogen that may be derived from ammonia (NH_3) by replacing one or more of the hydrogen atoms with an alkyl group.
- **Ammonia (NH_3)** – the different directions fifth highest volume of chemical produced in the United States that exists as a colorless gas or liquid with a sharp, intensely irritating odor; the inhalation of concentrated fumes may be fatal, and it has a TLV of 25 ppm in air and explosive limits of 16–25% in air.
- **Ampere** – a unit of electrical current, equal to 1 C/s.
- **Amplitude** – the peak, or maximum value, in wave-type transmissions.

- **Anaerobe** – an organism that carries out microbial reactions in the absence of oxygen.
- **Anaerobic digestion** – the degradation of organic matter by anaerobic organisms in the absence of oxygen to produce primarily carbon dioxide and methane.
- **Ancillary equipment** – any secondary equipment or device including piping, fittings, valves, pumps, etc.
- **Anhydrous** – an inorganic compound that does not contain water, either adsorbed on its surface or combined as water of crystallization.
- **Anion** – a negatively charged ion.
- **Anisotropic** – a term used to describe a material with different optical or other physical properties in different directions (e.g., quartz an graphite).
- **Annual energy production (AEP)** – the energy generated by turbine(s) over the course of 1 year.
- **Anode** – the positive electrode of an electrolytic cell to which negatively charged ions travel when an electric current passed through the cell.
- **Anticipated transient without scram (ATWS)** – for a nuclear reactor, an accident for which the initiating event is an anticipated operational occurrence and in which the system for fast shutdown of the reactor fails to function.
- **Applicant** – any person or organization applying to a regulatory body for authorization (or approval) to undertake specified activities.
- **Approval** – the granting of consent usually by a regulatory *body*.
- **Aqua** – a term used to designate water such as in a stream, a lake, or other sources.
- **Aqueous solution** – any solution dissolved in water.
- **Archimedes' principle** – the principle that the buoyant force on an object immersed in a fluid is equal to the weight of fluid displaced by it.
- **Aromatic** – a term used to describe a major group of unsaturated cyclic hydrocarbons containing one or more rings, typified by benzene which has a six-carbon ring containing three double bonds.
- **Arrangements (for operations)** – the integrated set of infrastructural elements necessary to provide the capability for performing a specified function or task required to carry out a specified operation.
- **Assessment** – the process, and the result, of analyzing systematically and evaluating the hazards and problems associated with sources and practices, and associated protection and safety measures for the purposes of protecting the public and environment against risks.
- **Atmosphere** – the gaseous envelope of fluid surrounding the Earth that comprises the troposphere, stratosphere, mesosphere, and thermosphere; the Earth's atmosphere consists of 79.1% nitrogen, 20.9% oxygen, about 0.03% carbon dioxide, a trace of the noble gases, water vapor, and traces of ammonia, organic matter, ozone, various salts, and suspended solid particles.
- **Atmospheric dispersion** – the mechanism of dilution of smoke or gaseous pollution in which the concentration is progressively decreased.
- **Atmospheric pressure** – the pressure exerted by the atmosphere at any given point.
- **Atmospheric stagnation** – the condition of very little or no air movement.
- **Atom** – the smallest unit of an element that retains all the properties of the element; it is composed of electrons, neutrons, and protons.
- **Atomic energy** – the energy, in the form of heat and radiation, that is derived in a controlled fashion from nuclear fission or fusion.
- **Atomic fission** – the breaking down of large atoms into smaller atoms or elements, involving the liberation of heat, gamma rays, alpha particles, and beta particles.

- **Atomic mass unit (amu)** – a unit equal to one-twelfth the mass of one neutral C^{12} atom.
- **Atomic weight** – the average weight or mass of all the isotopes of an element as determined by the proportion in which they are present in a given element; it is based on the mass of isotopes of carbon which is assigned a reference value of 12.000.
- **Attenuation** – the reduction in intensity of radiation passing through matter due to processes such as absorption and scattering.
- **Authorization** – the granting by a regulatory body or other governmental body of written permission for a person or organization to conduct specified activities. Authorization could include licensing (issuing a license), certification (issuing a certificate), or registration.
- **Autogenous combustion** – any burning without the addition of supplementary fuel.
- **Autoignition** – the starting of a fire without the addition of an external source such as a flame, spark, or heat.
- **Autoignition temperature (AIT)** – the temperature to which a closed or nearly closed container must be heated so that a flammable liquid, when introduced into the container, will ignite spontaneously or burn without the presence of a spark or flame.
- **Automatic control** – the maintenance of desired process conditions by means of sensing devices.
- **Available hydrogen** – naturally occurring liquid, gaseous, or solid hydrogen (or deuterium or tritium) on Earth often in combination with oxygen, carbon, or other elements.
- **Available oxygen** – the quantity of dissolved oxygen available for oxidation of organic matter.
- **Avogadro's number** – the number ofmolecules in 1 mol of a substance, 6.23×10^{23}.
- **Baffle** – a flow-regulating device consisting of a perforated metal plate placed horizontally to restrict or divert the passage of a fluid, usually used for the purpose of providing a uniformly dispersed flow.
- **Ball joint** – a flexible pipe joint formed in the shape of a ball or sphere.
- **Ball valve** – a nonreturn valve consisting of a ball resting on a cylindrical seat within a fluid passageway or pipe.
- **Barometric pressure** – the pressure of the ambient air in the atmosphere at a particular point or above the surface of the Earth.
- **Barrels of oil equivalent** – a unit of measure to quantify crude oil, natural gas liquids, and natural gas amounts using the same basis.
- **Base** – any compound that dissociates in an aqueous solution to yield hydroxyl ions; it is employed to neutralize acids.
- **Baseload** – the minimum level of demand on an electrical grid over a given period of time (usually over the year, or over a season or day); base load power sources are power stations that are capable of consistently generating the electrical power needed to cover this demand.
- **Baseload generation** – electric generation units that typically run all 24 hours a day.
- **Basic** – a term used to describe a compound that is alkaline, i.e., has a pH greater than 7.0.
- **Batch process** – a process that is not continuous; its operations are carried out with discrete quantities of material.
- **Battery** – a device consisting of one or more electrically connected electrochemical cells that is designed to receive, store, and deliver electric energy; it converts chemical energy to electrical energy.
- **Bernoulli's theorem** – a physical law of hydraulics that states that the energy per mass is constant at any point in a conduit; this energy comprises a velocity head, a pressure head, and a static head.
- **Beta (β) particle** – a charged particle emitted from a radioactive atomic nucleus; it has moderate penetrative power and can damage living tissue.

- **Bias** – a measure of the systematic error between an actual or true value and a prediction by a model or a measured mean value; the bias of a model represents the tendency of a model to overpredict or to underpredict.
- **Binary fission** – a process by which bacteria reproduce by splitting into equal parts.
- **Binary power plant** – a power plant that transfers heat from hot water to a working fluid that vaporizes at lower temperatures than water.
- **Biofuel** – a fuel produced from biomass or biomass used directly as a fuel.
- **Biomass fuel** – fuels derived from recently living organisms; sometimes also includes fuels derived from municipal solid waste (containing methane and carbon dioxide).
- **Biosphere** – that part of the environment normally inhabited by living organisms. In practice, the biosphere is not usually defined with great precision but is generally taken to include the atmosphere and the Earth's surface, including the soil and surface water bodies, seas and oceans, and their sediments.
- **Biotechnology** – the application of scientific and engineering principles to the processing of any organic or inorganic substance by biological agents to provide goods and services including energy.
- **Bituminous coal** – a coal between lignite and anthracite in rank, burning with a smoky, luminous flame.
- **Black hydrogen** – hydrogen produced from coal via coal gasification and extraction.
- **Blade** – the long fiberglass pieces designed to catch the wind at the top of the wind turbine, causing the turbine to turn. Three blades connect to the hub to create the rotor.
- **Blast-furnace gas** – a by-product gas from smelting iron ore obtained by the passage of hot air over the coke in blast furnaces.
- **Bleeding** – the gradual release of material and/or reduction of pressure from a system or process (e.g., by a value).
- **Blowdown** – the cyclic or constant removal of a portion of any process flow to maintain the constituents of the flow at a desired level.
- **Blower** – a fan employed to force or move air or gas under pressure.
- **Blue hydrogen** – this term is used for hydrogen produced using low-carbon processes.
- **Boiler** – a pressure vessel designed to produce vapor from liquid, usually water, by the applications of heat.
- **Boiler scale** – a covering, varying from a porous, friable crust to a dense, hard coating, that is deposited on boiler heating surfaces by precipitation of minerals from the water used.
- **Boiling point** – the temperature of a liquid at which its vapor pressure is equal to or varies slightly above that of the atmospheric pressure of the environment; it is 212°F for water at sea level.
- **Bond energy** – the amount of energy required to disrupt the bonds in 1 mol of a chemical compound.
- **Boyle's law** – a law that states that the volume of a gas is inversely proportional to the pressure if the temperature remains constant; this relation holds truc for ideal gases.
- **Brine** – a concentrated solution of salt and water that remains after the removal of a distilled product.
- **British Thermal Unit (Btu)** – the quantity of heat required to raise the temperature of 1 lb of water by 1°F at the temperature at which water has its greatest density (approximately 39°F).
- **Brown hydrogen** – hydrogen produced from lignite (see black hydrogen).
- **Brownian movement** – the constant, random movement of small, suspended particles due to collisions with other molecules.

- **Bubble point** – the temperature at which the first bubbles of gas appear in a vapor–liquid equilibrium system with two or more components.
- **Bubble tower** – a plate tower employed in distillation (primarily) and absorption; it contains plates with bubble caps to promote intimate vapor–liquid contact.
- **Buffer** – (1) a solution containing both a weak acid and its conjugate weak base, which is employed to stabilize the pH value in a solution, and (2) anything that acts to diminish and/or regulate change in a system or process.
- **Bulk density** – the mass per unit volume of the solid in a solid–gas mixture such as a packed bed or soil mass; unlike the real density, the pore pace is included in the volume.
- **Bulk sample** – a small portion of material that is collected and sent to a laboratory for analysis.
- **By-product** – a material that is not one of the primary products of a production process and is not solely or separately produced by the production process.
- **Bypass** – a device to inhibit, deliberately but temporarily, the functioning of a circuit or system by, for example, short circuiting the contacts of a relay.
- **Calcination** – the heating of a solid waste material to a temperature below its melting point to bring about a state of thermal decomposition or a phase transition other than melting.
- **Calibration** – the determination, checking, or adjustment of the accuracy of any instrument that gives quantitative measurements.
- **Calorimeter** – a device for determining the amount of heat evolved or absorbed in a chemical reaction.
- **Capacitance** – the property of a capacitor that determines how much charge can be stored in it for a given potential difference between its terminals; it is measured by the ratio of the charge stored to the potential difference.
- **Capacitor** – a device for accumulating and holding a charge of electricity, usually consisting of conducting surfaces separated by a dielectric.
- **Capacity** – the maximum electric power capability of a generating unit or transmission line measured in MW.
- **Capacity factor** (aka **load factor**) – the actual output of a generating unit over a specific period of time divided by the unit's total capacity multiplied by the total number of hours in the period.
- **Capital costs** – initial, one-time costs of planning, designing, and constructing a facility; it also refers to additional costs of long-term investment in facilities maintenance or upgrades.
- **Carbon neutral** – a process that has no net emissions of greenhouse gases into the atmosphere.
- **Carbon-free hydrogen** – hydrogen produced from renewable feedstock with emissions below 36.4 g CO_2eq/MJ H_2 e.g., by electrolysis using renewable electricity as feedstock.
- **Carnot cycle** – an ideal, closed heat cycle of reversible changes with which the performance of beat engines may be compared; it assumes no frictional losses.
- **Carryover** – the entrainment of liquid or solid particles in the vapor evolved by a boiling liquid or from a processing unit.
- **Catalysis** – a process in which the rate of a reaction is altered by the presence of an added substance that remains unchanged at the end of the reaction.
- **Catalytic combustion** – a process in which a catalyst is introduced into an exhaust gas stream to encourage the reaction of gaseous hydrocarbons at a lower temperature than would otherwise be possible.
- **Catalytic cracking** – the breaking of carbon–carbon bonds with the aid of a catalyst; it is an essential process in the refining of petroleum.

- **Cathode** – the negative electrode of an electrolytic cell to which positively charged ions travel when an electric current is passed through the cell.
- **Cation** – a positively charged ion.
- **Cavitation** – the formation of vapor bubbles in a liquid when subjected to localized low-pressure regions causing severe mechanical damage to the surface of the metal exposed to it.
- **Centrifugal fan** – a device for moving gases.
- **Centrifugal pump** – a device that increases the pressure of a liquid by using centrifugal force.
- **Certificate** – a legal document issued by the regulatory body stating the applicable conditions to be met for certification and certifying compliance with regulatory requirements if the conditions are met.
- **Certification** – the issuance of a certificate.
- **Channel** – an arrangement of interconnected components within a system that initiates a single output.
- **Charles' law** – the law which states that the volume of a confined gas at a constant pressure is proportional to its absolute temperature.
- **Check valve** – a one-way valve that prevents flow through a pipe in an undesired direction; it opens in the direction of the normal flow and closes with a reversal of flow.
- **Chemical agent** – an element, compound, or mixture that coagulates, dissolves, neutralizes, solubilizes, oxidizes, concentrates, makes a pollutant mass more rigid, or otherwise facilitates the lessening of harmful effects or the removal of a pollutant from a fluid.
- **Chemical bond** – an attractive force between atoms strong enough to permit the combined aggregate to function as a unit.
- **Chemical oxygen demand (COD)** – a measure of the oxygen that is needed for the decomposition of inorganic and organic waste matter present in wastewater.
- **Chemical reaction** – a process that involves changes in the structure and energy content of atoms, molecules, or ions but not their nuclei; a chemical change that may occur in several ways, including combination, replacement, decomposition, or some modification of these.
- **Chemical thermodynamics** – the aspects of thermodynamics concerned with the relationship of heat, work, and other forms of energy to equilibrium in chemical reactions and phase changes within a system.
- **Chemisorption** – the adsorption that occurs by the formation of bonds between the surface molecules of a metal and another substance in contact with it.
- **CHP** – an abbreviation that stands for combined heat and power. A term used to describe a technology that produces both heat and power for commercial uses.
- **Closed-loop** – a term used to describe an enclosed process.
- **Closure** – administrative and technical actions directed at a disposal facility at the end of its operating lifetime.
- **Co-firing** – the simultaneous use of multiple fuels to create thermal energy.
- **Coal** – a natural, solid, combustible material that is utilized as a fuel; it is formed from prehistoric plant life that occurs in layers or veins in sedimentary rock.
- **Coal gasification** – a process through which coal is deconstructed into a gas via a combination of high-pressure and high-temperature steam, and by external heat; this process transforms the complex hydrocarbons from a solid slate into a gaseous one.
- **Co-current** – a term condition used to describe a situation in which two or more controls or systems exist in an operated condition at the same time.
- **Coefficient of performance (COP)** – the ratio of the refrigeration effect to the work input.

- **Coke** – the porous, brittle, carbonaceous, solid residue of the destructive distillation of bituminous coal, petroleum, and coal–tar pitch; it can be utilized as a fuel.
- **Coke oven gas** – the mixture of permanent gases produced by the carbonization of coal in a coke oven at temperatures in excess of 1,000°C.
- **Combined-cycle turbine** – a device that uses both a gas turbine and a steam turbine to convert thermal energy to kinetic energy.
- **Combustible** – a material that can be ignited at a specific temperature in the presence of air or oxygen to release heat.
- **Cbustion** – a reaction at a high temperature with oxygen that produces carbon dioxide, water, and energy in the form of light and heat; it is a basic cause of air pollution.**Combustion air** – the air required in an incinerator or furnace for combustion.
- **Combustor** – the component of a power plant where fuel is burned.
- **Compaction** – the process of reducing the bulk of solid waste by rolling and packing down or compressing and crushing.
- **Composting** – an aerobic, microbial decomposition of solid organic wastes, both agricultural and municipal/industrial.
- **Compound** – a substance composed of atoms or ions of two or more elements in chemical combination where the constituents are united by bonds or valence forces; it is a homogeneous entity where the elements have definite proportions by mass and are represented by a chemical formula.
- **Compressed natural gas (CNG)** – natural gas, which is comprised primarily of methane, compressed to a pressure at or above 2,400 $lb/in.^2$ and stored in special high-pressure containers; it is used as a fuel for natural gas-powered vehicles.
- **Compressor** – any device employed to increase the pressure of gas.
- **Concentration** – the relative amount of a component in a given substance.
- **Condensate** – any liquid resulting from the cooling of a gas or vapor.
- **Condenser** – any device that cools gases or vapors to liquid.
- **Conditional probability value (CPV)** – the upper bound for the conditional probability that a particular type of event will cause unacceptable consequences. The term is used in the detailed event screening process for site evaluation.
- **Conduction** – (1) the transfer of heat from one substance to another when the two substances are in physical contact, and (2) the transfer of an electric current through a solid or liquid in a metal.
- **Conductive heat transfer** – a transfer of heat from one point to another within a body, or from one body to another if the bodies are in physical contact.
- **Conductivity** – the property of a substance or mixture that describes its ability to transfer heat or electricity.
- **Confidence interval** – an interval estimate with a specified level of confidence for a given set of data.
- **Confinement** – prevention or control of releases of radioactive material to the environment in operation or in accidents.
- **Congestion** – occurs when the level of electrical power that needs to be transported across a transmission line exceeds its transmission capacity.
- **Consequence assessment** – assessment of the radiological consequences (e.g., doses, activity concentrations) of normal operation and possible accidents associated with an authorized facility or part thereof.
- **Conservation of energy** – the principle that states that energy can never be created or destroyed, but only changed in form; this is also known as the First Law of Thermodynamics.

- **Conservation of mass** – the principle which states that mass can neither be created nor destroyed.
- **Conservation of momentum** – the law which states that the total momentum in a steady state flow process must remain constant.
- **Continuity equation** – a conservation of mass equation in fluid flow relating fluid density and flow velocity.
- **Convection** – the transfer of heat through a fluid by the movement of the fluid, and the vertical movement of air leading to cooling.
- **Conveyor** – any device for the transport of raw materials (usually solids), assembly units, or finished products.
- **Cooling tower** – a hollow, vertical structure with internal baffles to disperse water so it is cooled by flowing air and by evaporation at ambient temperature.
- **Corona** – an electrical discharge effect that causes ionization of oxygen and the formation of ozone; it is found in electrostatic precipitators.
- **Corrosion** – the deterioration or destruction of a material by chemical action.
- **Cost-effective** – said of an energy efficiency measure (such as a compact fluorescent light bulb, a refrigerator, insulation, or any other measure that makes a home more energy efficient) or of a program that delivers energy efficiency measures, including weatherization.
- **Coulomb** – a unit of electrical charge, equal to 6.281×10^{18} electronic charges.
- **Countercurrent** – a term used to describe the flow pattern in a piece of equipment in which two streams travel in opposite directions.
- **Countermeasure** – an action aimed at alleviating the consequences usually of an accident.
- **Cracking** – a refining process involving decomposition and molecular recombination of organic compounds to form molecules of smaller sizes that are suitable for fuels.
- **"Cradle-to-grave"** – an approach in which all the stages in the lifetime of a facility, activity, or product are taken into consideration.
- **Cradle-to-grave** – a waste producer's responsibility for the proper management and disposal of the material from the time of generation to the time of destruction.
- **Critical mass** – the minimum mass of a fissionable material needed to start and maintain a chain reaction.
- **Critical point** – (1) the transition point between the liquid and gaseous states of a substance; (2) the temperature above which a gas cannot be liquefied however high the pressure; and (3) the temperature at which internal changes take place within a metal.
- **Critical Reynolds number** – the value of the Reynolds number at which the flow makes a transition from laminar to turbulent flow as the velocity is increased.
- **Critical temperature** – the temperature above which a gas or vapor cannot be liquefied by an increase in pressure alone.
- **Cryogenics** – the production and utilization of extremely low temperatures.
- **Crystallization** – the change of state of a substance from a liquid to a solid by the phenomenon of crystal formation by nucleation and accretion (e.g., the freezing of water into ice).
- **Curie** – a quantitative measure of radioactivity equal to 3.7×10^{10} disintegrations per second.
- **Current** – the rate of motion of electricity through a substance.
- **Curtailment (of generation)** – curtailment is the reduction of electrical generation, used as a last resort to balance a system or solve congestion.
- **Dalton's law** – in a mixture of non-reacting gases, the total pressure exerted by the mixture is equal to the sum of the partial pressures of the individual gases in the mixture.
- **Damper** – a manual or automatically controlled valve or plate in a duct or stack that is employed to regulate the flow of air or other gases.

- **Decarbonized hydrogen** – hydrogen produced from nonrenewable feedstock with emissions below 36.4 g CO_2 eq/MJH_2. See carbon-free hydrogen.
- **Decarbonized power sector** – the reduction of the carbon intensity of the power sector by increased use of technologies that result in a lower concentration of fossil fuels released into the atmosphere during power generation.
- **Decommissioning** – administrative and technical actions taken to allow the removal of some or all regulatory controls from a facility. Decommissioning can entail activities that are similar to remediation (also an authorized process).
- **Decomposition** – a fundamental type of chemical change where one substance breaks down into simpler substances.
- **Deep well injection** – a method of ultimate (EPA's definition) disposal that involves depositing liquid waste into a deep well beneath the surface of the Earth for permanent storage.
- **Deflagration** – a surface phenomenon that involves the very rapid burning of the particles of an explosive; it is usually initiated by impact or friction.
- **Degrade** (as in energy) – the transformation of energy into a form in which it is less available for doing useful work.
- **Dehydration** – the chemical process where water in a chemical or material is removed.
- **Deionization** – a method for purifying water that involves converting the soluble salts into acids by passing them through a hydrogen exchanger, and their removal by an acid adsorbent or synthetic resin.
- **Demand-side management (DSM)** – controlling the use of energy on the customer's side of the meter; sometimes used interchangeably with "energy conservation" and "energy efficiency."
- **Demonstration power plant** – a pre-commercial project used to demonstrate the viability of a new technology.
- **Depletion** – the exhaustion of natural resources because of their removal.
- **Depressurization** – a reduction in pressure.
- **Desalination** – the extraction of freshwater from sea or other saltwater bodies by the removal of salts, usually by evaporation, distillation, or membrane separation.
- **Design basis** – the range of conditions and events taken explicitly into account in the design of structures, systems and components, and equipment of a facility, according to established criteria, such that the facility can withstand them without exceeding authorized limits.
- **Desorption** – a process of removing an adsorbed material from the solid on which it is adsorbed by increasing the temperature or reducing the pressure, or both.
- **Destructive distillation** – an operation in which a highly carbonaceous material is subjected to high temperature in the absence of air or oxygen resulting in the decomposition of solids, liquids, and gases.
- **Desulfurization** – the removal of sulfur from fossil fuels to reduce pollution.
- **Detention time (detention period)** – the average time that a unit volume of a fluid is retained in a unit during a flow process.
- **Deterministic analysis** – analysis using, for key parameters, single numerical values (taken to have a probability of 1), leading to a single value for the result.
- **Deuterium** – an isotope of hydrogen with an atomic weight of 2.0; unlike normal hydrogen, it has neutrons in the nucleus.
- **Deviation** – a departure, usually from specified conditions or requirements.
- **Dew point** – the temperature at which the first droplet of water forms on the progressive cooling of a mixture of air and water vapor; at the dew point, the air becomes saturated with water.
- **Diatomic** – a term used to describe a single molecule consisting of two atoms (e.g., O_2 H_2).

- **Dielectric** – a term used to describe a substance with very low electrical conductivity (i.e., an insulator).
- **Dielectric constant** – an index of the ability of a substance to impair the transmission of an electrostatic force from one charged body to another.
- **Diffuser** – a tube or plate that separates a gas stream into bubbles suitable for diffusion.
- **Digestion** – The process of breaking down a substance into smaller molecules.
- **Dipole** – a pair of equal and opposite electric charges separated by a small distance.
- **Direct combustion** – combustion of a material without first converting the material to another form.
- **Direct conversion** – the conversion of one form of energy directly into electrical energy, such as the conversion of sunlight into electricity.
- **Direct radiation** – the unscattered radiation that strikes a surface.
- **Distillate** – any overhead product of distillation, especially of petroleum.
- **Distillate oil** – the fuel oils that contain 0.05% nitrogen by weight or less and comply with the specifications for fuel oil numbers 1 and 2, as defined in the Standard Specifications for Fuel Oils.
- **Distillation** – a process of separating the constituents of a liquid mixture by means of vaporization of the mixture and separate recovery of vapor and residue.
- **Distributed energy resources (DERS)** – decentralized electrical generation and storage technologies that are located close to the load served.
- **Distributed generation (embedded generation)** – energy produced near the point of consumption rather than at a distance, usually small plants that provide power onsite.
- **Distribution grid** – the distribution grid is the group of power lines and associated equipment for the transport of electric energy at a lower voltage to most customers.
- **Distribution system** – a powerline system that comprises the parts of an electric power system between the sub-transmission system and the consumers' service switches.
- **Distribution** – the delivery of electricity to homes and businesses through low-voltage power lines.
- **Donor** – an atom that furnishes a pair of electrons to form a bond or linkage with another atom called the acceptor.
- **Dose assessment** – assessment of the dose(s) of hazardous compounds or radiation to an individual or group of people.
- **Dose rate** – the dose per unit time of a hazardous compound or radiation.
- **Downcomer** – a pipe or flue that conveys gases, vapors, or condensate downward in blast furnaces, distillation towers, and refineries.
- **Downtime** – the lost production time during which a place of equipment is not operating correctly due to a breakdown, maintenance, or power failure.
- **Draft** – the difference between the pressure within a unit and that in the atmosphere.
- **Dry bulb** – thermometer bulb maintained dry; it is used with a wet bulb to provide a measure of air humidity.
- **Dry-bulb temperature** – the air temperature where the thermometer is shielded from direct radiation and moisture; it is a measure of the heat content of the air.
- **Dry ice** – the solid form of carbon dioxide, subliming at temperatures of −72°C and above; it is used as a convenient portable refrigerant.
- **Dry injection** – the injection of a dry reagent to control acid emissions from incinerators.
- **Dry rock formation** – an underground location where rocks are heated to high temperatures by the heat of the Earth but where little or no water exists within the rock.
- **Dry steam** – steam that does not contain much, if any, liquid.

- **Duct** – a round or rectangular conduit, usually metal or fiberglass, employed to transport fluids.
- **Dynamic pressure** – the pressure exerted by a fluid in motion against a surface or body.
- **Earth's crust** – the outermost solid layer of the Earth.
- **Earth's mantle** – a solid layer of the Earth, approximately 2,300 kin thick, located between the Earth's crust and the Earth's core.
- **Economic conservation** – the management of natural resources, or the environment, to sustain a regular yield of a commodity at the highest feasible level.
- **Economic growth** – the annual rate of change in the Gross National Product, usually expressed as the percentage difference from the previous year.
- **Economizer** – a device that uses the heat of combustion of a gas to raise the temperature of boiler feedwater prior to entry into the boiler.
- **Efficient energy** – the use of a relatively small amount of energy for a given task, purpose, or service; achieving a specific output with less energy input.
- **Effluent** – any fluid emerging from a pipe or similar outlet that enters the environment.
- **Ejector** – a device for moving a fluid or solid by entraining it in a high-velocity stream of air or water jet.
- **Elbow** – a pipe fitting that connects two pipes at a 90° angle.
- **Electric dipole moment** – the condition in which one part of a molecule has a positive electrical charge while the other part has a negative electrical charge.
- **Electrochemistry** – the branch of chemistry concerned with the relationship between electrical forces and chemical reactions.
- **Electrode** – a conductor that establishes electrical contact with an electrolyte; a pair of electrodes is a necessary component of an electrolytic cell.
- **Electrolysis** – the splitting of an electrolyte by the application of electric current; this process can be employed to remove ions from a waste stream.
- **Electrolyte** – a substance that dissociates into two or more ions when it is dissolved in water.
- **Electrolyzer** – a technology that converts water and electricity into hydrogen, oxygen, and heat.
- **Electromotive force** – the force that causes a flow of electric current.
- **Electron** – a negatively charged fundamental unit of matter with a negligible mass that circles the positively charged nucleus of an atom.
- **Electron valence** – the number of electrons needed to be gained or lost by a material for it to gain stability.
- **Electronegative** – a term used to describe a substance that attracts electrons to bond to another substance.
- **Electrostatic field** – a region in which a stationary electrically-charged particle is subjected to a force of attraction or repulsion because of another stationary electric charge.
- **Electrostatic precipitator (ESP)** – a device that separates particles from a gas stream by passing the carrier gas between two electrodes across which a high voltage is applied.
- **Element** – any substance that cannot be broken down into smaller substances while still keeping all its properties; it is one of the presently known types of substances that comprise all matter.
- **Emergency plan** – a description of the objectives, policy, and concept of operations for the response to an emergency and of the structure, authorities, and responsibilities for a systematic, coordinated, and effective response; the emergency plan serves as the basis for the development of other plans, procedures, and checklists.
- **Emergency preparedness** – the capability to take actions that will effectively mitigate the consequences of an emergency to human life, health, property, and the environment.

- **Emergency procedures** – a set of instructions describing in detail the actions to be taken by emergency workers in an emergency.
- **Emergency response** – performance of actions to mitigate the consequences of an emergency to human life, health, property, and the environment; also provides the basis for the resumption of normal social and economic activity after the emergency.
- **Emissivity** – the ratio of the radiating power of a real surface to a blackbody surface.
- **Empirical** – anything derived from experimentation or observation and not from theory.
- **Endothermic** – a term used to describe a process or change that occurs with an absorption of heat.
- **Energy** – a quantity of electricity; often measured in MWh. It is the capacity to do work.
- **Energy balance** – an inventory of the energy of a process system; it includes energy crossing the system boundary and energy generated, consumed, or accumulated within the system.
- **Energy carrier** – a source of energy that has been subject to human-induced energy transfers or transformations; examples include hydrogen fuel and electricity.
- **Energy conversion** – the conversion of one form of energy to another form, usually for convenience of use.
- **Energy converter** – a device that transforms one energy carrier into another or into an energy service.
- **Energy resource** – a form of energy that can be extracted from the environment and placed into useful services in the economy; fossil fuels, sunlight, wind currents, and uranium represent energy resources, while electricity and hydrogen are energy carriers, not resources.
- **Energy service** – the forms of energy demanded by consumers and industry. These generally include (but are not limited to) heating, cooling, lighting, mechanical power, and information.
- **Energy transition** – a particularly significant set of changes to the patterns of energy use in a society, potentially affecting resources, carriers, converters, and services.
- **Enforcement** – the application by a regulatory body of sanctions against an operator, intended to correct and, as appropriate, penalize noncompliance with conditions of an authorization.
- **Engineering notation** – a method of representing very large or small numbers by factoring the number into two parts, one, a number between 1 and 10, the other, the number 10 raised to an integer power, e.g., 0.0000089 may be represented by 8.9×10^{-6}.
- **Enhanced Geothermal Systems (EGS)** – geothermal power plants that use dry rock as the heat source.
- **Enthalpy** – a measure of the heat content of a body or system.
- **Entrainment** – the carryover of drops of liquid during a process such as distillation, absorption, scrubbing, or evaporation.
- **Entropy** – a measure of the randomness of a system; it is also a measure of the ability of a system to perform useful work.
- **Environmental impact assessment (EIA)** – a procedure evaluating the effects on the environment of an infrastructure project.
- **Environmental justice** – a term that is used to describe environmental crimes and offenses committed against minorities and low-income individuals; although industry has received the brunt of these allegations, it almost always can be attributed to politicians and the political process.
- **Epicenter** – the point on the Earth's surface directly above the focus (i.e., hypocenter) of an earthquake.
- **Equation of state** – an equation relating the pressure, temperature, and volume of a system.
- **Equilibrium** – a state in which there is no tendency toward spontaneous change.

- **Essential services** – the supply of resources, including electricity, gas, water, compressed air, fuel, and lubricants, necessary to maintain systems operational at all times.
- **Estuary** – a region of interaction between rivers and near-shore ocean water where tidal action and river flow create a mixing of fresh and salt water.
- **Evacuation** – the rapid, temporary removal of people from an area to avoid or reduce emergency; evacuation may be taken as a precautionary urgent protective action based on observable conditions or plant conditions.
- **Evaporation** – the conversion of a liquid to vapor.
- **Evaporator** – a process unit in which a solution is converted to a vapor and a more consternated solution.
- **Event tree analysis** – an inductive technique that starts by hypothesizing the occurrence of basic postulated initiating events and proceeds through their logical propagation to system failure events.
- **Excursion** – an unintentional occurrence, such as a discharge of pollutants above the permitted amount, due to reasons beyond human control.
- **Exhaust** – a duct for the escape of gases, fumes, and odors from an enclosure.
- **Exothermic** – a term used to describe a reaction or process during which heat is released.
- **Expansion joint** – a joint installed in a structure to provide for changes (often in length) during expansion or contraction resulting from changes in temperature, without distortion of the structure.
- **Explosive** – a term used to describe a substance that undergoes a rapid chemical change on heating or detonation, with the evolution of significant heat and a large volume of gas.
- **External combustion engine** – an engine that converts thermal energy to kinetic energy with combustion occurring outside the chamber (cylinder or turbine).
- **Extract** – the solvent phase stream leaving a liquid–liquid extraction process.
- **Extraction** – the dissolving and separating out of specific chemical constituents.
- **Fabric collector** – a dust collector that operates by removing particles from airstreams through straining, interception, diffusion, filtration, or electrostatic charge.
- **Fahrenheit** – a relative temperature scale on which the boiling and freezing points of water at atmospheric pressure are 212 and 32°F, respectively.
- **Failure** – loss of the ability of an energy structure, system, or component to function within acceptance criteria.
- **Fallout** – the radioactive debris or material that settles to the Earth after a nuclear explosion.
- **Fatigue** – the incremental weakening of a material because of repeated cycles of stresses that are far lower than its breaking load; the end result is failure.
- **Fault tree analysis** – a deductive technique that starts by hypothesizing and defining failure events and systematically deduces the events or combinations of events that caused the failure events to occur. The fault tree is the diagrammatic illustration of the event sequence.
- **Feedstock** – raw material to supply or fuel a power plant, machine, or industrial process.
- **Federal Energy Regulatory Commission (FERC)** – the federal agency charged with regulating rates, terms, and conditions of interstate and wholesale power sales and transmission services.
- **Fermentation** – a biochemical reaction accompanied by microorganisms that are supplied with nutrients and organic substrates and carried out at critical conditions of heat, pressure, and light that are specific to the reaction at hand; the reaction leads to reduced organic compounds rather than complete oxidized end products.
- **Fiberglass** – a fiber created from the spinning of glass, possessing high resistance to chemical reactions and heat.

- **Fill** – any material deposited by an agent to fill a channel, valley, sink, or other depression.
- **Filter** – a porous material through which a gas or liquid is passed to remove suspended particles of dust.
- **Filter cake** – a concentrated solid or semisolid material that is separated from a liquid and remains on the filter after pressure filtration.
- **First Law of Thermodynamics** – the law which states that energy can be neither created nor destroyed, only transformed.
- **First-order reaction** – a reaction in which the rate of disappearance of a chemical is directly proportional to the concentration of the chemical and is not a function of the concentration of any other chemicals present in the reaction mixture.
- **First responders** – the first members of an emergency service to respond at the site of an emergency.
- **Fission** – the splitting of atomic nuclei into smaller nuclei, accompanied by the release of significant quantities of energy; it is induced by bombardment with neutrons from an external source and continued by the neutrons that are released.
- **Fixation** – the process of causing an unreactive element to combine into a chemical compound, as in ammonia synthesis.
- **Fixed carbon** – the ash-free, carbonaceous material that remains after volatile matter is driven off a dry waste sample.
- **Fixed-bed operation** – an operation in which the additive material (e.g., catalyst, absorbent, and filter media) remains stationary in a chemical reactor.
- **Flammability** – the ease with which a material (gas, liquid, or solid) will ignite spontaneously autoignition) from exposure to a high-temperature environment, or from a spark or open flame.
- **Flammable** – any solid, liquid, vapor, or gas that will ignite easily and burn rapidly.
- **Flared** – gas disposed of by burning in flares usually at the production sites or at gas processing plants.
- **Flash distillation** – a separation process in which an appreciable portion of a liquid is quickly converted to vapor by changing the temperature and/or pressure; this method is widely employed for the desalination of seawater.
- **Flash point** – the minimum temperature at which a liquid or solid gives off enough vapor to form an ignitable mixture with the air near the surface of the liquid or solid.
- **Flashing** – the conversion of a portion of a hot fluid under pressure into vapor by the release of the pressure.
- **Flocculant** – a substance (e.g., lime, and alum, ferric chloride) that induces flocculation, or the agglomeration of colloidal particles suspended in water through charge neutralization for improved separation of particles during water treatment.
- **Flotation** – a treatment process using air, injected into a stream, that adheres to the solids and causes them to rise, allowing for their removal.
- **Flow diagram** – a chart or line drawing employed by engineers to indicate successive steps in a process; it generally includes materials input and output, by-products, wastes, and other relevant data.
- **Flue gas** – the air and pollutants emitted into the atmosphere after a production process or after combustion occurs.
- **Fluidization** – a technique in which a finely divided solid is caused to behave like a fluid by suspending it in a moving gas or liquid.
- **Flux** – the rate of flow of a fluid across an area.
- **Fly ash** – the noncombustible residual particles from the combustion process that are carried by the flue gas.

- **Foot-pound (force)** – the standard unit of work (energy) in the American Engineering (AE) system.
- **Force** – the push or pull that tends to impart motion to a body at rest, to increase or diminish speed, or to change the direction of the body already in motion.
- **Forced draft** – the positive pressure created by the action of a fan or blower.
- **Formula** – a written, symbolic representation of a chemical entity or mathematical relationship.
- **Fossil** – any material that results from an animal or vegetable source from past geologic ages that has been buried and naturally compressed in the Earth.
- **Fossil fuel power** – electricity generated (produced) by coal, oil, or natural gas.
- **Fossil fuels** – fuel formed from biomass by a process taking millions of years or longer to occur; major fossil fuels include coal, petroleum, and natural gas, minor ones include peat and shale.
- **Free convection** – the motion and mixing of a fluid caused solely by density differences within a fluid.
- **Free electron** – an electron that is not associated with any one atom.
- **Free energy** – a thermodynamic quantity employed to predict the maximum work obtainable from the spontaneous transformation of a given reaction.
- **Free radical** – a molecular fragment with one or more unpaired electrons; it is usually short-lived and highly reactive.
- **Friction head** –the pressure differential required to overcome the resistance to flow.
- **Fuel cell** – an electrochemical device that employs continuous oxidation–reduction reactions to produce electricity.
- **Fusion** – the process in which the nuclei of two light elements (chiefly the hydrogen isotopes) combine to form the nucleus of a heavier element resulting in the release of substantial amounts of energy.
- **Fusion point** – the temperature at which a particular substance can flow.
- **Fusion reactor** – a nuclear reactor that devices energy from the fusion of two atoms (deuterium, tritium, or lithium) to one helium atom, with the release of energy.
- **Galvanization** – the coating of steel or iron with zinc, either by immersion in a bath of molten zinc or by electrolytic deposition from a solution of zinc sulfate, to give protection against corrosion.
- **Gamma decay** – a nuclear reaction in which the nucleus of an atom emits a pulse of energy in the form of gamma radiation.
- **Gas turbine** – a device that converts the thermal energy of hot air and hot fuel to kinetic energy of a spinning shaft.
- **Gasification** – the conversion of solid or liquid fuels into gas for use as a convenient and clean source of energy.
- **Gate valve** – a valve in which the closing element consists of a disk that slides over an opening or cross-sectional area through which a fluid passes; it is suitable for on-off control but not for throttling.
- **Gauge** – an instrument for measuring and indicating process variables such as pressure, liquid level, thickness.
- **Gaussian distribution (normal distribution)** – a distribution that represents the maximum number of occurrences at or near a center or mean point, a progressive decrease in occurrences with increasing distance from the center, and symmetrical distribution of occurrences on both sides of the center.
- **Generation** – the amount of electricity a generator has produced over a specific period of time.
- **Generator** – any machine that produces electrical power.

- **Geopressure** – an unusually high underground pressure caused by a subsurface formation.
- **Geothermal energy** – energy stored in the form of heat below the Earth's surface.
- **Glacier** – a large body of ice formed on land by the compaction and recrystallization of snow and showing evidence of present or past movement.
- **Global warming** – the warming of the Earth's atmosphere due to the greenhouse effect.
- **Grab sample** – a single sample that is collected at a time and place to be most representative of a total discharge.
- **Gradient** – the rate of change of a quantity or variable.
- **Gravimetric** – any measurement by weight.
- **Gray hydrogen** – hydrogen produced with emissions of more than 36.4 g CO_2eq/MJ H_2, e.g., by methane gas reformation without carbon capture technology, using natural gas as feedstock.
- **Gray water** – domestic wastewater composed of water from sinks, washing machines and tubs not containing liquid or solid human waste.
- **Green hydrogen** – hydrogen produced from 100% renewable sources.
- **Greenhouse effect** – the rise in the average global temperature due to the absorption of infrared radiation by increasing amounts of carbon dioxide, methane, nitrous oxide, and other greenhouse gases in the air which retard the dissipation of heat from the Earth's surface.
- **Grid** – a system of interconnected high-power transmission lines.
- **Gross national product (GNP)** – the total monetary value of all goods and services produced within a country during a period of time (usually a year), making no allowance for depreciation of stock or other consumption of capital, but including investments.
- **Guide vane** – a fixed or adjustable device intended to direct fluid flow in a conduit or channel.
- **Half-life** – (1) the time required for the concentration of a material to be reduced to one-half its initial value, and (2) the time required for an unstable element or nuclide to lose one-half of its radioactive intensity in the form of alpha, beta, or gamma radiation.
- **Hard water** – an alkaline water containing dissolved salts (divalent cations) that interfere with some industrial processes and prevent soap from lathering.
- **Hazard** – the potential for harm or other detriment; a factor or condition that might operate against safety.
- **Head** – the height of the free surface of fluid above any point in a hydraulic system; it is a measure of the pressure or force exerted by a fluid.
- **Head loss** – the drop in the sum of pressure head and static head between two points along the path of a flowing fluid due to energy loss such as fluid friction.
- **Heat** – a form of energy (thermal) associated with and proportional to molecular motion.
- **Heat capacity** – the amount of thermal energy or heat a substance can store, expressed as the heat required to raise the temperature of a certain amount of substance by 1°.
- **Heat engine** – a device that converts thermal energy to kinetic energy.
- **Heat exchanger** – a unit or vessel in which a hot fluid stream transfers part of its energy to a cooler fluid stream.
- **Heat of combustion** – the heat released by the oxidation of all carbon and hydrogen (usually in a fuel) to carbon dioxide and water.
- **Heat of fusion** – the heat necessary to convert a solid to its liquid form.
- **Heat of neutralization** – the heat liberated in the reaction of acidic solutions with basic solutions.
- **Heat of reaction** – the heat absorbed or evolved by a chemical reaction.
- **Heat of vaporization** – the heat required to convert a liquid or solid to a vapor.

- **Heat recovery** – the transfer of heat from a substance that has been heated during a process to the substance about to be treated to conserve energy.
- **Heat recovery steam generator** – a device employed in cogeneration to recover the heat from a flue gas and transfer it to water to produce steam and/or a hot water heat sink, i.e., a unit designed to absorb heat.
- **Heat transfer** – the transfer of energy between two materials as heat through conduction, convection, or radiation.
- **Heavy hydrogen** – see deuterium.
- **Heavy water** – water consisting of deuterium oxide, or water in which the hydrogen has been replaced by deuterium.
- **Helium (He)** – the first element in the noble gas group with atomic number 2 and atomic weight 4.0; its nuclei are alpha particles.
- **Helper in an emergency** – member of the public who willingly and voluntarily helps in the response to an emergency.
- **Hertz (Hz)** –an electrical unit of frequency equal to one cycle per second.
- **Heterogeneous** – a term used to describe a mixture of different phases (e.g., liquid–vapor, liquid–vapor–solid).
- **High-level radioactive waste** – the waste that is generated from the fuel of a nuclear reactor or found at nuclear reactors or nuclear fuel reprocessing plants.
- **Holdup** – a volume of material held or contained in a process vessel or line.
- **Holocene** – the most recent epoch of the geological Quaternary period, defined as the interval from 10,000 years before the present to the present.
- **Homogeneity** – the even distribution of species or characteristics so that a sample taken at any one point resembles one taken at any other.
- **Homogeneous reaction** – a chemical reaction in which the reacting substances are in the same state, i.e., solid, liquid, or gas.
- **Horsepower** – a unit of power (rate of work) in the AE system, equal to 550 ft-lb (force) per second.
- **Hot well** – reservoir of a condensing unit receiving condensate from a barometric condenser.
- **Humidification** – a process for increasing the water content of air or other gases.
- **Humidity** – the measure of the amount of water vapor in the air at any given time.
- **Hybrid** – (1) an offspring of two organisms that differ in at least one gene, and (2) any equipment that differs from another in at least one respect or accomplishes two or more objectives in a process.
- **Hydrate** – a compound formed by the joining of water with another molecule.
- **Hydrides** – a hydrogen storage technology that can absorb hydrogen into differing solids, including certain metallic compounds and porous nanoparticles. The hydrogen stored in this form can then be released back via changes in pressure, decomposition over a catalyst, or an increase in heat.
- **Hydrocarbon** – a chemical compound composed of only hydrogen and carbon.
- **Hydrocarbon cracking** – the decomposition of petroleum or other hydrocarbons by heat to give materials employed as gasoline, kerosene, other motor fuels, domestic fuel oil, and other products.
- **Hydroelectric power** – the electricity produced from converting the elevation head of water to kinetic energy of moving water to drive a turbine.
- **Hydrogen bond** – an attractive force, or bridge, occurring in highly polar compounds (e.g., covalent compounds containing fluorine, oxygen, nitrogen, or chlorine atoms, such as water) in which a hydrogen atom of one molecule is attracted to an unshared electron pair of another molecule.

- **Hydrogen** – a nonmetallic element with atomic number 1 and atomic weight 1.0 that is highly flammable and explosive; it is used in the production of ammonia, ethanol, and aniline. The lightest element in the periodic table and the most common in the universe.
- **Hydrogenation** – any reaction of hydrogen with an organic compound.
- **Hydrogenolysis (destructive hydrogenation)** – a type of hydrogenation reaction in which molecular cleavage of an organic compound occurs with the addition of hydrogen to each portion.
- **Hydrogeology** – the geology of groundwater with particular emphasis on the chemistry and movement of water.
- **Hydrologic cycle** – the constant cycling of water by evaporation, condensation, and precipitation.
- **Hydrology** – the science and engineering disciplines dealing with the physical properties, distribution, and circulation of water.
- **Hydrolysis** – the reaction of a salt with water to form an acid and a base.
- **Hydrometer** – a device for measuring the density of liquids.
- **Hydrometry** – the measurement and analysis of the flow of water.
- **Hydrophilic** – a substance with an affinity for or high solubility in water.
- **Hydrophobic** – a substance that has low solubility in water and an affinity for organic compounds.
- **Hydropower** – a process to use moving water (kinetic energy) to produce electricity. See hydroelectric power.
- **Hydrostatic level** – the elevation to which the top of a column of water would rise from a conduit under pressure.
- **Hydrothermal** – a geological process involving heated or superheated water.
- **Hydroxide** – an inorganic compound that contains an OH group.
- **Hygrometer** – a device used to measure the water vapor content of the atmosphere.
- **Ideal flow** – fluid flow that is incompressible, two-dimensional, irrotational, steady, and nonviscous.
- **Ideal fluid** – a hypothetical gas or liquid that offers no resistance to external forces and therefore has zero viscosity (i.e., a perfect fluid).
- **Ideal gas** – a hypothetical gas in which there is a complete absence of cohesive forces between the component molecules; the behavior of such a gas can be predicted accurately by the Ideal Gas Law.
- **Ideal Gas law** – the equation of state of an ideal gas which is a good approximation to real gases at ambient to high temperature and low pressure. It states that $PV = nRT$, where P is the absolute pressure, V is the volume, n is the number of moles of gas, R is the universal gas constant, and T is absolute temperature.
- **Igneous rock** – rock that has formed from volcanic magma.
- **Ignition temperature** – the lowest temperature of a fuel at which combustion becomes self-sustaining.
- **Impactor** – a sampling device that employs impingement for the measurement of particles in a gas.
- **Impoundment** – a stored body of water behind a dam, also called a reservoir.
- **Inches of water** – a unit of pressure, equal to 0.0361 lb (force) per square inch.
- **Incident** – any unintended event, including operating errors, equipment failures, initiating events, accident precursors, near misses or other mishaps, or unauthorized act, malicious or non-malicious, the consequences or potential consequences of which are not negligible from the point of view of protection and safety.

- **Independent system operator (ISO)** – the entity that dispatches electricity (directs it from where it is generated to where it is needed) within a grid.
- **Induced draft** – the negative pressure created by the action of a fan, blower, or other gas-moving device.
- **Industrial hygiene** – the art and science of anticipating, recognizing, evaluating, and controlling occupational and environmental health hazards in a work area and the surrounding community.
- **Inertia** – the force exerted by a body in opposition to a change in its motion.
- **Inertial separator** – a device that employs centrifugal force to separate particles from a gas or liquid stream.
- **Initiating event** – an identified event that leads to anticipated operational occurrences or accident conditions.
- **In situ** – (1) a term used to describe any reaction or analysis occurring in place; and (2) a term used to describe a fossil, mineral, or rock found in its original place of deposition, growth, or formation.
- **Insolation** – the rate at which heat is delivered to any place on the surface of the Earth directly from the sun.
- **Inspection** – an examination, observation, surveillance, measurement, or test undertaken to assess structures, systems, components, and materials, as well as operational activities, technical processes, organizational processes, procedures, and personnel competence.
- **Insulating brick** – a firebrick with a low thermal conductivity that is suitable for insulating industrial incinerators.
- **Interacting event** – an event or a sequence of associated events that, interacting with a facility, affect site personnel or items important to safety in a manner that could adversely influence safety.
- **Interconnector** – transmission lines that allow the transfer of electricity across international borders.
- **Interested party** – a person, company, etc., with a concern or interest in the activities and performance of an organization, business, system, etc.
- **Interlock** – a device designed to ensure that operations (or motions) occur in proper sequence.
- **Internal energy** – the energy contained within a body due to the motion of its constituent atoms and molecules.
- **Internal work** – the work done by a system against internal forces, or between its parts, or on the system by these forces.
- **Inversion** – an atmospheric condition in which the air temperature increases, rather than decreases, with an increase in altitude, due to a layer of warm air that prevents the rise of cooler air trapped beneath it; this prevents the dispersion of pollutants within the lower atmosphere.
- **Investor-owned utility (IOU)** – the type of utility that delivers most of the power in the United States; other types are municipally owned, cooperatives (often rural), and federal power companies.
- **Ion** – a charged atom or molecule that has gained or lost one or more electrons; the migration of an ion affects the transport of electricity through an electrolyte or, to a certain extent, through a gas.
- **Ion electrode** – an electrode that develops and measures an electrical potential in response to the activity of an ion in solution.
- **Isentropic process** – a process that occurs without any increase or decrease in entropy.
- **Isobar** – (1) a line drawn on a map connecting points of equal surface atmospheric pressure, and (2) a line of constant pressure on a graph.
- **Isobaric** – a process that occurs at constant pressure.

- **Isokinetic sampling** – a technique for collecting air particles so that the gas entering a sampling nozzle is at the same speed as the surrounding air.
- **Isopleth** – a line drawn on a map through points of equal value for a given meteorological variable.
- **Isotherm** – a line connecting points of equal temperature that is employed on climatic maps or graphs of thermodynamic relations.
- **Isothermal** – a process that exhibits no change in temperature.
- **Isotropic** – anything that exhibits uniform properties throughout and in all directions.
- **Iterative process** – a process for calculating a desired result by means of a repeated cycle of operations in which each result becomes closer and closer to the desired result.
- **Joule Thompson effect** – the cooling that is achieved when a highly pressured gas is allowed to expand without performing work.
- **Kiln** – a large, cylindrical mechanized type of furnace or incinerator.
- **Kilowatt (kW)** – A unit of power equivalent to 1,000 W.
- **Kilowatt-hour (kWh)** – one thousand Watt-hours, which is the basic measure of electricity usage.
- **Kinetic energy** – the energy possessed by a body or material stream because of its motion.
- **Laminar flow** – one of the two types of flow that occur in fluids (the other being turbulent) in which a fluid follows a smooth well-defined path; this type of flow occurs at low Reynolds numbers.
- **Lapse rate** – the rate at which temperature decreases in the atmosphere as height above the surface increases; the average adiabatic lapse rate is approximately −6 to −8°C/km rise.
- **Latent heat** – the heat released or absorbed by a change of state.
- **Latent heat of vaporization** – the amount of heat required to convert a unit mass of a substance from a liquid to a gas at its melting point.
- **Leaching** – the process by which soluble material dissolves and is carried through a solid in a solution.
- **Leakage** – an undesired and gradual escape or entry of a material from a system or reactor.
- **License** – a legal document issued by a regulatory body granting authorization to perform specified activities relating to a facility.
- **Life-cycle** – the phases, changes, or stages that an organism, process, system, facility, or asset passes through during its lifetime.
- **Life-cycle analysis (LCA)** – (also called Life Cycle Assessment) a technique to assess environmental impacts associated with all the stages of an asset's life, from cradle-to-grave, from raw material extraction, materials processing, manufacture, distribution, use, repair and maintenance, and disposal or recycling.
- **Life-cycle management** – lifetime management in which due recognition is given to the fact that at all stages in a lifetime, there may be effects that need to be taken into consideration and may be altered to extend the useful life of the product or asset.
- **Life, lifetime** – the period of time during which a component, facility, or asset is expected to perform according to the technical specifications to which it was produced.
- **Light water** – common water, as opposed to heavy water.
- **Liquefied natural gas (LNG)** – natural gas that has been liquefied for easier shipment and use. Natural gas, primarily methane, has been liquefied by reducing its temperature to −260°F at atmospheric pressure.
- **Liquefied petroleum gas (LPG)** – light gases, such as butane and propane, that can be maintained as liquids while under pressure.
- **Liquid trap** – the sumps, well cellars, and other traps employed in association with oil and gas production, gathering, and extraction operations for the purpose of collecting oil, water, and other liquids.

- **Liquid–liquid extraction** – the removal of a solute from one liquid by bringing the first liquid into intimate contact with a second liquid with which it is immiscible and in which the solute has higher solubility.
- **Load (also demand)** – (1) the level of electricity consumption by a consumer, and (2) a unit that consumes electrical power as opposed to a generator that provides it.
- **Load curve** – a chart illustrating the variation in demand/electrical load over a specific time.
- **Load factor** – the ratio of the average power supplied during a designated period to the peak or maximum load occurring in that period.
- **Local distribution company (LDC)** – A legal entity engaged in the retail sale and/or delivery of natural gas through a distribution system that includes mainlines and laterals.
- **Local exhaust system** – a system for capturing and exhausting contaminants from the air at the point where the contaminant is produced.
- **Low NO_x burner** – one of several combustion technologies employed to reduce emissions of NO_x (oxides of nitrogen).
- **Lower explosive limit (LEL)** – the lowest concentration (usually on a mole or volume percent or fraction basis) of a compound in a mixture that will propagate a flame in the air at 25°C and atmospheric pressure.
- **Maintenance** – the organized activity, both administrative and technical, of keeping structures, systems, and components in good operating condition, including both preventive and corrective (or repair) aspects.
- **Make-up solvent** – the solvent introduced into a process that compensates for solvent loss during a process.
- **Malfunction** – a failure in the normal operation of a system or process.
- **Management system** – a set of interrelated or interacting elements (a system) for establishing policies and objectives and enabling the objectives to be achieved in an efficient and effective manner.
- **Manifest** – the document employed for identifying the quantity, composition, origin, routing, and destination of hazardous wastes during disposal, treatment, or storage.
- **Manifold** – a pipe fitting with numerous branches to convey fluids among a large pipe and several smaller pipes or to permit the choice of diverting flow from one of several sources to one of the many discharge points.
- **Manometer** – an instrument for measuring pressure that consists of a U–shaped tube containing a liquid.
- **Manufactured gas** – a gas obtained by destructive distillation of coal, or by the thermal decomposition of oil, or by the reaction of steam passing through a bed of heated coal or coke.
- **Market barrier** – anything that interferes with people being able to participate in a market; usually used to explain why energy efficiency is not the standard for energy usage.
- **Mass** – the fundamental measure of the quantity of matter; it differs from weight in that it does not depend on gravitational force.
- **Mass balance** – an inventory of the mass of a chemical in a process or system; it includes mass crossing the system boundary and mass generated, consumed, or accumulated within the system.
- **Mass velocity** – the mass flow rate divided by the cross-sectional area through which the mass is flowing.
- **Material recovery** – the separating and processing of waste materials for beneficial use or reuse.
- **Mathematical model** – a mathematical representation of a real process, expressed as a set of equations or algorithms.
- **Matter** – anything that has a mass or occupies space; it can be solid, liquid, or gas.

- **Maxwell's rule** – a changing magnetic field produces a changing electric field, and this changing electric field produces another changing magnetic field. Thus, the cycle continues, and an electromagnetic wave is made and propagates through space.
- **Mechanical draft tower** – a tower through which the airflow is maintained by fans.
- **Mechanical power** – energy service provided by motors and engines for transportation and other physical manipulation of materials.
- **Megawatt (MW)** – a unit of power equal to 1 million Watts.
- **megawatt-hour (MWh)** – a unit of power equal to 1,000 kilowatt-hours (kWh). It is equal to 1,000 kW of electricity used for 1 hour.
- **Metastable** – an intermediate state in between stability and instability.
- **Metric system** – a system of measurement, exclusively using multiples of 10, that is based on the meter as its unit of length.
- **Metric ton** – a unit of mass equal to 1,000 kg = 2,204 lb (mass).
- **Minimization** – a decision criterion that selects the alternative that has the lowest value to the decision-maker.
- **Miscibility** – the ability of a liquid or gas to dissolve uniformly in another liquid or gas.
- **Mist eliminator** – a control device employed to remove water/liquid droplets entrained in a gas stream.
- **Model** – an analytical or physical representation or quantification of a real system and the ways in which phenomena occur within that system, used to predict or assess the behavior of the real system under specified (often hypothetical) conditions.
- **Modeling** – refers to the development and use of a computer model that represents a physical system by means of mathematical equations.
- **Model validation** – the process of determining whether a model is an adequate representation of a real system being modeled by comparing the predictions of the model with observations of the real system.
- **Molar volume at STP** – the volume that is occupied by 1 mol of an ideal gas at standard temperature and pressure.
- **Mole (mol)** – a measure of the amount of a substance; 1 mol of a compound has a mass equal to its molecular weight.
- **Mole fraction** – a ratio expressing concentrations of solutions and mixtures; defined as the number of moles of that component divided by the total number of moles in a system.
- **Molecular distillation** – distillation at low pressure on the order of 0.001 mmHg.
- **Molecular weight (MW)** – the sum of the atomic weights of the atoms in a molecule.
- **Molecule** – a chemical unit composed of one or more atoms combined to form a single unit; it is the smallest unit of a substance that still possesses the properties of that substance.
- **Momentum** – the product of a body's mass and its velocity.
- **Monomer** – a molecule that is capable of conversion to polymers, synthetic resins, or elastomers by combination with itself or other similar molecules; it usually contains carbon and is of relatively low molecular weight.
- **Monte Carlo method** – a method that constructs a stochastic model of a process on which sampling experiments are performed; it usually involves the use of random numbers in its calculations.
- **Multicyclone collector** – a dust collector consisting of several cyclone collectors that operate in parallel.
- **Multiple-effect evaporator** – a series of single-effect evaporators connected so that the vapor from one effect is the heating medium for the next.

- **Nano** – a prefix for units in the SI system indicating 10^{-9}.
- **Natural draft** – the negative pressure created by the height of a stack or chimney and/or the difference in temperature between flue gases and the atmosphere.
- **Natural gas** – a natural fuel that exists as a mixture of gases, containing primarily methane and some ethane; it is found in certain geological formations.
- **Natural gas liquids (NGLs)** – separated from natural gas, these include ethane, propane, butane, and natural gasoline.
- **Negative pressure** – a pressure difference that is negative.
- **Neutral** – (1) a particle without an electric charge, and (2) a solution that is neither acidic nor basic but with $pH \approx 7.0$.
- **Neutron** – the fundamental unit of matter with no electric charge and about the same mass as a proton; it is present in all atomic nuclei except that of ordinary hydrogen.
- **Node** – a significant point in a grid where several elements, such as loads, generation, storage, transmission lines, or external grids are connected.
- **Nonconservative** – a substance that can react.
- **Nonferrous** – any metal that contains no iron.
- **Normal boiling point** – the boiling point of a liquid when the ambient pressure is 1 atm.
- **Normal operation** – operations carried out within specified operational limits and conditions.
- **Notification** – a document submitted to the regulatory body by a person or organization to notify an intention to carry out a practice or other use of a source.
- **Nuclear energy** – the energy obtained from the fission of uranium or other heavy elements, or the fusion of lighter elements.
- **Nuclear power plant** – a facility that converts nuclear energy into usable power.
- **Nuclear reaction** – a reaction that involves a change in the nucleus of an atom, as opposed to a chemical reaction in which only the electrons take part; a reaction, as in fission, fusion, or radioactive decay that alters the energy, composition, or structure of an atomic nucleus.
- **Nuclear reactor** – a device for generating energy by the controlled fission of atoms or by the fusion of light atoms.
- **Nucleate boiling** – a method of vaporization where high rates of heat transfer are obtained because of bubble formation at a liquid–solid interface.
- **Nucleus** – (1) the small, positively charged central mass of an atom that contains essentially the entire mass of the atom in the form of protons and neutrons, (2) the central portion of a living cell that controls its functions, and (3) any small particle that can serve as the basis for crystal growth.
- **Nuclide** – a particular species of atom that is characterized by the mass, charge, and energy content of its nucleus.
- **Octane number** – a parameter defining the quality of petroleum; a fuel with a high-octane number has better antiknock properties than one with a lower octane number.
- **Off-gas** – a gaseous product from the chemical decomposition of a material.
- **Oil sands** – a naturally occurring mixture of *bitumen* (a heavy, viscous form of crude oil), water, sand, and clay. Using hydro-processing technology, bitumen can be refined to yield synthetic oil.
- **Oil shale** – a black, brown, or green shale impregnated with kerogen, and from which gaseous and liquid petroleum can be obtained by destructive distillation at 300–400°C.
- **Oil-equivalent gas** – the volume of natural gas needed to generate the equivalent amount of heat as a barrel of crude oil. Approximately 6,000 ft^3 of natural gas is equivalent to 1 barrel of crude oil.
- **Operations and maintenance (O&M)** – (1) the actions taken after a facility is constructed to ensure the proper operation, maintenance, and management to achieve intended efficiency and

production levels in an ongoing and optimum manner, and (2) activities required to properly and efficiently operate a system or process.

- **Operating personnel** – individual workers engaged in the operation of facility or the conduct of an activity.
- **Operating pressure** – the system pressure at which a process is operating.
- **Operation** – all activities performed to achieve the purpose for which a facility was constructed.
- **Operational states** – states defined under normal operation and anticipated operational occurrences.
- **Organic** – a term used to describe anything that contains carbon and is thus derived from living organisms.
- **Orifice** – an opening with a closed perimeter in a plate, wall, or partition, through which fluid flows; it is generally employed for the purpose of measuring or controlling a fluid.
- **Orifice meter** – an instrument that measures fluid flow by recording the differential pressure across an orifice plate placed in the flow steam.
- **Orsat method** – a portable mechanism for the analysis of mixtures of carbon dioxide, carbon monoxide, and oxygen in a gas stream.
- **Outflow** – the flow of a product stream out of a process facility.
- **Overfire** – the high-velocity jets of air employed to promote combustion in a furnace.
- **Overload** – the operation of equipment in excess of normal or full capacity.
- **Oxidation** – (1) the addition of oxygen, accompanied by a transfer of electrons, that breaks down organic waste or chemicals such as cyanides, phenols, and organic sulfur compounds in sewage by bacterial and chemical means; and (2) a general term describing a reaction with oxygen.
- **Oxygen (O)** – a colorless odorless, tasteless, nonmetallic, diatomic gaseous element with atomic number 8 and atomic weight 16.0; it is an oxidizing agent and a moderate fire risk as a gas, explosive as a liquid when in contact with heat and oxidizable materials, and an irritant to skin and tissue.
- **Oxygen demand** – the quantity of oxygen utilized in the biochemical or chemical oxidation of organic material in a specified time, at a specified temperature and under other specified conditions.
- **Packing** – (1) a collar or gasket employed to seal mechanical devices to prevent leakage of liquid, and (2) the inert material employed in absorption towers and distillation columns.
- **Parameter** – a quantitative or characteristic property that describes the physical, chemical, or biological conditions of a system.
- **Partial condensation** – the cooling of a saturated vapor until part of it is condensed as liquid.
- **Partial pressure** – the pressure exerted by one of the several components of a gaseous or vapor mixture (see also Dalton's law).
- **Partial vacuum** – a condition in which the pressure is less than atmospheric.
- **Partial volume (pure component volume)** – the volume that would be occupied by one component of a gaseous mixture if it were alone and under the total pressure and temperature of a mixture.
- **Particle-size distribution** – an equation, graph, or table that quantifies, usually by mass percent, the various sizes of particulates found in a sample.
- **Parts per billion (ppb)** – the fraction (ppb_m for mass fraction and ppb_v for volume fraction) multiplied by 10^{-9}; it is a unit used to measure extremely small quantities of a substance.
- **Parts per million (ppm)** – the fraction (ppm_m for mass fraction and ppm_v for volume fraction) multiplied by 10^{-6}; it is a unit used to measure extremely small quantities of a substance.

- **Peak load** – the highest demand placed on a grid by a load or group of loads over a specified period of time.
- **Peaking unit** – a unit that runs only during hours when loads are at their highest point in the day or year.
- **Periodic** – phenomena that repeat after regular intervals of time.
- **Periodic law** – the law which states that elements in the periodic table exhibit regular and predictable changes in their properties with increasing atomic numbers.
- **Persistence** – the length of time a compound or material, once introduced into the environment, remains there.
- **Petrochemicals** – compounds derived from petroleum.
- **Phase** – a physically homogenous component of a system.
- **Phase transition** – an abrupt change in physical properties as temperature and/or pressure are changed continuously.
- **Photochemical air pollution** – brownish–gray haze caused by the photochemical oxidation by solar ultraviolet radiation of an atmosphere containing hydrocarbons and oxides of nitrogen.
- **Photovoltaic** – the means to convert solar radiation directly to electricity using solar cells that are made up of semiconducting materials that are similar to computer and television chips.
- **Photovoltaic cell** – a solid-state electronic device that converts light directly into electricity.
- **Physical model** – the simulation of a real process by a physical experiment that models the important features of the original process that are the object of study.
- **Physical property** – a property of a substance that can change without involving a change in chemical composition, e.g., density, heat capacity.
- **Physical treatment** – a treatment process that alters physical structure or that uses physical processes (sedimentation, filtration and screening) to treat a waste stream.
- **Pilot plant** – a trial assembly of small-scale reactions and processing equipment used to test new processes or reactions.
- **Pilot scale testing** – a small-scale study designed to provide information about a process before it is scaled up to full production size.
- **Pipeline** – continuous pipe conduit, complete with such equipment as valves, compressor stations, communications systems, and meters, for transporting natural and/or supplemental fluids from one point to another, usually from a point in or beyond the producing field or processing plant to another pipeline or to points of utilization.
- **Pitot tube** – a device that is inserted into a pipe to measure the velocity of a flowing fluid.
- **Planned maintenance** – a form of preventive maintenance consisting of refurbishment or replacement that is scheduled and performed prior to unacceptable degradation of a structure, system, or component.
- **Plate column** – a mass transfer (e.g., distillation) column consisting of a number of perforated, equally spaced, horizontal plates on which a layer of liquid is maintained and through which gas flows.
- **Plugging** – the act or process of stopping the flow of water (e.g., the stopping of a leak and sealing off a pipe or hose).
- **Plume** – a visible or measurable discharge of a contaminant from a given point of origin, or the shape it takes.
- **Plutonium (Pu)** – a radioactive element with atomic number 94 and atomic weight 239.1; it is one of the most toxic substances known to humans.
- **Pneumatic** – a term used to describe anything of or containing air, or gases.

- **Political** – of, relating to, or dealing with the structure or affairs of government, politics, or the state.
- **Polymerization** – a chemical reaction, usually carried out with a catalyst, heat, and often under high pressure, in which a large number of relatively simple molecules combine to form chain-like macromolecules.
- **Pore** – any interstice or open space, generally one that allows the passage or adsorption of liquid or gas in a material.
- **Porosity** – the ratio of the volume of open space in a bed to the total bed volume.
- **Porous** – any material with small passages that are permeable to fluids.
- **Positive head** – the energy possessed per unit mass of fluid due to its elevation above some datum point.
- **Potential energy** – the energy possessed by a body due to its position in a force field (e.g., under gravitational influence).
- **Potential head** – the height above a specified datum point or position.
- **Power** – (1) the rate at which work is performed, and (2) a source of physical or mechanical force or energy expressed in watts, kilowatts, or megawatts.
- **Power plant** – a unit constructed for the conversion of stored energy (usually in fossil fuels) into another desired form of energy (usually electrical).
- **Precipitate** – a solid that separates out from a liquid due to some physical or chemical change in the liquid.
- **Preheater** – a unit employed to heat air needed for combustion by absorbing heat from the hot flue gases.
- **Pressure disk** – a thin piece of metal between flanges that breaks at a certain pressure as a safety pressure relief device.
- **Pressure drop** – a measure of the difference in pressures measured immediately upstream and downstream of a unit or process.
- **Pressure relief valve** – a valve that opens automatically to a sufficient area when the pressure reaches an assigned limit in order to relieve the stress on a pipeline or vessel.
- **Pretreatment** – any or all of the operations prior to treatment, such as collection, segregation, chemical neutralization, and decontamination.
- **Preventive maintenance** – actions that detect, preclude, or mitigate degradation of a functional structure, system, or component to sustain or extend its useful life by controlling deterioration and failures to an acceptable level.
- **Proton exchange membrane (PEM)** – a chemical solution used for the electrolyte in either a fuel cell or an electrolyzer that shares the name.
- **Primary air** – admitted to a reactor or incinerator during the first stage of a combustion process (i.e., with the fuel).
- **Primary energy source** – a source of energy found in nature that has not been subject to any human-induced energy transfers or transformations. Examples include fossil fuels, solar, wind, and hydropower.
- **Primary pollutant** – a pollutant emitted directly from a process stack.
- **Prime mover** – a component that converts energy into action when commanded by an actuation device.
- **Probe** – (1) a tube employed to measure pressure at a distance from the actual measuring equipment, and (2) a general term to describe a sampling device.
- **Procedure** – a series of specified actions conducted in a certain order or manner; the set of actions to be taken to conduct an activity or to perform a process is typically specified in a set of instructions.

- **Process** – a manufacturing procedure or the equipment employed in the chemical, pharmaceutical, food, metalworking, or other industry.
- **Process control** – the manipulation of the condition of a process to effect a desired change in the output of the process.
- **Process vent** – any open-ended pipe in a process or stack that is vented to the atmosphere.
- **Processing** – any method, system, or other treatment designed to change the physical form or chemical content of a substance.
- **Product** – the end result of a process reaction.
- **protocol** – the plan and procedures that are to be followed in conducting a test or project.
- **Proton** – the positively charged fundamental unit of matter with about the same mass as a neutron; it is present in all atomic nuclei.
- **Proximate analysis** – analysis, on a mass basis, of a fuel to determine how much moisture, volatile matter, and ash the sample contains, in addition to its heating value.
- **Psychrometer** – an instrument employed to calculate dew points and relative humidity through the use of dry-bulb and a wet-bulb thermometer.
- **Psychrometric** chart – a chart employed to determine the properties of moist air and humidity.
- **Pump** – any mechanical device designed to cause an entering fluid to have an increased velocity and/or pressure at its exit.
- **Purging** – a cleansing or removal of impurities, foreign matter, or undesirable elements.
- ***p*-value** – exceedance probabilities for a given variable, i.e., a p_{90} energy production value denotes the production level predicted to be exceeded 90% of the time.
- **Pyrolysis** – the chemical decomposition of organic matter through the application of heat in the absence of oxygen.
- **qualified expert** – an individual who, by virtue of certification by appropriate boards or societies, professional license, or academic qualifications and experience, is duly recognized as having expertise in a relevant field of specialization.
- **qualitative analysis** – the analysis of a gas, liquid, or solid sample or mixture to identify the elements or compounds present in a sample.
- **quality assurance/quality control (QA/QC)** – a system of procedures, checks, audits, and corrective actions to ensure that all research, design and performance, environmental monitoring and sampling, and other technical and reporting activities are of the highest achievable quality.
- **quantitative analysis** – measurement of the presence and quantities of particular elements or compounds present in a sample.
- **quantum** – (1) any quantity with its magnitude restricted to a discrete set of values and a multiple of a definite unit, and (2) the smallest indivisible quantity of energy.
- **Quark** – the smallest known quantity of matter.
- **Quenching** – a rapid cooling of metals or alloys by immersion in cold water or oil.
- **Quiescent** – a system at rest.
- **Radiant energy** – energy traveling in the form of electromagnetic waves, e.g., light, heat, or electricity.
- **Radiation source** – a radiation generator or a radioactive source or other radioactive material outside the nuclear fuel cycles of research and power reactors.
- **Radiation** – the emission or transmission of energy in the form of waves or particles through space or through a material medium.
- **Radioactive** – (1) the release of energy from the decay of the nuclei of certain kinds of atoms and isotopes either naturally or as a result of intentional manipulation, and (2) exhibiting radioactivity; emitting or relating to the emission of ionizing radiation or particles (adjective).

- **Radioactive half-life** – the time required for the radioactivity to decrease, by a radioactive decay process, by half.
- **Radioactive material** – material designated in national law or by a regulatory body as being subject to regulatory control because of its radioactivity.
- **Radioactive nuclide** – any atom that disintegrates by emitting radiation.
- **Radioactive waste** – for legal and regulatory purposes, material for which no further use is foreseen that contains, or is contaminated with, radionuclides at activity concentrations greater than clearance levels as established by the regulatory body.
- **Radioactive waste management** – all administrative and operational activities involved in the handling, pretreatment, treatment, conditioning, transport, storage, and disposal of radioactive waste.
- **Radioactivity** – the phenomenon whereby atoms undergo spontaneous random disintegration, usually accompanied by the emission of radiation.
- **Radium (Ra)** – a radioactive element with atomic number 88 and atomic weight 226.0 that exists as a brilliant-white solid; it is highly toxic, radioactive, and destructive to living tissue.
- **Radon daughter** – a radioactive decay product of radon.
- **Rankine** – an absolute temperature scale on which the freezing and boiling points of water at atmospheric pressure are 459.7°R and 639.7°R, respectively.
- **Raoult's law** – the law that states that in a vapor–liquid equilibrium mixture, the partial pressure of any component in the vapor is equal to the product of the vapor pressure of the pure component and the liquid mole fraction of that component.
- **Rapping** – a process to remove dust off the plates in an electrostatic precipitator.
- **Rare Earth element** – one of the groups of 15 chemically related elements in group IIIB of the periodic table with atomic numbers from 58 to 71; this group is referred to as the lanthanide series.
- **Raschig ring** – a short section of a metal tube that is employed in mass transfer operations; it is similar in shape to the pasta called ziti.
- **Reaction** – a chemical change in which a substance decomposes or combines with other substances to form a new substance.
- **Reaction order** – the sum of all the numerical orders for each reactant as given by the rate equation (e.g., a first-order reactant plus a second-order reactant gives a third-order reaction).
- **Reaction velocity** – the rate at which a chemical reaction occurs as measured by the rate of formation of the products or the rate of disappearance of reactants.
- **Reactor** – a vessel in which a reaction is carried out.
- **Real gas** – a gas that deviates from the ideal gas law as a result of the interaction of gas molecules.
- **Reboiler** – an auxiliary heating unit for a distillation column designed to supply additional heat to the lower portion of the column.
- **Recirculation** – the repeated flow of a fluid around a closed system.
- **Rectification** – the enrichment or purification of vapor during the distillation process by contact or interaction with a countercurrent stream of liquid condensed from the vapor.
- **Recycled** – any used, reused, or reclaimed material that is employed to produce a new material.
- **Recycling** – (1) an operation in which a substance is passed through the same series of processes more than once, and (2) processes used to separate waste components and concentrate them for return to the raw materials market.
- **Reducing agent** – any material that adds hydrogen or electrons to an element or compound.
- **Reduction** – a chemical reaction in which an atom or ion gains an electron.
- **Refinery gas** – noncondensate gas collected in petroleum refineries.

- **Refining** – a separation process in which undesirable components are removed from various types of mixtures to yield a more purified product.
- **Reflux** – the liquid that has condensed from a rising vapor and is allowed to flow back down a column in a distillation process.
- **Reflux ratio** – the quantity of liquid reflux (return) per unit quantity of product removed from a processing unit.
- **Refractory** – an inert, ceramic material with the ability to retain its physical shape and chemical identity when subjected to extremely high temperatures.
- **Refrigerant** – a substance that is suitable as the working medium of a cycle of operations producing refrigeration (e.g., liquid ammonia, and Freon).
- **Regulatory body** – an authority or a system of authorities designated by the government as having legal authority for conducting the regulatory process, including issuing authorizations.
- **Reinforced plastic** – a composite structure strengthened by glass fibers or other materials.
- **Relative humidity** – the ratio of the amount of water vapor in air to the maximum amount of water vapor the air could hold at a specific temperature.
- **Relief valve** – a pressure relief device such as pressure valves, rupture disks, and other pressure relief systems employed to protect process components from possible explosions or other damage due to overpressure.
- **Rem** – the unit dose of ionizing radiation that gives the same biological effect as that due to 1 Roentgen of X-rays.
- **Renewable energy** – energy obtained from sources that are virtually inexhaustible (defined in terms of comparison to the lifetime of the Sun) and replenish naturally over small-time scales relative to the human life span.
- **Renewable energy resource** or **renewables** – electricity generated (produced) by sources that are sustainable, like the sun (solar or photovoltaic power), wind, geothermal (uses the heat beneath the ground), biomass (trees or other crops grown to be burned in a power plant), wave, or tidal power.
- **Reproducibility** – the ability to repeat an experiment, reaction, or process and produce the same results.
- **Research** – a systematic investigation, including the development, testing, and evaluation that is designed to develop or contribute to knowledge.
- **Reservoir** – a place where a supply or store of something is kept or located.
- **Residual** – the portion of a pollutant remaining in the environment after a natural or technological process has taken place.
- **Residue** – the solid or semisolid material (e.g., ash, ceramics, glass, and metal) that remains after completion of a chemical or physical process such as combustion, distillation, evaporation, or filtration.
- **Retrofit** – the addition or removal of a piece of equipment, or a required adjustment, connection, or disconnection of an existing piece of equipment for the purpose of reducing emissions or updating the process associated with it.
- **Reverse flow** – a flow in a direction opposite to normal flow.
- **Reversible process** – an ideal thermodynamic process that can be exactly reversed.
- **Reversible reaction** – a chemical reaction that can proceed in either direction.
- **Reynolds number** – a dimensionless number used in fluid flow calculations to describe the state of flow in the system, i.e., turbulent or laminar.
- **Risk assessment** – the qualitative and quantitative evaluation of the risk posed to human health and the environment by the presence of specific pollutants, potential natural, or man-made risks associated with normal operation or possible accidents involving facilities and activities.

- **Risk characterization** – the final step of a risk assessment that describes the nature and often the magnitude of human risk, including attendant uncertainty.
- **Rotameter** – a device composed of a float inside a tapered glass tube employed for measuring fluid flow.
- **Rotary dryer** – a long steel cylinder, slowly revolving, with a slightly inclined axis, through which a material passes to be dried by hot air.
- **Rotary valve** – a value consisting of a casing, more or less spherical in shape, and a gate that turns through 90° when opening or closing; they have a cylindrical opening of the same diameter as that of the pipe it serves.
- **Rotor** – the blades and the hub together on a wind turbine.
- **Safety assessment (safety analysis)** – evaluation of the potential hazards associated with the activities of a facility or the conduct of an activity.
- **Safety data sheet (SDS)** – a compilation of information required under the OSHA communication standard on the identity of hazardous chemicals, health and physical hazards, exposure limits, and precautions for safe handling and spill clean-up.
- **Safety valve** – a valve that automatically opens when prescribed conditions, usually of pressure, are executed in a pipeline or other closed receptacle containing liquids and/or gases.
- **Salinity** – the amount of salt in water.
- **Saturated hydrocarbon** – a hydrocarbon molecule that contains only single bonds and hence cannot bond with more atoms.
- **Saturated steam** – steam that is in equilibrium with liquid water at a given temperature and pressure.
- **Saturation temperature** – the minimum temperature at which condensation will occur.
- **Scalar** – a quantity that possesses only a magnitude, but not direction.
- **Scale-up** – the calculation and planning involved in increasing operations from the pilot plant stage to the large-scale production stage.
- **Scheduled maintenance** – any periodic activity that is necessary to maintain the integrity or reliability of a system, which can be anticipated and scheduled in advance.
- **Scope of work** – a document detailing program requirements.
- **Scrubber** – a device that uses a liquid spray to remove particulates and gaseous pollutants from an air stream; the gases are removed by absorption, or a chemical reaction, and the particulates are removed through contact and capture by the liquid droplets.
- **Sealed source** – radioactive (or other energy) source in which the dangerous or radioactive material is permanently sealed in a capsule or closely bonded and in a solid form.
- **Second law of thermodynamics** – the law which states that heat will flow from a warmer body to a cooler body when brought into contact with each other.
- **Second-order reaction** – a chemical reaction in which the rate of reaction is proportional to a concentration term raised to the second power.
- **Seepage** – the passage of fluids through porous media, i.e., water through soil.
- **Semiconductor** – a substance that can conduct electricity under some conditions but not others.
- **Sensible heat** – the heat that, when added or removed, results in a change of temperature.
- **Sensitivity analysis** – the study of how the uncertainty in the output of a mathematical model or system can be apportioned to different sources of uncertainty in its inputs.
- **Serendipity** – an unexpected scientific discovery that turns out to be more important than the project being researched.
- **Service life** – the period of time during which a component of a system will be capable of performing its desired function.

- **Settling chamber** – any chamber designed to reduce the velocity of a gas stream in order to promote the settling of particulates from the gas stream before it is discharged into the environment.
- **Shield** – a wall that protects individuals and/or workers from harmful radiation.
- **Shielding** – the protection of personnel from the harmful effects of ionizing radiation and/or neutrons by enclosure of the radiating equipment.
- **Shutdown** – the cessation of operation of an affected facility for any purpose.
- **Side reaction** – a secondary reaction accompanying or following a primary reaction.
- **Significant figures** – the number of digits that reflect the accuracy with which a quantity is measured or known.
- **Simple distillation** – a distillation process in which no appreciable rectification of a vapor occurs.
- **Single-effect evaporator** – an evaporator in which the liquid is subjected to only one evaporation step.
- **Sink** – a receptacle for material moved through a system.
- **Site area** – area of a site over which the management of an authorized facility can be given some degree of authority.
- **Site boundary** – boundary of the site area.
- **Site inspection** – an on-site investigation to determine whether there is a release or potential release and the extent and severity of hazards posed by the release from the site.
- **Site personnel** – all persons working in the site area, either permanently or temporarily.
- **Site survey** – process of identifying candidate sites for a facility installation after the investigation of a large region and the rejection of unsuitable sites.
- **Siting** – process of selecting a suitable site for a facility, including appropriate assessment and definition of the related design basis.
- **Slurry** – a mixture of particulate matter and liquid.
- **Small hydro** – a hydro plant with a capacity between 100 kW and 30 MW.
- **Smelter** – a facility that melts or fuses ore.
- **Soft water** – a water with a low concentration of calcium and magnesium ions, as opposed to hard water.
- **Solar cell** – a battery-like device in which the radiant energy of the sun is converted to electrical energy by means of a semiconductor that is sensitive to the photovoltaic effect.
- **Solar farm** – a power utility, covering a considerable area, where large amounts of electrical energy would be generated from solar energy.
- **Solar heated system** – a system that uses focused solar radiation for the development of energy.
- **Solar radiation** – the electromagnetic waves emitted by the sun.
- **Solid oxide fuel cells (SOFCs)** – among the most efficient and longest-duration fuel cells commercially available.
- **Solubility** – the amount at which one substance can mix or blend uniformly with another.
- **Soluble** – a substance that is able to be dissolved in a liquid.
- **Solute** – the substance that is dissolved in a solvent to form a solution.
- **Solution** – a homogeneous mixture of two or more substances constituting a single phase.
- **Solvent** – the substance (usually liquid) in which a solute is dissolved to form a solution.
- **Sorption** – the action of adsorption. The interaction of an atom, molecule, or particle with a solid surface at a solid–solution or a solid–gas interface.
- **Source** – anything that may cause radiation exposure, such as by emitting ionizing radiation or by releasing radioactive substances or radioactive material and can be treated as a single entity for purposes of protection and safety.
- **Space velocity** – the volume of gas or liquid, measured at a specified temperature and pressure, passing through a unit volume per unit time.

- **Specific gravity** – (1) the ratio of the density of a liquid or solid at a particular temperature to the density of water at 4°C, and (2) the density of a gas to that of air at the same temperature and pressure.
- **Specific heat** – the ratio of the heat capacity of a substance to the heat capacity of water at a particular temperature.
- **Specific volume** – the volume of a unit mass of a substance; it is the reciprocal of density.
- **Spectrum** – the radiant energy emitted by a substance as a characteristic band of wavelengths by which it can be identified.
- **Spent** – any material that has been used and, because of contamination, can no longer serve the purpose for which it was produced without additional processing.
- **Spent source** – a source that is no longer suitable for its intended purpose.
- **Stability** – a measure of the degree of atmospheric turbulence commonly defined in terms of the vertical temperature gradient referenced to the adiabatic lapse rate. See lapse rate, inversion.
- **Stable** – a term used to describe a chemical with the ability to resist changes in its composition despite exposure to water, air, pressure, heat, or shock; see also stable air; stable isotope.
- **Stable isotope** – an isotope of an element that is not radioactive.
- **Stack** – a chimney or any vertical conduit through which gaseous products are released into the atmosphere.
- **Stack sampling** – the collecting of representative samples of gaseous and particulate matter that flows through a duct or stack.
- **Stage** – (1) a part of a series of operations in a piece of equipment, and (2) a subdivision of a series, which is a subdivision of a system, as applied to geological strata.
- **Stakeholder** – an individual, a group of individuals, a company, or an institution that has a stake in a plan, project, or any other planning-related matter.
- **Stand-alone system** – an independent system or process.
- **Standard condition** – any reference temperature and pressure (e.g., a temperature of 70°F and a pressure of 14.7 psia).
- **Standard cubic feet per minute (scfm)** – a flow rate corrected to standard pressure and temperature.
- **Standard enthalpy of reaction** – the change in enthalpy of a given chemical reaction with all reactants and products in their reference states; the reference state is usually 25°C and 1 atm.
- **Standard heat of formation** – the heat of formation of a compound from its elements at 1 atm and 25°C.
- **Standard temperature and pressure (STP)** – standard conditions, typically 0°C, 25°C or 60°F and 1 atm.
- **State** – one of the three forms that matter can take; these are solid, liquid, and gas.
- **Static electricity** – electricity that is at rest, not flowing.
- **Static head** – the pressure in a fluid due to the height of fluid above a given datum point.
- **Steady flow** – flow that does not vary with time; the mass flow rate is constant and all other quantities (e.g., temperature, pressure, and velocity) are independent of time.
- **Steam** – the gaseous form of water.
- **Steam drum** – a vessel in a boiler in which saturated steam is separated from the steam-water mixture and into which feedwater is introduced.
- **Steam methane reforming (SMR)** – a process by which hydrogen is extracted from natural gas or methane.
- **Steam plume** – a dense white cloud that results from the rapid cooling of moisture-carrying air below its saturation temperature.

- **Steam quality** – a measure of the amount of liquid water in steam.
- **Steam turbine** – a device that converts the thermal energy of steam to the kinetic energy of a spinning shaft.
- **Still** – an apparatus for distilling that is employed to purify liquids (e.g., to prepare alcoholic beverages, distilled water.)
- **Stoichiometric** – (1) the elements of a compound in exactly the proportion represented by the compound's chemical formula, and (2) the minimum amount of a chemical necessary to completely react with another reactant.
- **Stoichiometric combustion** – a combustion condition that is characterized by the exact amount of oxygen required for complete combustion of the chemicals in the influent gas stream.
- **Stoichiometry** – the branch of chemistry and chemical engineering that deals with the quantities of substances that enter into and are produced by chemical reactions.
- **Stoker** – a mechanical device employed to feed solid fuel or solid waste to a furnace.
- **Stoker boilers** – a mechanical system that feeds solid fuel into the furnace of a steam boiler.
- **Stop valve** – a large valve installed in a pipeline to shut off flow for the purpose of inspection or repair.
- **Storage** – an element or system that stores energy; it allows the capture of electrical energy produced at one time for use at a later time by converting it to a different form of energy, e.g., chemical energy in batteries or potential energy in pumped storage systems.
- **Strip mining** – a process that uses machines to scrape soil or rock away from mineral deposits just under the Earth's surface.
- **Stripper** – a column in which one or more components of a liquid stream are removed by being transferred into a gas stream.
- **Strontium (Sr)** – a pale-yellow, soft, metallic element with atomic number 38 and atomic weight 87.6; it is spontaneously flammable in powdered form, ignites when heated above its melting point, and reacts with water to evolve hydrogen.
- **Subatomic particle** – any particle smaller than an atom.
- **Subterranean stream** – a stream flowing through large interstices or openings in rock, such as caves and caverns.
- **Suction pump** – a pump set above the surface of the liquid that is supplied to the pump.
- **Sump** – a pit or tank that catches liquid runoff for drainage or disposal.
- **Supercooled liquid** – a liquid cooled below its normal freezing point without solidification; it is at nonequilibrium, and therefore is a temporary, condition.
- **Superficial mass velocity** – the quantity obtained when the mass rate of flow is divided by the total cross-sectional area, regardless of the presence of any obstruction to flow.
- **Superheated steam** – steam at a temperature above its boiling point.
- **Supersaturation** – an unstable condition in which a solvent contains more dissolved matter than is present in a saturated solution of the same components at an equivalent temperature.
- **Surface tension** – the attractive force exerted by molecules below the surface on those at the surface of a liquid.
- **Surge** – a momentary increase in flow or pressure that passes along the length of a conduit due to sudden changes in velocity.
- **Surveillance** – a type of inspection to verify the integrity of a facility or structure.
- **Sustainable** – able to be maintained at a steady level without exhausting natural resources or causing severe ecological damage, as in a behavior or practice.
- **Synergism** – the cooperative interaction of two or more chemicals or processes producing a greater total effect than the sum of their individual effects.

- **Synthesis** – (1) the combination of parts to form a whole, and (2) creation of a substance that either duplicates a natural product or is a unique material not found in nature.
- **Synthetic gas** – a gas consisting of hydrogen, carbon monoxide, and often carbon dioxide that is produced through a human-controlled process. Also called synthesis gas.
- **Synthetic natural gas (SNG)** – a manufactured product, chemically similar in most respects to natural gas, resulting from the conversion or reforming of hydrocarbons that may easily be substituted for or interchanged with pipeline-quality natural gas.
- **System** – a set of connected components or parts forming a complex whole.
- **Systemic** – something that affects the entire system or body.
- **Systems-based approach** – emphasizes the interdependence and interactive nature of elements within and external to events, processes, and phenomena.
- **Tariff** – the document approved by a public utility commission that lists a utility's rates, terms, and conditions of service.
- **Temperature** – a measure of the average kinetic energy of the molecules of a substance; it is generally measured by the change of volume it causes in a liquid, as in a thermometer.
- **Temperature gradient** – the change in temperature with location or position.
- **Temperature inversion** – an increase in temperature with altitude that is caused by a warm air mass covering a colder one. See inversion, lapse rate.
- **Theoretical air** – the quantity of air, calculated from the chemical composition of a waste that is required to combust a waste/feedstock completely so that no oxygen remains.
- **Thermal conductivity** – a property that quantifies the heat transfer ability of a material as the specific rate of heat flow per unit area per unit temperature gradient.
- **Thermal decomposition** – the decomposition of a material by heat.
- **thermal energy** – also known as heat energy; the energy associated with atomic and molecular vibration within a substance.
- **Thermal pollution** – the discharge of heated water from industrial processes and power plants that can adversely affect the life processes of aquatic plants and animals.
- **Thermal storage** – technology that stores thermal (heat) energy in a reservoir.
- **Thermodynamics** – a rigorously mathematical analysis of energy relationships.
- **Third Law of Thermodynamics** – the law which states that absolute zero can never be attained; it also sets the entropy of all pure crystalline substances to be zero at absolute zero temperature.
- **Three-phase equilibrium** – the state of equilibrium among the vapor, liquid, and solid phases of a pure substance; it occurs at a definite pressure and temperature known as the "triple point."
- **Throat velocity** – the gas velocity through the throat of a venturi.
- **Throttle valve** – a valve designed to control the rate of fluid flow.
- **Throttling** – the control of the rate of fluid flow using a throttle valve.
- **Tidal power** – the mechanical power that is generated by the rise and fall of ocean tides which may be converted to electrical power.
- **Ton (U.S., short)** – a unit of mass equal to 2,000 lb (mass).
- **Total head** – the sum of the static head, velocity head, and pressure head.
- **Total pressure** – the pressure given by the sum of static pressure and velocity pressure.
- **Trace element** – an element present in extremely small amounts.
- **Transformer** – a piece of electrical equipment used to step up or step-down voltage; a device used to change voltage on an AC electrical system.
- **Transient** – a term used to describe anything that changes with time.
- **Transmission** – the electric grid that connects electricity generators to distribution substations.

- **transmission bottlenecks** – transmission lines that are frequently subject to congestion.
- **Transmission system** – set of devices that deliver electricity over high-voltage lines from generators to the interconnection with the distribution system.
- **Transmission** – the movement of electricity from a generator to local distribution lines over high-powered lines; regulated by the FERC.
- **Triple bond** – a highly unsaturated linkage between two atoms; it occurs between two carbon atoms of acetylenic compounds.
- **Triple point** – the temperature and pressure at which all three states (i.e., solid, liquid, and gas) of a substance exist together.
- **Tsunami** – a sea wave caused by an underwater seismic disturbance such as sudden faulting, landslides, or volcanic activity.
- **Turbine** – a device for converting the flow of a fluid (air, steam, water, or hot gases) into mechanical motion by passing the fluid through fan-like blades, causing them to rotate and turning the rotation into mechanical energy that can be used to produce electricity.
- **Turbulence** – the fluid property that is characterized by irregular variations in the speed and direction of the movement of individual particles or elements of flow.
- **Ultimate analysis** – chemical analysis on a mass basis of a solid, liquid, or gaseous fuel in terms of its elemental composition and concentration.
- **Ultra-low-carbon hydrogen** – hydrogen produced with emissions below 36.4 g CO_2eq/MJ H_2 and encompasses both decarbonized hydrogen as well as and carbon-free hydrogen.
- **Ultraviolet (UV)** – the invisible, electromagnetic radiation with a very short wavelength in the range between visible light waves and X-rays.
- **Uncertainty** – anything that is unknown and uncontrollable or not perfectly known or controllable.
- **Uncertainty analysis** – an analysis of uncertainty and error bounds of the estimates involved in, and results from, the solution of a problem.
- **Unit operation** – one of the many physical operations employed in chemical engineering in the industrial production of various chemicals that do not involve a chemical reaction.
- **Unit process** – one of many chemical, thermal, or biological processes in which a chemical reaction takes place between two or more chemicals which results in another chemical or chemicals being formed.
- **Unsealed source** – a radioactive source in which the radioactive material is neither permanently sealed in a capsule nor closely bonded and in a solid form.
- **Unstable** – a chemical that tends to move toward decomposition or other unwanted chemical changes during normal handling or storage.
- **Unsteady flow** – a flow that is characterized by a mass flow rate and/or other quantities that vary with time.
- **Upper explosive limit (UEL)** – the concentration of a combustible substance above which combustion will not occur in the presence of a flame or spark.
- **Upper flammable limit (UFL)** – see upper explosive limit (UEL).
- **Upset** – a disturbance in the functioning, fulfillment, or completion of a process or material.
- **Upstream** – the direction from which a stream is flowing.
- **Uptake** – the act of taking up, drawing up, or absorbing.
- **Uranium-235 (U–235)** – a heavy, radioactive, metal element employed in nuclear reactors and in the production of nuclear weapons.
- **Uranium-238 (U–238)** – a heavy, radioactive, metal element employed in nuclear reactors and in the production of nuclear weapons; it is the most abundant uranium isotope on Earth.

- **Useful life** – the period of time during which a piece of equipment or a process operates at its design performance level.
- **Vacuum distillation** – the distillation at a pressure below atmospheric but not so low that it would be classified as molecular distillation.
- **Valence** – a whole number that is a measure of an element's capacity for combining with other elements as defined in balanced chemical equations.
- **Valence electrons** – the electrons in the outer shell of an atom; they determine the chemical properties of the atom.
- **Validation** – the process of determining whether a product or service is adequate to perform its intended function satisfactorily.
- **Valve** – a device for controlling the flow of a fluid through a pipe or tube.
- **Valve stem** – the rod employed to open or close a valve.
- **Vane** – a device that pivots on a rooftop or other elevated location for the purpose of determining the wind direction.
- **Vapor** – the gaseous state of a substance that is below its critical pressure.
- **Vapor pressure** – the pressure exerted by a vapor in equilibrium with a liquid at a given temperature.
- **Vaporization** – the process by which a substance changes from the liquid state to the gaseous state.
- **Velocity** – the speed and direction of a moving body.
- **Velocity head** – the head equivalent of the kinetic energy per unit mass of a flowing stream.
- **Velocity pressure** – the kinetic pressure in the direction of flow necessary to cause a fluid at rest to flow at a given velocity; it is always positive.
- **Vena contracta** – the most contracted surface area of a stream passing through or over an orifice or weir.
- **Vendor** – a design, contracting, or manufacturing organization supplying a service, component, or facility.
- **Vent** – an opening through which a fluid is ejected from a system.
- **Vented gas** – gas released into the atmosphere on a production site or at processing plants.
- **Ventilation** – the act of admitting fresh air into a space to replace stale or contaminated air.
- **Venturi meter** – a device for measuring the flow of water or other fluid through a closed conduit or pipe; it consists of a venturi tube and a flow registering device.
- **Venturi scrubber** – a unit using a liquid, usually water, to remove pollutants from an air stream.
- **Verification** – the process of determining whether the quality or performance of a product or service is as stated, as intended, or as required.
- **Violet rays** – the shortest rays of the visible spectrum of electromagnetic radiation.
- **Virgin material** – a raw, unused material.
- **Viscometer** – a device for measuring the viscosity of a liquid.
- **Viscosity** – the internal resistance of a fluid (liquid or gas) to a change in shape, or movement of neighboring portions of fluid relative to one another.
- **Volatility** – the tendency of a solid or liquid material to pass into the vapor state at a given temperature.
- **Voltage** – a quantitative expression of the potential difference in charge between two points in an electrical field; the greater the voltage, the greater the flow of electrical current.
- **Volume reduction** – a treatment method that decreases physical volume of a waste stream.
- **Wash solvent** – a liquid added to a liquid–liquid extraction process to wash or enrich the purity of the solute used in the extraction.
- **Waste energy** – energy that is not usefully transferred or transformed.

- **Waste minimization** – the prevention or restriction of waste generation at its source by redesigned products or process modification.
- **Water** – a colorless, odorless, tasteless liquid composed of the elements, hydrogen and oxygen.
- **Water softening** – a treatment process designed to remove hardness-producing ions completely or partially from water.
- **Watershed** – the area surrounding a stream that supplies it with runoff.
- **Watt** – a unit of power (rate of work) in the SI system, equal to 1 J/s.
- **Weir box** – a fabricated channel into which a weir plate has been installed to direct open channel flow for measurement purposes.
- **Wet bulb** – a thermometer bulb maintained wet with distilled water; it is employed with a dry bulb to allow the determination of humidity in an air sample.
- **Wet-bulb temperature** – the temperature of a parcel of air cooled to saturation (100% relative humidity) by the evaporation of water into it, with the latent heat of evaporation supplied by the parcel or air.
- **Wholesale competition** – sales to utility distribution companies and other retailers, not to end-use customers, and regulated by the FERC.
- **Wind power** – mechanical or electrical power generated by a windmill using the kinetic energy of the wind as the primary energy source.
- **Wind speed** – the speed of the wind usually denoted in meters per second (m/s) or miles per hour (mph).
- **Wind vane** – an instrument employed to determine wind direction.
- **Working capacity** – see working charge.
- **Working charge** – the net amount of pollutant that is adsorbed during a complete cycle in an adsorber.
- **X-ray** – electromagnetic radiation of extremely short wavelength; overexposure can permanently damage cells and tissue structures, and its effect is cumulative.
- **Zeolite** – any of the hydrated aluminum complex silicates, either natural or synthetic, which possesses cation exchange properties; used in water and wastewater treatment.
- **Zero-order reaction** – a reaction in which the rate of disappearance of a chemical is independent of the concentration of the chemical or the concentration of any other chemical present in the reaction mixture.
- **Zeroth Law of Thermodynamics** – this law states that two objects are in thermal equilibrium when no heat transfers occur between them when they are placed in contact with each other.

1.4 Symbols and Acronyms

A short list of symbols and acronyms employed in the energy industry is provided below.

$^\circ$	degree
Δ	change
ΔG	Gibbs free energy of reaction
ΔH	enthalpy of reaction
ΔH_f°	standard heat of formation
Å	Angstrom

abs	absolute
acfm	actual cubic feet per minute
AE	alkaline electrolyzer
AFC	alkaline fuel cell
AFV	alternate fuel vehicle
ALK	alkaline electrolysis
amb	ambient
ANL	Argonne National Lab
ANSI	American National Standards Institute
ASME	American Society of Mechanical Engineers
ASTM	American Society for Testing and Materials, ASTM International
atm	atmosphere, atmospheric
bbl	barrel
BEV	battery-electric vehicle
BF	blast furnace
BOF	basic oxygen furnace
b.p.	boiling point
Btu	British thermal unit
°C	degree Celsius
CAA	Clean Air Act
CAD	computer-aided design
cal	calorie
CARB	California Air Resource Board
CBP	Customs and Border Patrol
CCS	carbon capture and storage
CCUS	carbon capture, utilization, and storage
CERCLA	Comprehensive Environmental Response, Compensation, and Liability Act
cfm	cubic feet per minute
cfs	cubic feet per second
cgs	cm-g-sec
CH_4	methane
CHP	combined heat and power
CHHP	combined heat, hydrogen, and power
cf	cubic foot
cm	centimeter
cm^3	cubic cm
CNG	compressed natural gas
CO_2	carbon dioxide
cP	centipoise
Cp	heat capacity
CWA	Clean Water Act
d	day
D_2	deuterium
DC	direct current
DoD	Department of Defense
DOE	Department of Energy
DOI	Department of the Interior

DOT	Department of Transportation
DP	dew point
EA	environmental assessment
EIA	US Energy Information Administration
EMF	electromagnetic field
EPA	US Environmental Protection Agency
EPCRA	Emergency Planning and Community Right-to-Know Act
ES	energy storage
ETA	event tree analysis
EU	European Union
EV	electric vehicle
°F	degree Fahrenheit
FAA	Federal Aviation Administration
FC	fuel cell
FCEB	fuel cell electric bus
FCEV	fuel cell electric vehicle
FE	DOE's Office of Fossil Energy
FERC	Federal Energy Regulatory Commission
f.p.	freezing point
fpm	foot per minutes
fps	foot per seconds
ft	foot
ft-lb	foot pound
FY	fiscal year
g	gram
g/kW	gram per kilowatt
g/L	gram per liter
GHG	greenhouse gas
GJ	gigajoules
gr	grain
GW	gigawatt
$\mathbf{H_2}$	hydrogen
He	helium
HEV	hybrid electric vehicles
$\mathbf{H_2O}$	water
$\mathbf{H_2O_2}$	hydrogen peroxide
hp	horsepower
HVAC	heating, ventilation, and cooling
ICE	internal combustion engine
IEA	International Energy Agency
IEEE	Institute of Electrical and Electronics Engineers
I-O	input–output
IRENA	International Renewable Energy Agency
ISO	International Organization for Standardization
hr	hour
in	inch
K	Kelvin

kcal	kilocalorie
kcal/mol	kilocalories per mole
kg	kilogram
km	kilometer
kW	kilowatt
kWh	kilowatt-hour
kWh/L	kilowatt-hour per liter
L	liter
lb	pound
L/D	length to diameter ratio
LEL	lower explosive limit
LFG	landfill gas
LH_2	liquid hydrogen
LHV	lower heating value
ln	natural logarithm
LNG	liquified natural gas
log	logarithm Base 10
LOHC	liquid organic hydrogen carrier
m	meter
m^3	cubic meter
Mcf	million cubic feet
meq	milliequivalents
mg	milligram
MHz	megahertz
mi	mile
min	minute
MJ	megajoule
mL	milliliter
mm	millimeter
mol	mole
mol wt.	molecular weight
m.p.	melting point
mpg	miles per gallon
mph	miles per hour
Mt	megaton
MW	megawatt
MWh	megawatt-hour
NAS	National Academy of Sciences
NASA	National Aeronautics and Space Administration
NEPA	National Environmental Policy Act
ng	nanogram
NH_3	ammonia
NO_x	nitrogen oxide
μm	micrometer, micron
O&M	operation and maintenance
OSHA	Occupational Safety and Health Administration
oz	ounce

PFD process flow diagram
pH log concentration of the hydrogen ion
PHA process hazard analysis
PM particulate matter
ppb parts per billion
ppm parts per million
ppm_v parts per million by volume
ppm_w parts per million by weight
PSA pressure swing adsorption
PSD particle size distribution, pore size distribution
psi pounds per square in
psia pounds per square in absolute
psig pounds per square in gauge
PV photovoltaic
°R degrees Rankine
RCRA Resource Conservation and Recovery Act
RFP Request for Proposals
R&D research and development
R&I research and innovation
RMP EPA's CAA Risk Management Plan
RMS root mean square
ROI return on investment
rpm revolution per min
s second
scf standard cubic feet
scfh standard cubic feet per hour
scfm standard cubic feet per min
SDWA Safe Drinking Water Act
SMR steam methane reforming
SNR synthetic natural gas
SOE solid oxide electrolysis
SOFC solid oxide fuel cell
SO_x sulfur oxide
sp.gr. specific gravity
sp.ht. specific heat
STP standard temperature and pressure
Tcf trillion cubic foot
temp temperature
TWh terawatt-hour
UEL upper explosive limit
UFL upper flammability limit
UV ultraviolet
Wh waat-hour
WHO World Health Organization
yr year

References

Dupont, R.R., Ganesan, K., and Theodore, L. (2016). *Pollution Prevention, Industrial Ecology, Green Science and Engineering, and Sustainability*. Boca Raton, FL: CRC Press/Taylor and Francis Group.

Holmes, G., Singh, B., and Theodore, L. (1993). *Handbook of Environmental Management and Technology*. New York: Wiley–Interscience.

Skipka, K. and Theodore, L. (2014). *Energy Resources: Availability, Management and Environmental Impacts*. Boca Raton, FL: CRC Press/Taylor and Francis Group.

Theodore, M.K. and Theodore, L. (2021). *Introduction to Environmental Management*, 2nd Edition. Boca Raton, FL: CRC Press/Taylor and Francis Group.

Theodore, L., Reynolds, J., and Morris, K. (1997). *Concise Dictionary of Environmental Terms*. Amsterdam, The Netherlands: Gordon and Breach Science Publishers.

Theodore, L., Ricci, F., and VanVlet, T. (2007). *Thermodynamics for the Practicing Engineer*. Hoboken, NJ: John Wiley & Sons.

2

Introduction to Energy and Energy Issues

2.1 Introduction

In terms of introduction, energy may be defined in general terms as having the capacity to do work. It is a quantity that is conserved, although it may be exchanged among bodies and transformed from one form to another, converted between heat and work, or even interconverted with mass. Energy associated with motion is known as kinetic energy, and energy-related to its position is defined as potential energy. All forms of energy are interconvertible by appropriate processes. In the process of transformation, either kinetic or potential energy may be lost or gained but the total of the two remains essentially the same. In addition, energy and mass are interconvertible. Thus, the conservation of energy and of mass are unified (see also Chapter 5 in this part).

Over the past half-century, there has arisen among informed leaders of different industries, government, and the environmental movement, an acute awareness of energy as a problem of impending critical magnitude on the local, national, and international scenes (Gordon and Suzuki 1991). The energy crisis has been created by historical increases in demand for energy and the continuing lack of viable energy management policies. This situation has resulted in two issues that are fast becoming pervasive concerns: one is how to provide an adequate, reliable supply of all forms of energy in both the developed and underdeveloped countries of the world to meet the demands of a growing population. The second issue is the growing concern with the environmental and social consequences of producing and distributing usable energy (catalog.libraries.psu.edu).

The solution to some of the energy problems is conservation as well as the development of new, less destructive/consumptive energy forms. Energy conservation has helped reduce the waste of resources that have caused some of the problems resulting from the exploitation of energy resources. More conservation efforts could help slow the rapid growth in the energy demand and cause energy to be used much more efficiently. At the same time, new sources of energy must be developed to take the place of exhaustible resources and to ensure the availability of adequate, long-term energy supplies. Further expansion is required of solar, wind, tidal, geothermal, and nuclear carbon-free renewable energy. Fusion and other so-called unconventional sources of energy must continue to be investigated until a truly viable renewable or unlimited source of energy is discovered. (Skipka and Theodore 2014) Hydrogen as an alternative fuel could play a major role as a carbon-free fuel to power fuel cells, combustion engines, etc.

The availability of energy which for many years was just a thermodynamic term to engineers and applied scientists has now become a familiar matter to many others in the United States and elsewhere in the world. Because of demand and the need to address global warming, energy

Hydrogen Energy: Principles and Applications, First Edition. Vincent J. DelGatto, Louis Theodore, R. Ryan Dupont, and Matthew C. Ogwu.

availability has become a matter of concern to virtually every household, institution, business, and industry on the planet. Three major factors are: increasing energy requirements brought about by technological advances, especially in developing countries, international socio-political-economic instability, and transition to a drastically lower-carbon energy economy.

It should also be noted that humans have been benefiting from the energy of the Sun since the birth of dawn. Much of this energy is stored in organic materials on the surface of the Earth. It was absorbed over large areas and after billions of years, this solar energy was transformed, converting carbon into fossil fuels (Ponting 1991). Solar energy continues to be transformed, converting carbon into fossil fuels. Solar energy also continues to be transformed into the kinetic energy of air and water.

Section titles for this chapter "Introduction to Energy and Energy Issues" are presented below:

2.2 Early History of Energy

Energy as well as environmental problems have bedeviled humanity since the first person discovered fire. The earliest reported humans appear to have inhabited a variety of locales within a tropical and semitropical belt stretching from Ethiopia to Southern Africa approximately 2 million years ago. Certain energy sources found only in localized areas, were also used in ancient times, e.g., asphalt, coal, and peat from surface deposits and oil from seepages of underground deposits. The situation changed when wood began to be used during the Middle Ages to make charcoal, which in turn was used to reduce ores to metals (Skipka and Theodore 2014).

Early in the 19th century, an awesome new force was gathering strength in Europe. The term *Industrial Revolution* was coined by the French economist, August Blanqui in 1837 to describe the connection between technology and the great political revolution of modern times. Soon exported to the United States, the Industrial Revolution swept away any visions of America being an agrarian society. In 1880, agricultural workers outnumbered nonagricultural workers three to one, but by 1920 the numbers were approximately equal because of the Industrial Revolution taking place in the United States (Hirschman and Mogford 2009). The steam engine, the railroad, the mechanical thresher, and hundreds of other ingenious artifacts that increased man's ability to transform the natural world and put it to use would soon become an integral part of the nation. The new machines not only swiftly accelerated the consumption of raw materials from farms, forests, and mines but also increased the nation's energy demands.

The pace and nature of energy innovations led to dramatic transformations in human life and the economy, as well as major environmental impacts. In the 19th century, some developed agrarian economies dependent on wood and animal power became industrial economies driven by coal. In the 20th century, many economies underwent another transformation, relying on a variety of fuels to power new internal combustion engines and electricity generators. As the demand for energy services has changed, so have the fuels and conversion technologies employed to meet those demands (worldenergysupply.blogspot.com).

Although petroleum had been used in small quantities for centuries for purposes as diverse as medicine and building, the modern petroleum era began when a commercial well was brought into production in Pennsylvania in 1859. The American oil industry expanded rapidly as refineries sprang up to make numerous products from crude oil. It was John D. Rockefeller who formed the Standard Oil Company for producing, processing, refining, transporting, and marketing oil and with it, the modern oil industry was officially born.

2.3 Later History of Energy

Energy consumption in the United States shifted from approximately 70% wood in 1870, to 70% coal in 1900, to 70% oil and gas around 1960. These changes referred to by some as energy transformations were accompanied by changes in energy technologies and the provision of energy services such as heating, cooling, lighting, and mechanical power. As populations have risen and economic activity has expanded, energy consumption has increased, primarily through the use of fossil fuels.

The major energy "revolution" was the ability to generate electricity and transmit it over large distances. During the first half of the 19th century, British physicist Michael Faraday demonstrated that electricity would flow in a wire exposed to a changing magnetic field, now known as Faraday's Law (repository.futiminna.edu.ng:8080). Humans then understood how to generate electricity. In the 1880s, alternating currents (AC) motors and transformers made long-distance transmission of electricity possible. Humans could now generate electricity on a large scale, at a single location, and then transmit that electricity efficiently to many different locations. Electricity generated at Niagara Falls, for example, could be used by customers all over the region (www.energy.gov).

Most of the energy utilized by mankind until recently has originated from the sun. This solar energy is absorbed and stored in plants by photosynthesis and provides the energy found in foodstuffs. The energy of coal and oil is also derived from sunlight. These fuels occurred from decomposed plants that lived millions of years ago. In recent times society has begun to fully utilize the resource of these fossil fuels. Today (at the time of the preparation of the chapter), benchmark oil is sold for $70–$100 per barrel. Supplies cannot be fully guaranteed due to potential geopolitical instabilities. The present major sources of energy available in the United States and elsewhere are coal, oil, natural gas, nuclear, oil shale, solar, hydroelectric, and geothermal.

2.4 Energy "Emergencies"

Energy is a basic economic commodity; it is available to users to the extent that they are willing to pay the price that is asked for it by the seller. The relevant questions are how much customers will demand at a given price and at a given time and how much businesses can afford to spend to meet the energy demand. At some point, answers to these questions match, and the level of supply and demand meet, a simple example of Economics 101 (Dorf 1978). However, this can be altered by regulatory and other government controls.

The last half-century has produced rising prices, shifting relationships among energy suppliers, uncertain supply, and political influences on fuel supply. Superimposed on these happenings is the problem of overcoming the threat of the aforementioned excessive damage to the environment due to the extensive use of fossil fuels. The process of compensating for these environmental effects usually consumes more energy and often requires new approaches. Recycling of materials, rapid transit, sewage treatment, and environmental control equipment often consume additional energy and thus increase the total use of energy as environmental impacts are attempted to be mitigated (baadalsg.inflibnet.ac.in).

Energy "emergencies" did not originate in 1973, with the Arab oil embargo; they appeared much earlier. Examples include gas utilities refusing to connect service for new residential customers, enforced electrical voltage reductions, and the electrical blackout in the northeastern United States in 1965. Recently the rise in gasoline prices and the intermittent interruptions of natural gas have brought home the energy crisis to most Americans. Some Americans in the past were searching for a villain who could be blamed for energy "emergencies" and higher prices. Some blamed the oil companies, others the federal government, while others blamed the Arab governments or the environmentalists. The three most convenient targets were the energy industry, the federal government, and the environmentalists. As seen by the adversaries, the first conspired the second bungled, and the third obstructed, respectively. In effect, the first is a knave, the second a fool, and the third a dreamer (Dorf 1978). Commonly blamed "villains" of today are the Biden Administration, Putin, and (once again) the profiteering oil companies. Overarching all the finger-pointing, a robust global economy with lower-carbon emissions is the goal.

During the past half-century, there has been an increasing tendency for industry and transportation to become more energy-intensive. Most industries responded to increased demand for production by becoming more energy-intensive and less labor-intensive. Fossil fuel energy has been useful in increasing one's ability to accomplish work, move humans and their goods, and provide physical comfort. Energy will continue to be needed in the future. The challenge that remains is to supply energy in a manner that is compatible with the environment and the Earth's limited resource pool.

It should also be noted that both World Wars I and II imposed enormous military demands for fuel and resulted in a shortage of fuels for industry and private use at home. During World War II, rationing of gasoline to the public and industry was instituted to conserve fuel for use by the military. Immediately following World War II, an energy shortage occurred because of a rapid increase in nonmilitary demand for all forms of energy. In Great Britain, the rationing of gasoline remained in force for several years after World War II. Shortages of energy and energy "emergencies" are not new – they are a fact of life. Still, most people would prefer to live with an unlimited supply of energy and go about their daily lives without consideration for energy supply and demand constraints. This will be an increasing challenge as we move to provide more energy with much less carbon dioxide emitting fuel. Innovation amongst the low-carbon alternatives is a priority. Hopefully, innovation will provide a pathway to a low-carbon lower cost hydrogen economy.

2.5 Net Energy Analysis

No discussion on energy issues would be complete without addressing net energy analysis (Skipka and Theodore 2014), presented earlier in the work of Theodore (2011, personal notes). His work is briefly summarized below.

How much energy does it actually take to produce usable energy or materials? The term *energy analysis* represents a broad field of study dealing with the development and use of all aspects of energy in human society and the environment. *Net energy analysis*, a more limited field of study, deals with the analysis of the energy made available to society by energy production processes *after* accounting for energy lost to society/environment as a result of these processes. This subject can also include the energy analysis of materials production (i.e., how much energy must be invested in the total system for the production of a unit of material). Net energy analysis is a topic that will be addressed several times in this book on hydrogen energy.

Net energy analysis differs significantly from traditional engineering efficiency studies. First, net energy analysis is concerned with the total system of production, starting with resources in the ground. Second, it is concerned with the total quantity of energy throughout society that must be input into the construction and operation of an energy or material production system up to the point where the produced energy is utilized. In effect, it is a cradle-to-grave energy analysis.

The objectives of a net energy analysis are the following:

1) Provide reliable, objective, credible information to the government and industry on the net energy inputs and outputs of energy systems.
2) Provide a workable methodology that can be used in subsequent expanded net energy studies.
3) Provide the best possible documentation of data related to net energy use.
4) Discuss and describe the usefulness and limitations of net energy studies and their potential values in all aspects of decision-making.
5) Discuss the philosophy of and issues about net energy studies.

Four major concerns or issues to which the general subject of net energy analysis might apply include the following:

1) How much energy is required from the industrial component of society to *drive* or establish and operate an energy production process relative to the energy yield of the process?
2) In extracting, processing, and moving a resource (if applicable) to provide energy to end users, what final yields are obtained relative to losses of the total energy of the recovered (fuel) resources and of the industrial energy needed to establish and operate the energy production systems?
3) For a given output of energy for end use, what total amounts of the gross (fuel) resources and industrial energies are necessary to establish and operate the system?
4) The issues of the finiteness of (fuel/energy) resources and the rate of depletion are also of concern to society.

It is important to realize that energy must be expended when a material is extracted from its source, processed, or transported. As material moves *downstream* through a series of processing steps and/or operations; there is an accumulation of energy expenditures. This energy expenditure is embodied in the material as a result of a process defined as *sequestered* energy. A petroleum-derived chemical has such an energy value. Thus, the energy requirements of end products include fuel values in some cases, and expended processing energy in all cases, to represent the total sequestered energy.

Net energy analysis should not be used as the primary factor in decisions regarding the processes being evaluated. In effect, other factors may generally carry more weight.

In terms of net energy analysis, consider the life-cycle example of using Middle East heating oil that one of the authors uses to heat their home during the winter (Theodore 2011, personal notes). In this case,

1) Gunky oil was extracted from below the Earth's surface somewhere in the Middle East.
2) The oil is sent to a holding tank.
3) The oil is transported to a larger holding tank near the Persian Gulf.
4) The oil is loaded onto a tanker.
5) The tanker travels approximately 8,000 miles to a port along the New Jersey coast.

6) The oil is unloaded into a holding tank.
7) The oil is transferred to a smaller holding tank at a nearby refinery.
8) The oil is processed at the refinery.
9) Part of the product is the author's heating oil.
10) The healing oil is sent to a holding tank.
11) The heating is transferred to a central holding tank.
12) The heating oil is fed to a delivery truck.
13) The heating oil is fed to a smaller holding tank.
14) The heating oil is transferred to a smaller delivery truck.
15) The delivery deposits the oil in the author's underground 1,500-gallon heating oil tank.
16) The heating oil is combusted, releasing energy that is employed to heat the house.
17) The flue gas produced can contain air pollutants.
18) The flue gas can possess usable sensible energy (Theodore, Ricci, and VanVleet 2007)

The reader should now consider how much energy was expended from Step 1 to Step 15 even before the energy was generated at the author's home to keep him warm in the winter. Concerns should also include the depletion of an irreplaceable natural resource.

Net energy analysis is not a panacea for energy planning but is worth utilizing as a framework for energy analysis in examining a variety of issues. This is addressed in more detail later in Part III of this book.

2.6 Hydrogen as an Energy/Fuel

Free hydrogen is found only in very small traces in the atmosphere. It is widely distributed on Earth in combination with other elements, where the most important and abundant compound of hydrogen is water, H_2O. It is a component of all the constituents of living matter as well as of many minerals. It also forms an essential part of all hydrocarbons and a vast variety of other organic substances. In addition, all acids contain hydrogen; the distinguishing characteristic of an acid is its dissociation upon going into the solution to yield hydrogen ions (studyres.com).

Hydrogen is also the most abundant gas in the universe and the source of all the energy received from the sun. The sun is a giant ball consisting of both hydrogen and helium gases. In a process defined as fusion, four hydrogen atoms combine to form one helium atom, releasing energy as radiation. This radiant energy is Earth's most abundant energy source. It provides light and heat and sustains plant growth. It also causes the wind to blow and the rain to fall. And it is stored as chemical energy in fossil fuels. Most of the present sources of energy originally came from the sun (Skipka and Theodore 2014).

Hydrogen may well emerge as an important fuel in the near future. Since hydrogen is not a basic energy resource (except in the sun), it must be supplied by using some other basic energy resource to generate hydrogen from water or other hydrogen-containing chemical compounds, e.g., menthol. Unlike carbon-based fuels, hydrogen used directly as a fuel produces only water at low combustion temperatures. Unfortunately, at higher combustion temperatures, as in all combustion processes requiring air, nitrogen oxide NO_x can form. Hydrogen is also ideal for catalytic combustion at low temperatures. At these low temperatures, essentially no NO_x pollutants are formed. However, large scale adaptation of hydrogen-based technologies may well produce other problems in the future. For example, water electrolysis cells in the past used a toxic asbestos membrane to separate the hydrogen and oxygen components, so its adoption must proceed using a holistic, systems approach – similar to that presented earlier for net energy analysis – to avoid unanticipated and undesirable consequences.

Hydrogen is the lightest in weight of all gases and has been used for the inflation of balloons and dirigibles. It ignites very easily – a small spark causing it to burn. Several dirigibles, including the infamous *Hindenburg*, may have been destroyed by hydrogen fires (Theodore and Dupont 2012). Today, helium, which has 92% of the lifting power of hydrogen and is not flammable, is used in place of hydrogen for these activities whenever possible.

In a flame of pure hydrogen gas (combusting in the air), the hydrogen (H_2) reacts with oxygen. It does not produce other chemical by-products, except as noted at elevated temperatures, $H_2 + (1/2)\ O_2 \rightarrow H_2O$, where a small number of (2.1) nitrogen oxides (from the air) can be produced. Hence, a key feature of hydrogen as a fuel is that it is relatively nonpolluting.

Pure hydrogen does not occur naturally, it takes energy to manufacture it. Once manufactured, it is an *energy carrier* (i.e., a store of energy first generated by other means). The energy is eventually delivered as heat when the hydrogen is burned. The heat in a hydrogen flame is a radiant emission from the newly formed water molecules. The water molecules are in an excited state on the initial formation and then transition to a ground state, with the transition unleashing thermal radiation. When burning in air, the temperature is roughly 3,630°F (≈2,000°C) (bmiet.net).

In recent years, hydrogen has been touted as the answer to many of society's energy problems. It is the perfect clean fuel. It has multiple applications through direct combustion and conversion to electricity in fuel cells.

Fuel cell technology in vehicles provides an opportunity to perhaps address and solve petroleum and urban air pollution in the future. Fuel cell technology is not new and is reviewed later in Chapter 6 of this part of the text. It has only been recently, however, that this technology has become a potentially viable alternative to low-cost fossil fuels. Fuel cells have already found numerous niche applications in highly selected areas such as the NASA Space Program and as an electrical backup system to protect sensitive electronic equipment should the power grid temporarily go down.

2.7 The Future

The United States is at a turning point in the use of its energy resources. The cost of energy will increase. Carbon reduction to control global warming will be a big factor. Energy will almost certainly cost more, much more, and be harder to obtain in the future. Fossil fuels (the most abundant energy resource in the United States) and their utilization will also play some role in filling both the demand and the price gap as the United States transitions its energy portfolio to continue as a growing lower-carbon industrialized nation (Skipka and Theodore 2014).

In the final analysis, one can either accept or reject the grim projections for the future obtained by extending the consumption patterns and trends of the past, which have established the basis for defining "energy demand." Once it has been determined that the demand exists, the choice among the various means of energy conversion systems, either available at present or in some stage of development for the future, can be made. This requires an evaluation of each means of power generation of the available (fuel) resources, the environmental implications of their development and use, the advances in waste energy recovery, power generation, and energy use technology improvements, and the relation of each fuel type to relevant economic and social issues and concerns. However, all of these considerations are themselves influenced by assumptions regarding future power demands, and these, too, must be reexamined (Skipka and Theodore 2014).

The demand for energy throughout the world obviously impacts the United States. Although it is both difficult to estimate and a strong function of time, there are rough estimates of world energy demands. The demand for energy could triple by 2050 as the global population rises and developing countries expand their economies. According to data from the United Nations, it is projected that the world population will increase from 7.0 billion to more than 9 billion by 2050. This increase, coupled with continued growth in per capita energy demand will cause a significant increase in the consumption of energy planetwide (ebin.pub).

Affordable energy resources will remain essential for economic and social development as well as sustainable food production, water supply availability, and healthy living. To avoid the long-term adverse and potentially irreversible impact of harvesting energy resources, the technical community must explore all aspects of energy production and consumption, including energy efficiency, clean energy, the global carbon cycle, carbon sources and sinks, and hydrogen energy, as well as their relationship to climate and natural resource issues to enable development of sustainable energy supplies for the future (idoc.pub).

Knowledge of how to generate, capture, and utilize energy has allowed humans to flourish in numbers unimaginable to earlier civilizations. It was the affordable oil that has powered the world economy since the 1950s. Engineers learned how to recover fossil fuels and harvest their energy for operating power plants, planes, trains, and automobiles, which has resulted in modifying the carbon cycle and generating additional greenhouse gas emissions. This has resulted in the debate on the future availability of fossil energy resources; the peak oil era, and timing for the anticipated end of the fossil fuel era; price and environmental impact versus various renewable resources and use; carbon footprint, emission, and control, including cap and trade; and the emergence of "green power" (epdf.pub).

The current energy consumption largely relies on oil and natural gas for mobile applications and coal, natural gas, nuclear, waterpower, solar and wind for stationary applications (US EIA 2023). To address energy issues comprehensively moving forward, it is vital to reconsider the complexity of energy usage. Any energy resource, including oil, gas, coal, geothermal, wind, hydrogen, etc., is one of many elements of a complex supply chain and all must be considered within a system of production through consumption. All the elements of the system are interrelated and interdependent. For example, the introduction of intermittent renewable generation with limited capacity factor restricts steady continuous production. Additional generation, process equipment (electrolyzers), and storage would be required to compensate for the intermittent nature of wind and solar.

The continued development and application of energy are essential to the sustainable advancement of society. Therefore, all aspects of each energy option must be considered, including performance against known criteria, basic economics and benefits, efficiency, processing and utilization requirements, infrastructure requirements, subsidies and credits, waste production and ecosystem impacts, as well as unintended consequences and impacts to natural resources and the environment. Additionally, one must include the overall changes to the emerging energy picture based on current and future efforts in renewable alternatives plus modified and enhanced fossil fuels to evaluate the energy return for the investments made and proposed (Glassley 2014).

More than any other time in history, this century will unquestionably be a turning point for human civilization. Human beings may be facing ecological disasters. However, the likelihood of existential scenarios is very unlikely. These crises will force society to reexamine the energy "system" and consumption that has governed their lives. Many humans have lost their feeling of connectedness to nature, resulting in a "we can manage the world" attitude. This attitude *might* lead to the decline of not only this country but also the planet if significant changes in energy "systems" and their use do not take place in the future (Skipka and Theodore 2014).

References

B.M. BMIET (n.d.). https://app.bmiet.net/.

Department of Energy. The Department of Energy's Energy.gov. (n.d.). http://www.energy.gov/.

Dorf, R. (1978). *Energy Resources and Policy*. Reading, MA: Addison-Wesley Publishing Company.

ebin.pub. (n.d.). *Brings together new 2024 papers*. https://ebin.pub/.

Epdf.pub.

Glassley, W.E. (2014). *Energy and the Environment*. Boca Raton, FL, CRC Press/Taylor and Francis.

Gordon, A. and Suzuki, D. (1991). *It's a Matter of Survival*. Cambridge, MA: Harvard University Press.

Hirschman, C. and Mogford, E. (2009). Immigration and the American industrial revolution from 1880 to 1920. *Social Science Research* 38(4): 897–920. https://doi.org/10.1016/j.ssresearch.2009.04.001.

INFLIBNET Centre, Gandhinagar (n.d.). *Inflibnet Centre Gandhinagar*. INFLIBNET Centre Gandhinagar. https://inflibnet.ac.in/.

Ponting, C. (1991). *A Green History of the World*. New York: St. Martin's Press.

Renewable Energy Blogs (2021). Renewable Energy World. https://www.renewableenergyworld.com/blogs/ (worldenergysupply.blogspot.com).

Skipka, K. and Theodore, L. (2014). *Energy Resources: Availability, Management, and Environmental Impacts*. Boca Raton, FL: CRC Press/Taylor & Francis Group.

Theodore, L. and Dupont, R. (2012). *Environmental Health Risk and Hazard Risk Assessment; Principles and Calculations*. Boca Raton, FL: CRC Press/Taylor & Francis Group.

Theodore, L., Ricci, F., and VanVleet, T., (2007), *Thermodynamics for the Practicing Engineer*, Hoboken, NJ, John Wiley & Sons.

U.S. Energy Information Administration (2023). What is the United States' share of World Energy Consumption? https://www.eia.gov/tools/faqs/faq.php?id=87&t=1#:~:text=The%20United%20States%27%20percentage%20share,energy%20consumption%20in%20the%20world (accessed 17 July 2023).

3

Energy Resources

3.1 Introduction

Interest in all forms and sources of energy resources is growing in the United States and globally. The energy industry is expanding its activities to incorporate nuclear, wind, geothermal, solar, and other energy resources into their energy portfolios. Former President Obama proposed to spend nearly $150 billion over a 10-year period on clean energy technology through the American Recovery and Reinvestment Act (Recovery Act, for short) (The White House 2016). That program was eventually put on hold during the Trump Administration. However, the Biden Administration committed to phasing out fossil fuels in the long term as a main source of energy to power the U.S. economy. In fact, most U.S. states now require that part of their electricity production come from renewable energy sources that is planned to increase in the future. The reader should note that this energy transition involves a particularly significant set of changes to the pattern of energy use or application in a society and is not generally a simple nor rapid transition because of many sociopolitical implications of such a significant societal change.

Improving access to all energy resources and services is considered by most to be a critical component for improving the quality of life. No discussion on this topic would be complete without an attempt to differentiate between the two major energy sources in use today: nonrenewable and renewable resources. Nonrenewable energy sources include coal, petroleum, natural gas, propane, and uranium. These energy sources are called nonrenewable because they cannot be replenished in a short period of time. Renewable energy sources include biomass, geothermal, hydropower, tidal, solar, and wind because these supplies can be replenished within a short period of time or represent sources of potential power driven by the Sun or planetary forces. Renewable energy is sustainable energy source that is continually replenished by nature – the Sun, wind, water, the Earth's heat, and plants. Renewable energy technologies turn these fuels into usable forms of energy – most often electricity, but also heat, chemicals, or mechanical power.

This chapter primarily provides information on the four major fossil fuels: coal, oil, natural gas, and shale oil/tar sands. This chapter also reviews a host of renewable energy resources. The section list provided is not complete. The last section in this chapter attempts to address the remaining energy sources that may someday become important in the energy management field. Skipka and Theodore (2014) provide details for all the above sources on availability/distribution and characterization, extraction, processing, transportation/transmission, environmental issues, and prospects and concerns. A host of other references on this subject are available in the literature.

Hydrogen Energy: Principles and Applications, First Edition. Vincent J. DelGatto, Louis Theodore, R. Ryan Dupont, and Matthew C. Ogwu.

It should be noted that although humans have found many different sources of energy to power their endeavors, fossil fuels remain the major source by a wide margin. Oil has been the major fuel supply for industrialized society since the middle of the 20th century and provides more of the energy used by humans than any other source. Coal is second on this list, followed closely by natural gas. Together they accounted for 82% of the world's energy use in 2022 (Global Change Data Lab 2023) despite our growing understanding of the human health effects of air pollution and despite our growing recognition of the major impact of climate change on our planet-driven largely by the use of these fossil fuels.

Finally, an expanded definition of the term fuel may be in order. Fuel is any substance that reacts chemically with another to produce heat or that produces heat by nuclear processes. The term *fuel* is generally limited to those substances that burn readily in air or oxygen, emitting large quantities of heat. Fuels are used for heating; to produce steam for heating, manufacturing, and power purposes; for powering internal-combustion engines; and as a direct source of power in jet and rocket propulsion.

Section titles for this chapter on energy resources are presented below:

3.1 Introduction
3.2 Coal
3.3 Oil
3.4 Natural Gas
3.5 Shale Oil/Tar Sands
3.6 Solar Energy
3.7 Nuclear Energy
3.8 Geothermal Energy
3.9 Wind Energy
3.10 Hydrokinetic Energy
3.11 Biomass-Based Fuels
References

Note that the bulk of the material in this chapter has been adopted from the work of Skipka and Theodore (2014).

3.2 Coal

Coal is a combustible black or brownish-black sedimentary rock normally occurring in layers or veins referred to as *coal beds* or *coal seams*. It is composed primarily of elemental carbon along with variable quantities of other elements – chiefly hydrogen, with smaller quantities of nitrogen, sulfur, oxygen, and chlorine depending on the type of coal.

Coal was formed from the energy that plants absorbed from the Sun millions of years ago. These plants converted solar energy to living plant materials through a process known as photosynthesis. This stored energy is usually released when the plants decay. Under conditions favorable to coal formation, however, this decay process is interrupted, preventing the release of the plant's stored solar energy; in effect, the energy becomes locked into the coal.

Over time, the dead plant matter was deposited into swampy water and a thick layer of dead plants decayed at the bottom of the swamps. As the surface and climate of the Earth changed, more swamps seeped above the plants, halting the decay process. The mass of the top layer of swamp water compressed the lower layers of dead pants. Under both heat and increased pressure, the plant matter underwent chemical reactions and physical changes. This resulted in leaving rich hydrocarbon deposits; in effect, the plants gradually turned into coal. As the preceding physical and chemical processes continued, the decaying plant matter was successively transformed into peat, lignite, sub-bituminous coal, and anthracite. The chemical and physical process referred to here that transformed plant debris into coal is referred to as *coalification*. The first product of this process was peat. However, peat was progressively transformed to lignite, which eventually became anthracite, etc., depending on the progression of geological changes. These chemical changes affected an increase in carbon content plus loss of water, oxygen, and hydrogen. The degree of coalification was determined by:

1) The varying types of vegetation from which the coal originated
2) The depth of burial
3) Temperatures and pressures at those depths
4) The length of time the coal has been forming in the deposit

Additional details on the rank of coal are provided in the literature (Skipka and Theodore 2014).

The United States' coal reserves are vast and the largest in the world. These coal resources are mainly in six states – Montana (25%), Illinois (22%), Wyoming (12%), West Virginia (6%), Kentucky (6%), and Pennsylvania (5%) while another 25 states hold the remaining (23%) (U.S. EIA 2022). These reserves are estimated to range from 500 billion to more than 1 trillion tons. Such a bounty truly represents a major domestic energy supply for decades into the future, providing the technical and environmental sustainability problems with using this fuel can be overcome in a reasonable time and at an acceptable cost.

Coal became *the* energy resource of the 19th century. People used coal to manufacture goods and to power both steamships and railroad engines. By the mid-1800s, people also used coal in the iron and steel production process. Industries expanded and people even used coal to make electricity by the end of the 1800s. World coal production increased from 150 million metric tons annually in the mid-19th century to approximately 3,000 million metric tons today. The annual production rate in the United States increased from 10 to approximately 1,500 million metric tons during the same period. Coal became an energy mainstay for the United States in the 20th century. Today, approximately 90% of coal use in the United States is for the production of electrical energy; however, coal use is currently in decline in the United States due to regulatory constraints brought about by significant environmental and climate impact concerns with the conventional uses of coal for energy production.

3.3 Oil

The terms oil and petroleum have been used interchangeably by industry since the early 1900s. However, some refer to petroleum as crude oil before it is subjected to any treatment/refining, and oil is reserved as "any of the various kinds of greasy, combustible substances obtained from animal, vegetable, and mineral sources, and liquid at ordinary temperature." Both terms are employed interchangeably in this chapter as a matter of convenience. The typical petroleum reservoirs are mostly sandstone or limestone formations containing oil. The viscosity of the oil in these formations may be as thin as gasoline or as thick as tar. It may be almost clear or black. Petroleum created by the decay of biological materials like those creating coal, is called a nonrenewable energy source because it takes millions of years to form.

Unlike coal, in which the stored energy takes the form of rock, oil is an entirely separate substance from the rock in which it was formed. Oil, in its liquid form, is very mobile and tends to move upward as the heat in the Earth's crust causes it to expand. Rising oil frequently penetrates surrounding rock formations and continues its upward journey until it meets a layer of rock that is so tightly compressed that passage is not possible. Such a barrier might be produced by movements in the Earth's crust that have caused the hard, impenetrable rock to abut the porous layers through which oil is moving. When this happens, the oil becomes trapped in a formation that is defined today as an oilfield.

Water has also been trapped underground in the same layers of rock and some of the hydrogen and carbon atoms from the oil have combined to form a mixture of gases. Because the gases are lighter than the oil, they gradually separate and gather at the top of the oil reservoir. Similarly, oil is lighter than water and will float into the pore spaces above it. However, in most oilfields, there is not enough space for all three substances as they expand in the high temperatures deep within the Earth's crust. This explains why pressure builds up and drives oil to the surface when a well is drilled.

Major uses of oil in the United States are for transportation and industry. Approximately 42% of each barrel of oil is refined into gasoline. Transportation consumed approximately 27 quads of energy in 2021, 90% of which comes from petroleum, representing 37% of total U.S. energy consumption (U.S. EIA 2023a).

In 2020, the United States became a net exporter of petroleum for the first time since at least 1949. In 2022, total petroleum exports were about 9.58 million bbl/d and total petroleum imports were about 8.32 million bbl/d, making the United States an annual net total petroleum exporter for the third year in a row. Total petroleum net exports were about 1.26 million bbl/d in 2022. Also in 2022, the United States produced about 19.99 million bbl/d of petroleum and consumed about 20.28 million bbl/d. Although U.S. annual total petroleum exports were greater than total petroleum imports in 2020, 2021, and 2022, the United States still imported some crude oil and petroleum products from other countries to help supply domestic demand for petroleum and to supply international markets (U.S. EIA 2023b).

Since the 1970s the United States has maintained a Strategic Petroleum Reserve as an emergency stockpile in mostly underground facilities. The U.S.'s reserve is held in California, Louisiana, New Mexico, Texas, and the Gulf of Mexico and currently (in 2023) holds over 300 million barrels. The greatest portion of the world's oil reserves are located in OPEC member countries and stand at an estimated 1.2 billion barrels accounting for 80.4% of the world's reserves. Venezuela, Saudi Arabia, Iran, Iraq, United Arab Emirates, and Kuwait hold the greatest amount (between 8% and 24% each) whereas Libya, Nigeria, Algeria, Angola, Gabon, Congo, and Equatorial Guinea hold between 0.1% to 3% each.

3.4 Natural Gas

Natural gas is a nonrenewable fossil fuel because it was formed from the remains of tiny sea animals and plants that died over 300–400 million years ago. As noted earlier, when these animals and plants died, they sank to the bottom of the oceans where they were buried by layers of sediment that turned into rock. Over the millennia, these layers increased in height to several thousands of feet, subjecting the remains to high pressure, which, when combined with high temperatures of the Earth's crust, converted this decaying biomass into not only oil but also natural gas. Over time, high concentrations of natural gas became trapped in the layers of rock.

As with other fossil fuels, natural gas was formed in the Earth's crust, possessing the ability to rise toward the Earth's surface. It built up in pockets when its progress was blocked by an upper layer of solid impervious rock. This process has been described as analogous to trapping water with a sponge. These pockets of gas accumulation can be reached by drilling wells in a manner similar to how oil is recovered from an oil field. It should be mentioned that natural gas exists on all continents except Antarctica.

Natural gas is often found with oil. The oil fields of the Middle East once contained about one-fifth of the world's natural gas reserves. In the past, large quantities of natural gas recovered along with oil were burned because there was no local use for the energy. Today, this gas is recovered and liquefied for transport to user nations. The leading producers of natural gas are the United States, Russia, Iran, China, Qatar, and Canada. Before the commercialization of natural gas, fuel gas was manufactured from coal and oil. However, this manufactured gas had only about half the heating value of natural gas.

Natural gas consists of a mixture of different gases. The main component is methane (CH_4, about 87–96%) which is odorless, colorless, and tasteless. Companies add a chemical odorant called mercaptan (with an odor like rotten eggs) so that any escaping gas can be detected. This process is employed as a safety measure. Other hydrocarbon constituents of natural gas are ethane (1.8–5.1%), propane (0.1–1.5%), and butane (0.02–0.6%), and as well as nitrogen (1.3–5.6%), and carbon dioxide (0.1–1.0%) (NAESB 2023).

The full commercialization and application of natural gas required the development of a safe, economical method of piping (Abulencia and Theodore 2009). This occurred with the development of seamless steel pipes in the 1920s. Thus, the combination of improved drilling techniques, pipelines to deliver the gas, and machines and appliances that run on gas resulted in new viable fossil fuel for use in industries and other commercial and residential applications.

Natural gas ranks second, after oil, in energy consumption in the United States, and approximately 32% of the energy consumed is supplied by this fossil fuel. The best estimates of natural gas consumption by sector in 2022 (total = 32.31 Tcf) are as follows (U.S. EIA 2023c):

1) Electricity generation: 38%
2) Industry: 32%
3) Residential use: 15%
4) Commercial use: 11%
5) Transportation: 4%

The choice of power generation technology in the United States is heavily influenced by the cost of fuel. Historically, the use of fossil fuels has provided some of the lowest prices for generating electricity. As a result, fossil fuels (coal, petroleum, and natural gas) have historically accounted for more than 70% of electricity generation through 2008. With the concerns regarding the environmental and climate impacts of coal use and the growth of more economic renewable energy sources, this share of fossil fuel electricity generation in the United States has been declining. In 2022 natural gas (40%), coal (20%) and petroleum (1%) accounted for only 60% of electricity generation capacity in the United States (U.S. EIA 2023d).

Combustion of natural gas at a power plant causes air pollution – mainly in the form of carbon monoxide and nitrogen oxides – and results in thermal discharges to water. However, natural gas is considered by far the least environmentally damaging of the fossil fuel alternatives. There is essentially no water pollution other than thermal discharges and the amounts of solid wastes generated are insignificant. Unlike coal or oil, combustion of natural gas produces no sulfur oxides or other hazardous air pollutants that would require pollution controls, except for NO_x. By contrast, coal combustion is the leading source of sulfur dioxide emissions and produces about 20 hazardous air pollutants including mercury. Combustion of natural gas also produces fewer emissions of greenhouse gases (primarily CO_2) than other fossil fuels – about half the amount generated by coal in producing an equivalent amount of heat or power. However, the primary component of natural gas, methane, is a more potent greenhouse gas than CO_2 ($\approx$ 25 times more warming potential). Thus, concerns have arisen regarding the release or leaks of methane from the production, transmission, and processing of natural gas. As the use of natural gas continues to expand in the United States, reducing emissions of methane has become an increasingly significant concern for policymakers and the natural gas industry.

The United States became a net exporter for the first time in nearly a century in 2017. The United States continues to be a net natural gas exporter, and in 2022, the total annual U.S. natural gas exports were 6.90 Tcf –the highest on record (U.S. EIA 2023e).

3.5 Shale Oil/Tar Sands

3.5.1 Shale Oil

Shales are geologic formations composed of thinly layered sedimentary rocks and are widely dispersed on the Earth. In certain shale deposits, organic material has been laid down and intermixed with inorganic material. This organic material has partially decomposed and metamorphosed over the ages until it now exists as a mixture of organic substances locked within the shale known as *kerogen*.

The shale itself contains no oil; however, if the shale is pretreated and fractionated at elevated temperatures, the kerogen decomposes and part of it is discharged as an oily vapor that can be condensed into a viscous liquid referred to as shale oil. Unlike typical crude oil, this shale oil contains large amounts of nitrogen and sulfur. These must be removed before a product similar to crude oil results.

Vast new energy resources are being discovered every year across North America. One major source of these "vast new resources" comes from the development of deep shale natural gas formations. Deep shale natural gas is typically found thousands of feet below the Earth's surface in tight, low-permeability shale formations. Experts have known for years that natural gas deposits existed in deep shale formations, but until recently, the vast quantities of natural gas in these formations were thought to be unrecoverable. Today, through the use of a technique called hydraulic fracking, combined with sophisticated horizontal drilling techniques, extraordinary amounts of natural gas from deep shale formations across the United States are being produced, making it the predominant source of U.S. natural gas production since approximately 2010 (Mantell 2009).

The United States is fortunate to have a large remaining natural gas resource base in the form of oil shale. However, because some of the higher quality, lower cost portions of this resource base have already been produced, continued improvements and advances in extraction technologies are essential to expand future production. Fortunately, geoscientists and engineers determined that oil could be produced from these tight (low permeability) rocks. The application of wells with longer horizontal laterals, and the use of intensive multi-stage hydraulic fracturing stimulations to create many more fractures per foot of wellbore, enabled shale oil wells to become economically viable in many settings – such as in the Bakken Shale of Montana and North Dakota, the Eagle Ford Shale in South Texas, the Cana-Woodford Shale in the Anadarko Basin of Oklahoma, and the giant Wolfcamp Shale in the Permian Basin of West Texas and New Mexico. After borrowing and then adapting horizontal well drilling and hydraulic stimulation technology developed for unlocking shale gas, shale oil development has provided massive volumes of associated gas from shale oil wells, particularly in the fields of West Texas and North Dakota, enabling total shale gas production to continue to grow despite declines in production from some of the original shale gas fields (U.S. DOE 2022a).

Continuing research and development investments in advanced shale oil recovery methods and technologies, such as injecting carbon dioxide or other gases, will enable a much larger portion of the shale oil resource to become recoverable. This will expand domestic oil supplies and help keep marginal producers profitable during the energy transition.

3.5.2 Tar Sands

Tar sands (also alternately referred to as bituminous or oil sands) are located throughout the world but are found in vast quantities in Alberta, Canada, and Venezuela. They contain a somewhat dense, viscous form of petroleum, with the assigned technical term of bitumen. The term tar sands were first employed around the turn of the 20th century. The interstices between the grains of sand are largely filled with a thick tarry hydrocarbon mixture that can be extracted at a recovery rate of approximately 15%, with approximately 2 T of tar sands required to produce 1 bbl of oil. This extract can be processed to yield 65–75% synthetic crude oil. The sand was initially mined and transported to a separation plant. Only open-pit mining was practical, and since most of the deposits were buried too deeply for this activity, there were also relatively few places where commercial, large-scale operations were possible at that time. One of many of these accessible near-surface deposits is the Athabasca Tar Sands, located in northern Alberta, Canada, along both sides of the Athabasca River. This Alberta oil sand reserve is estimated to be between 1.7 and 2.5 T bbl, possibly the largest single oil reserve in the world.

While this fossil fuel source is abundant, its extraction and export have generated a significant amount of controversy and criticism from the environmental community. Greenhouse gas emissions from the production of gasoline and diesel from tar sands bitumen were found to be approximately 20% higher than from conventional U.S. crude oils (Cai et al. 2015), and pressure is increasing for shale oil producers to reduce the carbon footprint of their product to meet increasingly stringent global climate goals (Brower 2022). Fossil fuel demand, reflected by crude oil prices, continues to support expansion of new fossil fuel resources, and it appears that tar sands oil production will continue to grow, at least in the near future as world economies transition away from fossil fuels to more sustainable energy sources (Brower 2022).

3.6 Solar Energy

Solar energy is a renewable resource that cannot be depleted (at least for millions of years). Solar energy is radiant energy that is produced by the Sun. Every day the Sun radiates an enormous amount of energy that maintains planet Earth at habitable temperatures.

Where does this energy come from? The Sun's extremely high pressure and temperature cause nuclear fusion. Four hydrogen nuclei fuse to become one helium atom, a process that is accompanied by a loss in mass. This loss in mass results in the emission of radiant energy. It takes millions of years for the energy in the Sun's core to make its way to the solar surface, and then just in a little over 8 minutes, it travels the 93 million miles to Earth. Solar energy travels to the Earth at a speed of 186,000 mi/s, the speed of light.

Only a small portion of the energy radiated into space from the Sun strikes the Earth –approximately one part in 2 billion, or 0.5×10^{-7}%, yet this amount of energy is enormous. Enough solar energy strikes the United States every day to supply the nation's energy needs for 1.5 years. Approximately 15% of the energy that strikes the Earth is reflected back into space. Another 30% evaporates water, and some is absorbed by plants, the land, and the oceans. The remainder is available to supply society's energy needs.

Almost all the radiant energy impacting the surface of the Earth can be attributed to solar energy. This section covers what has been traditionally defined as solar energy. Solar energy is not a single energy technology, but rather a term that covers a diverse set of renewable energy technologies. Solar energy applications can be divided into two categories – heating and cooling, and electricity generation. The main problems lie in both utilizing and storing this energy day and night during the entire year.

In terms of solar energy, it should be noted that, for most points on Earth, the Sun does not shine 24 hours a day. The rest of the time solar energy is unavailable and cannot represent base-load power of a conventional power plant that is not subjected to such constraints. The angle of incidence of radiation, and therefore its intensity, also depends on the motion of the Sun and upon a specific geographic position of the Earth. The angle of radiation incidence becomes more pronounced the farther away a receptor site is located from the position where the Sun is directly overhead.

In terms of energy strength, the Sun constantly delivers 1.36 kW (1,360 J/s) of power per square meter (430 Btu/h-ft^2 or 1,324 W/m^2) to the Earth (in one solar cycle [i.e., an 11-year period], the Sun's emitted energy varies by only 0.1%). However, some of this energy is absorbed by the atmosphere so that, even under ideal conditions, a receptor on Earth may receive less than 1,000 W/m^2 or 0.1 W/cm^2. The total annual radiation reaching the Earth's atmosphere is 1.5×10^{18} kWh.

Interestingly, solar energy provided about 4.5% of the world's electricity in 2021. It should also be noted that solar is the only primary energy source that can generate electricity since particles of light provide the energy for the flow of electrons directly.

3.6.1 Passive Solar Lighting and Heating

People have used the Sun to heat and light their homes for centuries. Ancient Native Americans built their dwellings directly into south-facing cliff walls because they knew the Sun travels low across the southern sky in the Northern Hemisphere during the winter. They also knew the massive rock of the cliff face would absorb heat in winter and protect against wind and snow. At the same time, the cliff-dwelling design blocked sunlight during the summer, when the Sun was higher in the sky, keeping their dwellings cool. The modern version of this Sun-welcoming design is called passive solar because no pumps, fans, or other mechanical devices are used, nor are necessary. Its most basic features include large, south-facing windows that fill the home with natural sunlight, and dark tile or brick floors that store the Sun's heat and release it back into the home at night. In the summer, when the Sun is higher in the sky, window overhangs block direct sunlight, which keeps the house cool. Tile and brick floors also remain cool during the summer.

There are three types of solar collectors for heating water using *solar thermal* systems: flat plate, evacuated tube, and concentrating. The most common type, a *flat plate collector*, is an insulated, weatherproof box containing a dark absorber plate under a transparent cover. *Evacuated tube collectors* are made up of rows of parallel, transparent glass tubes. Each tube consists of a glass outer tube and an inner tube, or absorber, covered with a coating that absorbs solar energy but inhibits heat loss. *Concentrating collectors* for residential applications are usually parabolic-shaped mirrors (like a trough) that concentrate the Sun's energy on an absorber tube called a receiver that runs along the axis of the mirrored trough and contains a heat-transfer fluid. It is the transfer of collected solar thermal energy to receivers by all types of collectors, and transfer of this heat to water in a water storage tank, that results in solar water heating.

Passive solar water-heating systems use natural convection or household water pressure to circulate water through a solar collector to a storage tank. They have no electrical components that could break, a feature that generally makes them more reliable, easier to maintain, and possibly longer lasting than active systems. An *active* system uses an electric pump to circulate water or nontoxic antifreeze through the system. Active systems are usually more expensive than passive systems, but they are also more efficient. Active systems also can be easier to retrofit than passive systems because their storage tanks do not need to be installed above or close to the collectors. Also, the moving water in the active systems will not freeze in cold climates. However because these systems use electricity, they will not function in a power outage. That's why many active systems are now combined with a small solar panel to power a pump independently from the main power grid.

3.6.2 Solar Electricity Production

Solar electricity or photovoltaic (PV) technology (commonly known as solar panels) converts sunlight directly into electricity. Solar electricity has been a prime source of power for space vehicles since the inception of the space program. It has also been used to power small electronics and rural and agricultural applications for four decades. During the last decade, a strong solar electric market has emerged for powering urban grid-connected homes and buildings as a result of advances in solar technology along with global changes in electric industry restructuring.

Although many types of solar PV systems are available today, they all consist of basically three main items: *modules* that convert sunlight into electricity; *inverters* that convert that electricity (direct current) into alternating current so it can be used by most household appliances; and more commonly now *batteries* that store excess electricity produced by the system during daylight hours for use during nighttime periods. The remainder of the system is comprised of equipment such as wiring, circuit breakers, and support structures as well as internet connections to allow continuous system monitoring maintenance and performance evaluation.

The main problem with solar energy is the chemical process needed to convert silicate into the silicon used in the panels for solar energy capture and conversion to electricity. Making sure enough silicon for solar PV applications requires processing with hydrochloric acid, sulfuric acid, nitric acid, hydrogen fluoride, trichloroethane, and acetone. In addition, PV panels also require gallium, arsenide, copper-indium-gallium-diselenide, and cadmium-telluride, which are also highly toxic. In addition, silicon dust is a hazard to workers, and panel recycling requires significant improvement in safety and cost-effectiveness.

Global solar PV manufacturing capacity has increasingly moved from Europe, Japan, and the United States to China over the last decade. As China has invested over $50 billion in new PV supply capacity – 10 times more than Europe, China's share in all the manufacturing stages of solar panels exceeds 80%. In addition, the country is home to the world's 10 top suppliers of solar PV manufacturing equipment. This level of geographical concentration in global supply chains creates potential geopolitical challenges that governments need to address, and the International Energy Agency (IEA 2022) has provided a number of suggestions to ensure PV market security and sustainability to support the global green energy transition of the future:

1) Diversify manufacturing and raw material supplies around the world to reduce supply chain vulnerabilities.
2) De-risk investment in solar manufacturing and solar R&D.

3) Ensure environmental and social sustainability to help improve working conditions and ensure PV manufacturing facilities adopt low-carbon and material-efficient manufacturing practices.
4) Continue to foster innovation for further improving solar cell conversion efficiency and reducing raw material use, especially for critical rare Earth elements and other materials such as silver and copper.
5) Develop and strengthen recycling capabilities, especially in technology development efforts that improve solar PV panel design for recycling, reusability, and greater durability.

3.7 Nuclear Energy

Nuclear energy is energy that is obtained from the nucleus of an atom. The energy is released from an atom through one of two processes: (1) nuclear *fusion* or (2) nuclear *fission*. Fission is the splitting of a heavy nucleus into two approximately equal parts, accompanied by the release of a relatively large amount of energy and generally one or more neutrons. Fusion can occur spontaneously but usually is caused by nuclear absorption of neutrons or other particles. Fusion is defined as the combining of atomic nuclei of very light elements by collision at high speed to form new and heavier elements, resulting in the release of energy. As noted above, energy is released in nuclear fusion when atoms are combined or fused (fusion). This is how energy is generated in the Sun. In nuclear fission, energy is released when atoms are split apart. Nuclear fission is the only method currently used by nuclear plants to generate electricity. However, nuclear fusion and nuclear fission both represent a means of generating nuclear energy involving changes in (nuclear) mass.

Uranium is the heaviest of the 92 naturally occurring elements and was discovered by Martin Heinrich Klaproth in 1789. It has an unstable nucleus and is in constant decay in an attempt to reach a stable arrangement. Since it is also one of the few elements that are easily fissionable, it is the fuel of choice used by today's nuclear power plants.

Uranium was formed when the Earth was created and is usually found in rocks. Rocks that contain significant quantities of uranium are referred to as uranium ore or *pitchblende*. Two forms (isotopes) of uranium are found in these rocks: uranium-235 and uranium-238. The three-digit numbers refer to the number of neutrons and protons in each atom. Uranium-235 is the form normally used for energy production because, unlike uranium-238, the nucleus splits more easily when bombarded by a neutron. During fission, the uranium-235 atom absorbs a bombarding neutron, causing the nucleus to split apart into two atoms of lighter weight. The fission reaction simultaneously releases energy as both heat and radiation. It also releases more neutrons. These released neutrons go on to bombard other uranium atoms, and the process repeats itself. This repetitive process is referred to as a *chain reaction*.

When a uranium atom is split apart (fission), the mass of the fragments is less than the mass of the original atom. The energy corresponding to this loss of mass is defined as fission energy and as an example, the heat energy originating from the Sun is a result of hydrogen changing into helium (Eq. 3.1).

$$2H_2 \rightarrow He + (\text{mass converted to energy}) \tag{3.1}$$

Note that the atomic mass of hydrogen may be taken as 1.00813 and that of helium as 4.00389. Therefore, when 1 g atom of helium is formed from hydrogen, the loss of mass amounts to (4)

(1.00813) − 4.00389 = 0.02863 g. According to the Einstein relation between mass and energy (Eq. 3.2), this unaccounted-for mass appears as energy (Theodore, Ricci, and VanVlet 2007; Theodore 2014).

$$\Delta E = \Delta m c^2 \tag{3.2}$$

where ΔE = the amount of energy released, J; Δm = loss in mass, g; and c = velocity of light 3×10^{10} cm/s.

One can show that each kilogram of mass converted liberates approximately 10^{17} J. This thermal energy is enough to run a large turbine generator (a 1 GW unit) for a year at 35% thermal efficiency and is equivalent to the energy possessed by 2.5 million tons of coal.

When commercial use of nuclear power was first developed (in the late 1950s), many thought it would become the great savior of human civilization, providing a source of clean, limitless electricity that would replace fossil fuels in the future. Unfortunately, that has not come to fruition; simply put, nuclear power has yet to achieve that promise and concerns to be discussed later have severely hampered the development of this industry.

Nuclear power supplied approximately 18% of the U.S. electrical generation in 2022 (U.S. EIA 2023d). Despite this, the nuclear industry has faced significant obstacles in the past, and many continue to argue that the risks of nuclear waste disposal and the cost of the technology outweigh its benefits. A significant concern with the technology is the long-term management and disposal of spent nuclear fuel containing highly radioactive fission products. A long-term federal spent fuel storage facility, the Yucca Mountain repository in Nevada, has been proposed to serve as a spent nuclear fuel and high-level radioactive waste repository. If constructed, it would use a tunnel complex approximately 1,000 ft below the top of Yucca Mountain and about 1,000 ft above the aquifer underlying the repository. The basic idea of geologic disposal is to place carefully packaged radioactive materials in tunnels deep underground. To achieve this, the Yucca Mountain repository would utilize a mixture of natural and engineered barriers to isolate the waste from the surrounding environment (U.S. EPA 2022). This site has been embroiled in political and environmental controversy since 2002 when the site was officially selected to store the nation's spent fuel and high-level radioactive waste. After $8 billion in DOE studies funding ceased at the site in 2016, and its fate is currently uncertain, as is the long-term solution for the disposal of spent fuel waste from the nuclear power industry (Yucca http://Mountain.org 2022).

Other problems that are proving to be an almost intractable concern are the possibility of catastrophic accidents in these plants, i.e., Three Mile Island accident, U.S. (1979); the Chernobyl disaster, Ukraine (1986); Fukushima nuclear disaster, Japan (2011); spent fuel transportation risks, and the possibility of sabotage and theft of nuclear materials (Theodore and Dupont 2013). It seems clear that while some of the problems created (thermal discharges and radioactive releases) by a rapidly expanding nuclear industry are on their way to the solution, others will be compounded by planned developments. Many engineering difficulties are encountered in the construction and operation of nuclear reactors. Construction materials having low neutron absorption must be used. Steel and aluminum cannot withstand the high temperatures involved but stainless steel and chromium have been employed as materials of construction. Molten metals, such as lead, sodium, or potassium, have been found suitable as heat-transfer media while gases, including helium and air, have also been considered as possible materials for removing the heat generated in the reactors.

Small Modular Reactors (SMRs) are advanced nuclear reactors that have been proposed to move nuclear power forward in response to these many expansion concerns. SMRs have a power capacity of up to 300 MW per unit, which is about one-third of the generating capacity of traditional nuclear power reactors (IAEA 2023). Many of the benefits of SMRs are inherently linked to the nature of their design – small and modular. Prefabricated units of SMRs can be manufactured and then shipped and installed on-site, making them more affordable to build than large power reactors. SMRs offer savings in cost and construction time, and they can be deployed incrementally to match increasing energy demand. Power plants based on SMRs may also require less frequent refueling, every 3 to 7 years, compared to between 1 and 2 years for conventional plants (IAEA 2023).

Ultimately, the future of humankind may depend upon the fusion process. Planet Earth will eventually run out of fossil fuels. Significant concern over harmful by-products of fission reactors suggests that fusion offers one of the few well-founded hopes for low-risk production of almost unlimited amounts of energy. A 2022 experiment achieved a key milestone on the path to viable fusion energy: it was the first experiment in which the fusion reaction produced more energy than the energy injected into it (U.S. GAO 2023). However, this technology faces basic science, materials, and engineering challenges before it can produce commercial electricity and still remains decades from commercial implementation.

3.8 Geothermal Energy

The word *geothermal* comes from the Greek words *geo* (Earth) and *therme* (heat). Thus, geothermal energy is heat from within the Earth that is generated in the Earth's core. Geothermal energy in a modern context refers primarily to the generation of electricity from the stored heat in the Earth's crust. Temperatures hotter than the Sun's surface are continuously produced inside the Earth by the slow decay of radioactive particles, a process that occurs in all rocks. This energy (heat) can be recovered and used to heat buildings or generate electricity. Other uses of geothermal energy for process or space heating are termed *direct heat*. Geothermal power may be natural (i.e., from emissions of steam and/or hot water). In addition, cold water may be injected into deep wells where it is heated and returned to the surface for use.

Geothermal energy (also referred to by some as *geothermics*) is virtually limitless and since the Earth is hotter the deeper one drills below the surface. The Earth's core lies almost 4,000 miles beneath the Earth's surface. The double-layered core is made up of very hot molten iron surrounding a solid iron center. Estimates of the temperature of the core range from 5,000°F to 11,000°F. Surrounding the Earth's core is the mantle, which is part rock and part magma. The mantle is about 1,800 miles thick. The outermost layer of the Earth is the insulating crust; it is not one continuous sheet of rock but is broken into pieces called *plates*. These plates of continents and ocean floors drift apart and push against each other at the rate of about 1 in/yr in a process called *plate tectonics*. This process can cause the crust to become *faulted* (cracked), *fractured*, or *thinned*, allowing plumes of magma to rise into the crust.

This magma can reach the surface and form volcanoes, but most remain underground where they can underlie regions as large as spacious mountain ranges. The magma can take from 1,000 to 1,000,000 years to cool as its heat is transferred to surrounding rocks by conduction. In areas where there is underground water, the magma can fill rock fractures and porous rocks. The water becomes heated and can circulate back to the surface to create hot springs and mud pots, or it can become trapped underground, forming deep geothermal reservoirs.

Geothermal steam or superheated water is produced when the Earth's heat energy is transferred to subsurface water from rocks in the Earth's crust. Recent explorations have revealed that the resource is more extensive than had been supposed. There is evidence now that reservoirs of steam and hot water are widespread in the Earth's crust. Where the pressure and temperature are adequate, the steam output may be used in turbines for conversion to electricity. However, more often than not, the temperatures and pressures at which the steam emerges are well below those used in traditional electric-generating plants, giving rise to inevitable inefficiencies in power production. In addition to electricity, geothermal steam or hot water can be applied to desalting seawater (Theodore 2024); heating houses, greenhouses, and swimming pools; and providing nonelectrical energy for refrigeration and air conditioning.

Geothermal power plants access underground steam or hot water from wells drilled a mile or more into the Earth. The steam or hot water is piped up from the well to drive a conventional steam turbine which powers an electric generator. Typically, the water is then returned to the ground to recharge the reservoir and complete the renewable energy cycle. Direct-use applications require geothermal temperatures between about 70°F and 302°F – lower than those required for electricity generation. The United States already has about 1,300 geothermal direct-use systems in operation; however, geothermal systems remain largely underutilized and represent less than 1% of the renewable electricity generating capacity in the United States in 2022 (U.S. EIA 2023d).

While air temperatures can vary widely through the seasons, the temperatures of the shallow ground only range from 50°F to 70°F depending on latitude. Geothermal heat pumps (GHPs) draw on this relatively stable temperature as a source for heating buildings in the winter and keeping them cool in the summer. Through underground piping, a GHP discharges heat from inside a building into the ground in the summer, much like a refrigerator uses electricity to keep its interior cool while releasing heat into your kitchen. In the winter, this process is reversed; the GHP extracts heat from the ground and releases it into a building. Because GHPs move heat between homes and the Earth, instead of burning fuels, they operate very cleanly and efficiently. They are at least three times more efficient than even the most energy-efficient furnaces on the market today.

The Earth's geothermal resources are theoretically more than adequate to supply humanity's energy needs, but only a very small fraction may be profitably exploited. Drilling and exploration for deep resources are expensive, but costs continue to decrease. Forecasts for the future of geothermal power depend on assumptions about technology, energy prices, existing subsidies, and the value of money (interest rates). The Geothermal Technologies Office of the U.S. Department of Energy has as its mission to increase geothermal energy deployment through research, development, and demonstration of innovative technologies that enhance exploration and production to realize its potential to provide clean energy for electricity generation as well as heating and cooling across the entire United States (U.S. DOE 2023a).

3.9 Wind Energy

A significant quantity of power is contained in the movement of air in the atmosphere in the form of wind. In a technical sense, wind can be defined as air in motion, and it is so secret that a significant amount of power can be derived from the motion of air. This air movement is produced by the uneven heating of the Earth's surface by energy from the Sun. Since the Earth's surface is made of very different types of land and water, it absorbs the Sun's radiant energy at different rates. Much of this radiant energy is converted into heat as it is absorbed by land areas, bodies of water, and the air over these formations.

As noted, earlier the Earth is not uniformly heated by the Sun. The poles receive less energy from the Sun than the equator. Along with this, dry land heats up (and cools down) more quickly than the seas. Solar insolation causes temperature difference driving forces between the equator and the poles. These temperature difference driving forces create a global atmospheric convection system reaching from the Earth's surface to the stratosphere, which acts as a virtual ceiling. On a local scale, the warm air over land expands and rises, and the heavier, cooler air over the water rushes in to take its place, creating winds.

Wind power is generated by the conversion of wind energy into a useful form of energy. This is accomplished by using wind turbines to make electricity, windmills for mechanical power, wind pumps for water pumping or drainage, or sails to propel ships. For example, planes traveling from west to east use jet stream winds to increase ground speed. Thus, transcontinental air travel times are usually shorter from west to east and longer from east to west. However, airstreams within a few hundred feet of the ground are of primary importance to ground-based wind energy systems. The total amount of usable power that could be harvested from the wind is estimated to be more than present human power use from all other sources.

The maximum amount of energy that can be extracted from wind is given by (Eq. 3.3):

$$P = 0.5c\rho Av^3 \tag{3.3}$$

where P is the power, watts; c is the coefficient of performance, unitless; ρ is the density of air, kg/m^3; A is the area of the turbine, m^2; and v is the wind speed, m/s.

Compared to the environmental impact of traditional energy sources, the environmental impact of wind power is viewed by some as relatively minor. Wind power generation consumes no fuel and emits no air pollution, unlike fossil fuel power sources. The energy consumed to manufacture and transport the materials used to build a wind turbine is recovered by energy produced by the plant within a few months. While a wind farm may cover a large area of land many land uses, such as agriculture are compatible, with only small areas of turbine foundations and infrastructure made unavailable for use.

Most wind power technologies, as presently envisioned, would have a benign effect on the environment. The specific effects of wind power systems will depend upon the method of wind energy extraction, the ancillary equipment used for energy conversion, energy storage, and transmission, and the ultimate method of their use. Some people regard visual (aesthetic) pollution caused by wind turbines and transmission lines as their severe environmental impact. To some, windmills on the horizon are an eyesore, while to others they are a green technology at work.

Wind power will not solve all of the energy problems of the United States. However, it might provide future cost-effective options for addressing peak energy requirements in some areas. Since 2008, wind turbine prices have declined sharply while performance has improved. In addition, federal tax incentives have helped both wind and solar power purchase agreement prices fall below the projected cost of burning natural gas in existing gas-fired combined cycle units (U.S. DOE 2023b). The average levelized cost of wind power purchase agreements signed in recent years has been 2–3 cents/kW-hr, depending on the wind resource and the project's financing. Because the electricity from wind farms is sold at a fixed price over a long period of time (e.g., 20 years) and its fuel is free, wind energy mitigates the price uncertainty that fuel costs add to traditional sources of energy (U.S. DOE 2023b).

Wind energy continues to receive government or other support to encourage its development. Wind power in the United States receives a tax credit for each kilowatt hour produced; the credit has a yearly inflationary adjustment. Another tax benefit is accelerated depreciation. Many states also provide incentives, such as exemption from property tax, mandated purchases, and additional markets for "green credits." Countries such as Canada and Germany also provide some credits. It is assumed that these will continue or increase in the future.

In 2021, wind turbines operating in all 50 states generated more than 9% of and rose to more than 10% of the country's total electricity generation in 2022. Wind power provides over 50% of the power production in Iowa and South Dakota, and over 30% in Kansas, Oklahoma, and North Dakota (U.S. DOE 2023c). Wind power was the second largest source of U.S. electric-generating capacity additions in 2021 (behind solar) with 13.4 GW of U.S. wind capacity installed, bringing the cumulative total to 136 GW. (U.S. DOE 2023b). The U.S. installed 13 GW of land-based wind energy in 2021 bringing the cumulative total to 136 GW installed. In 2021, the U.S. offshore wind energy pipeline increased 14% to a potential generating capacity of 40 GW. The U.S. distributed wind sector – which includes power from small wind turbines installed near where the power will be used – added 11.7 MW of new distributed wind energy capacity with 1,751 new wind turbines installed across 15 states in 2021. This brings the total installed capacity to 1.1 GW from more than 89,000 wind turbines across all 50 states, Puerto Rico, the U.S. Virgin Islands, and Guam.

As with all energy projects, wildlife impacts from wind project development vary by location. The wind industry incorporates pre- and post-development studies, educated siting, and other impact reduction tools to decrease wildlife impacts (U.S. DOE 2023c). Research shows that wind projects rank near the bottom of the list of human-related bird mortalities, resulting in far fewer annual deaths than those caused by house cats, building collisions, or vehicle impacts. Since 2000, the impact of wind development on birds has been greatly reduced by improvements in turbine design and particularly through improved project and turbine siting.

Although research to develop sound mitigation techniques is ongoing, global peer-reviewed scientific data and independent studies consistently conclude that sound from wind plants has no direct impact on physical human health (U.S. DOE 2023c). The sound level from wind turbines at common residential setbacks is not sufficient to cause hearing impairment or other direct adverse health effects.

Recent research from DOE's Lawrence Berkeley National Laboratory has actually helped quantify the health benefits of wind energy power plants (U.S. DOE 2022b) making the outlook for continued development of wind power in the United States even brighter in the future.

3.10 Hydrokinetic Energy

3.10.1 Hydropower

Hydropower is the third largest source of renewable power, producing 2% of the United States' total electricity but 19% of total renewable electricity generation in 2021 (U.S. EIA 2023a). The existing U.S. hydropower capacity was approximately 80 GW in 2022. Hydropower plants convert the energy in flowing water into electricity. The most common form of hydropower uses a dam on a river to retain a large reservoir of water. Water is released through turbines to generate power. It is worth noting that most dams in the United States were not constructed to primarily produce energy but mainly for irrigation and flood control. "Run of the river" systems divert water from a river and direct it through a pipeline to a turbine to generate power.

Washington State (Grand Coulee Dam) is home to the largest U.S. hydropower facility at 6,800 MW capacity. Other major hydropower plants are located in Virginia, New York, Oregon, Nevada, and Arizona. Globally, the Three Gorges Dam in China is the largest hydroelectric energy facility with a generating capacity of 22,500 MW followed by the Itaipu Dam (on the Parana River between Brazil and Paraguay) at 14,000 MW. The Alqueva Dam in Portugal is the largest in Western Europe at 518 MW capacity, while the Grand Ethiopian Renaissance Dam currently under construction in Ethiopia on the Nile River is the largest in Africa at 5,150 MW design capacity.

Hydropower plants produce no air emissions but can affect water quality and wildlife habitats. Therefore, hydropower plants are now being designed and operated to minimize impacts on the river. Some of them are diverting a portion of the flow around their dams to mimic the natural flow of the river and use fish ladders to permit fish migration into historic spawning grounds upstream of dams to maintain and recover impacted fish habitat.

3.10.2 Tidal Energy

Tidal energy is considered a renewable energy source. The power generated is free and nonpolluting and the plants are easy to maintain. The plants can affect the ecology and there can be aesthetic objections. The United States has no commercial tidal plants at this time and only a few sites where tidal energy could be produced economically are available in the United States

The tides offer a virtually inexhaustible natural source of energy that is essentially untapped. The idea of harnessing the rise and fall of oceans has received the attention of engineers, scientists, and inventors in the past. However, maximum tidal ranges do not occur daily. Tides are caused by the gravitational attraction of the Moon and the Sun on the waters of the Earth and on the Earth itself. The Moon has a greater effect. It "pulls" water away from the Earth, increasing water height on the side toward the Moon. It also draws the Earth away from the water, increasing the water height on the other side, creating two low tides and two high tides each day. The Sun acts similarly but has a reduced effect on the tide. When the Sun and Moon are "pulling" together (or opposite), their tides are in phases and result in a high tidal range. About a week later, when the Sun and Moon are "pulling" at right angles (half-moon), their tides tend to cancel each other, and the tidal range is smaller.

Tidal energy is the most promising source of ocean energy. A dam called a *barrage* is built across an inlet. The barrage has one-way gates that allow the incoming flood tide to pass into the inlet. When the tide turns, the water flows out of the inlet through turbines built into the barrage, producing electricity. Unfortunately, generating electricity in the middle of the ocean is simply not possible because of transportation problems; there is no one there to use it. One can only use the energy near shore, where people need it; in effect, the energy resources would only be of regional interest not an option for large-scale energy supply.

3.10.3 Ocean Thermal Energy

Ocean thermal energy is also referred to as ocean thermal energy conversion (OTEC). Vast amounts of energy are available from ocean thermal gradients. When two extensive currents of water, one warm and one cold, exist near one another, it is possible to operate a power plant utilizing this temperature differential. Theoretically, this energy may be extracted wherever a temperature difference driving force exists. The extraction of such energy becomes (as one might suppose) more difficult, more costly, and less efficient as the temperature difference between the high- and low-temperature reservoirs decreases (Theodore, Ricci, and VanVliet 2009; Theodore 2014). The economics are not currently competitive where the technology for this idea has been shown to work.

If this energy source were developed, however, there would be several environmental impacts to be considered. The large-scale mixing of warm and cold water could have significant impacts on the ocean, biota, and climate. The large surface areas in the heat exchangers would be continually subjected to the flow of corrosive seawater, and metallic elements would therefore be introduced into the seawater. Loss of working fluid (typically ammonia) might also be a problem if leaks in the system were significant, or if there were unexpected spills. Other problems include the impacts of techniques used to inhibit biofouling and corrosion, the impacts of coastal zone facilities associated with the operations of offshore plants, and the installation and operation of electrical distribution systems (Theodore 2014).

3.10.4 Wave Energy

There is also tremendous energy in waves. Waves are caused by the wind blowing over the surface of the ocean. In some areas of the world, the wind blows with enough intensity and force to produce large waves. The west coasts of the United States and Europe and the coasts of Japan and New Zealand are excellent candidates for harnessing wave energy. There are no large commercial wave-energy plants, but there are a few small ones. They have been used to power the lights and whistles on buoys. This resource might produce enough energy to power local communities. Interestingly, Japan, which must import almost all of its fuel, has an active wave-energy program.

3.11 Biomass-Based Fuels

Biomass is renewable organic material that comes from plants and animals. Biomass contains stored chemical energy from the Sun that is produced by plants through photosynthesis. Biomass can be burned directly for heat or converted to liquid and gaseous fuels through various processes. Biomass in the form of wood was the largest source of total annual U.S. energy consumption until the mid-1800s. In 2022, biomass accounted for nearly 5% of U.S. total primary energy consumption and 40% of the U.S. renewable energy portfolio (U.S. EIA 2023a). Biomass is used for heating and electricity generation and as a transportation fuel. Biomass remains an important fuel in many countries, especially for cooking and heating in developing countries.

Biomass sources for energy include (U.S. EIA 2023f):

1) Wood and wood processing waste – firewood, wood pellets, wood chips, lumber and furniture mill sawdust and waste, and black liquor from pulp and paper mills.
2) Agricultural crops and waste materials – corn, soybeans, sugar cane, switchgrass, woody plants, algae, and crop and food processing residues, mostly to produce biofuels.
3) Biogenic materials in municipal solid waste – paper products; cotton and wool products; and food, yard, and wood wastes.
4) Animal manure and human sewage for producing biogas (renewable natural gas).

Biomass is converted to energy through various processes, including direct combustion (burning) to produce heat; thermochemical conversion to produce solid, gaseous, and liquid fuels; chemical conversion to produce liquid fuels; and biological conversion to produce liquid and gaseous fuels.

Direct combustion is the most common method for converting biomass to useful energy. All biomass can be burned directly for heating buildings and water, for providing industrial process heat, and for generating electricity in steam turbines.

Thermochemical conversion of biomass includes *pyrolysis* and *gasification*. Both processes are thermal decomposition processes wherein biomass feedstock materials are heated in closed, pressurized vessels called gasifiers at high temperatures. The processes mainly differ in the temperatures and the amount of oxygen present during conversion. Pyrolysis entails heating organic materials to between 800°F and 900°F (400°C and 500°C) in the nearly complete absence of free oxygen. Biomass pyrolysis produces fuels such as charcoal, bio-oil, renewable diesel, methane, and hydrogen.

Gasification entails heating organic materials to between 1,400°F and 1,700F (800°C and 900°C) with injections of controlled amounts of free oxygen or steam into the vessel to produce a carbon monoxide- and hydrogen-rich gas called synthesis gas or *syngas*. Syngas can be used as a fuel for diesel engines, for heating, and for generating electricity in gas turbines. It can also be treated to separate the hydrogen from the gas, and the hydrogen can be burned or used in fuel cells. The syngas can be further processed to produce liquid fuels using the Fischer–Tropsch process (U.S. EIA 2023f).

A chemical conversion process known as transesterification is used for converting vegetable oils, animal fats, and greases into fatty acid methyl esters (FAME) to produce biodiesel.

Biological conversion of biomass includes fermentation to make ethanol and anaerobic digestion to produce biogas. Ethanol is used as a vehicle fuel and comprises about 10% by volume (E10) of nearly all motor gasoline now sold in the U.S. Biogas, also called biomethane or renewable natural gas, is produced in anaerobic digesters at sewage treatment plants and at dairy and livestock operations. It also forms in and may be captured from solid waste landfills. Properly treated renewable natural gas has the same uses as fossil fuel natural gas.

Of the biomass generated energy in 2022 (4,930 TBtu), various biofuels comprised 49% of the total, wood and wood waste comprised another 43%, and municipal solid waste, animal manure, and sewage combined to represent the remaining 8% of the total.

Finally, biomass can provide an array of benefits including greenhouse gas emission reductions, decreasing the U.S. dependence on foreign oil, and supporting the U.S. agricultural and forest product industries.

The use of biomass energy has the potential to greatly reduce greenhouse gas emissions, as burning biomass releases about the same amount of carbon dioxide as burning fossil fuels. Fossil fuels release carbon dioxide captured by photosynthesis millions of years ago, while biomass combustion releases carbon dioxide that is largely balanced by the carbon dioxide captured in its own growth, depending on how much energy was used to grow, harvest, and process the fuel. Studies have found that clearing forests to grow biomass results in a carbon penalty that takes decades to recoup, so it is best if biomass is grown on previously cleared land, such as under-utilized farmland to minimize its overall environmental footprint (NREL 2023). U.S. dependence on foreign oil is reduced through the production and use of U.S.-based biofuels because biofuels are the only renewable liquid transportation fuels available and can be directly substituted for fossil fuel-based fuels in transportation and power generation applications.

The main biomass feedstocks for power are paper mill residue, lumber mill scrap, and municipal waste. For biomass fuels, the most common feedstocks used today are corn grain (for ethanol) and soybeans (for biodiesel). In the near future agricultural residues such as corn stover (the stalks, leaves, and husks of the plant) and wheat straw will also be used. Long-term plans include growing and using dedicated energy crops, such as fast-growing trees, grasses, and algae. These feedstocks can grow sustainably on land that will not support intensive food crops. All these efforts support the reuse of waste materials from the agricultural and forest product industries and provide new markets and opportunities for their expansion to support the growth of sustainable energy systems into the future.

References

Abulencia, P. and Theodore, L. (2009). *Fluid Flow for the Practicing Chemical Engineer*. Hoboken, NJ: John Wiley & Sons.

Brower, D. (2022). Canada's oil sands: why some of the world's dirtiest fuel is now in hot demand. *Financial Times* (8 June 2022). https://www.ft.com/content/276ecc11-15cd-45ef-8e10-5f64dcde77da (accessed July 24, 2023).

Cai, H., Brandt, A.R., Yeh, S., Englander, J.G., Han, J., Elgowainy, A., and Wang, M.Q. (2015). Well-to-wheels greenhouse gas emissions of Canadian oil sands products: implications for U.S. petroleum fuels. *Environmental Science & Technology* 49 (13): 8219–8227. https://doi.org/10.1021/acs.est.5b01255.

Global Change Data Lab (2023). Our world in data. Energy Consumption by Source, World. https://ourworldindata.org/grapher/energy-consumption-by-source-and-country (accessed 18 July 2023).

International Atomic Energy Agency (2023). What are Small Modular Reactors (SMRs)? https://www.iaea.org/newscenter/news/what-are-small-modular-reactors-smrs (accessed 24 July 2023).

International Energy Agency (2022). *Sola PV Global Supply Chains.* Paris: IEA. https://www.iea.org/reports/solar-pv-global-supply-chains, License: CC BY 4.0.

Mantell, M. (2009). Deep shale natural gas: abundant, affordable and surprisingly water efficient. Water/Energy Sustainability Symposium, Ground Water Protection Council, Salt Lake City, UT.

National Energy Research Laboratory (2023). Biomass energy basics. https://www.nrel.gov/research/re-biomass.html (accessed 26 July 2023).

North American Energy Standards Board (2023). Natural Gas Specs Sheet. Chemical Composition of Natural Gas. https://www.naesb.org/pdf2/wgq_bps100605w2.pdf (accessed 21 July 2023).

Skipka, K., and Theodore, L. (2014). *Energy Resources: Availability, Management, and Environmental Impacts.* Boca Raton, FL: CRC Press/Taylor & Francis Group.

The White House (2016). *Transforming our Energy System, Creating Good Paying Jobs, and Saving Americans on their Energy Bills.* The White House, Office of the Press Secretary. https://obamawhitehouse.archives.gov/the-press-office/2016/02/25/fact-sheet-recovery-act-made-largest-single-investment-clean-energy (accessed 15 April 2023).

Theodore, L. (2014). *Chemical Engineering: The Essential Reference.* New York: McGraw-Hill.

Theodore, L. (2024). Patent number 17/736,239 "System and Process for Obtaining Potable Water Employing Geothermal Energy", East Williston NY, 2024.

Theodore, L. and Dupont, R.R. (2013). *Environmental Health Risk and Hazard Risk Assessment: Principles and Calculations.* Boca Raton, FL: CRC Press/Taylor & Francis Group.

Theodore, L., Ricci, F., and VanVlet, T. (2007). *Thermodynamics for the Practicing Engineer.* Hoboken, NJ: John Wiley & Sons.

U.S. Department of Energy (2022a). *U.S. Oil and Natural Gas: Providing Energy Security and Supporting Our Quality of Life.* Washington, DC: Office of Oil and Natural Gas. https://www.energy.gov/sites/prod/files/2020/10/f79/Natural%20Gas%20Benefits%20Report.pdf.

U.S. Department of Energy (2022b). Wind Energy Technologies Office. How Wind Energy Can Help Us Breathe Easier. https://www.energy.gov/eere/wind/articles/how-wind-energy-can-help-us-breathe-easier (accessed 25 July 2023).

U.S. Department of Energy (2023a). Geothermal Technologies Office. About. https://www.energy.gov/eere/geothermal/about (accessed 25 July 2023).

U.S. Department of Energy (2023b). Wind Energy Technologies Office. Wind Market Reports: 2022 Edition. https://www.energy.gov/eere/wind/wind-market-reports-2022-edition (accessed 25 July 2023).

U.S. Department of Energy (2023c). Wind Energy Technologies Office. Frequently Asked Questions about Wind Energy. https://www.energy.gov/eere/wind/frequently-asked-questions-about-wind-energy (accessed 25 July 2023).

U.S. Energy Information Administration (2022). Coal explained: how much coal is left? https://www.eia.gov/energyexplained/coal/how-much-coal-is-left.php (accessed 19 July 2023).

U.S. Energy Information Administration (2023a). US energy facts explained. https://www.eia.gov/energyexplained/us-energy-facts (accessed 20 July 2023).

U.S. Energy Information Administration (2023b). Oil and petroleum products explained. Oil imports and exports. Basics. https://www.eia.gov/energyexplained/oil-and-petroleum-products/imports-and-exports.php#:~:text=Total%20petroleum%20net%20exports%20were,about%2020.28%20 million%20b%2Fd (accessed 20 July 2023).

U.S. Energy Information Administration (2023c). Natural gas explained. Use of natural gas. Basics. https://www.eia.gov/energyexplained/natural-gas/use-of-natural-gas.php (accessed 21 July 2023).

U.S. Energy Information Administration (2023d). Electricity explained. Electricity generation, capacity, and sales in the United States. Basics. https://www.eia.gov/energyexplained/electricity/electricity-in-the-us-generation-capacity-and-sales.php (accessed 21 July 2023).

U.S. Energy Information Administration (2023e). Natural gas explained. Natural gas imports and exports. Basics. https://www.eia.gov/energyexplained/natural-gas/imports-and-exports.php#:~:text=Today%2C%20the%20United%20States%20is,and%20by%20ship%20as%20LNG (accessed 20 July 2023).

U.S. Energy Information Administration (2023f). Biomass explained. Basics. https://www.eia.gov/energyexplained/biomass (accessed 25 July 2023).

U.S. Environmental Protection Agency (2022). What is the Yucca Mountain repository? https://www.epa.gov/radiation/what-yucca-mountain-repository (accessed 24 July 2023).

U.S. Government Accountability Office (2023). Fusion energy: potentially transformative technology still faces fundamental challenges. GAO-23-105813. https://www.gao.gov/products/gao-23-105813#:~:text=A%202022%20experiment%20 achieved%20a,it%20can%20produce%20 commercial%20electricity (accessed 25 July 2023).

Yucca http://Mountain.org (2022). About Yucca Mountain. https://www.yuccamountain.org/faq.htm#status (accessed 24 July 2023).

4

Environmental Policy and Regulatory Considerations for Hydrogen Energy

*Marybeth Reynolds**

4.1 Introduction

In the face of the growing climate changes, it has become clear that the world must pursue clean, renewable, low-carbon energy sources to reduce greenhouse gas emissions. Nations around the world are setting goals for decarbonization and transitioning to net-zero emissions. To reach those goals, governments are exploring energy and industrial strategies that can help reduce or eliminate the use of fossil fuels and other carbon-based resources.

The use of hydrogen as an alternative fuel source has been researched for some time, but its use has been somewhat limited. Hydrogen has enormous potential as an abundant, non-extractive energy resource that emits no climate-warming greenhouse gas pollution. Now, the urgent need for sustainable, climate-friendly solutions is driving widespread interest and investment – including significant support from the United States federal government – into expanding the role of hydrogen in energy systems, industrial applications, and transportation.

The United States has set a goal of reducing its greenhouse gas emissions to 50% below 2005 levels by 2030 and reaching net-zero emissions by 2050 (The White House 2021). As part of a series of efforts to reach that goal, the United States has stepped up plans to bring hydrogen-based solutions out of the research phase and into widespread implementation. It is investing significant resources in projects designed to drive down the cost of clean hydrogen and to build out the infrastructure necessary to support its use.

In conjunction with these strategies to develop hydrogen technologies and processes, the federal government must also establish comprehensive regulatory frameworks to manage the new and expanded ways hydrogen will be deployed throughout the energy and industrial sectors. Policies should ensure these new technologies meet strict environmental, public health, and safety standards so that the investments of time, effort, and resources result in real, measurable benefits to the economy and the climate.

This chapter will explore the potential environmental benefits of hydrogen energy, how hydrogen fits into existing U.S. energy policy, the priorities and challenges the U.S. government faces in

* Technical Editor and Digital Content Provider.

Hydrogen Energy: Principles and Applications, First Edition. Vincent J. DelGatto, Louis Theodore, R. Ryan Dupont, and Matthew C. Ogwu.

ensuring that hydrogen is an effective part of a net-zero energy landscape, and global policies to promote a clean hydrogen market.

Section titles for this chapter on environmental policy and regulatory considerations for hydrogen energy are presented below:

4.2 Opportunities and Benefits for the Emerging Hydrogen Energy Industry

Environmental challenges and opportunities exist on both the supply and demand sides of the emerging hydrogen economy. Hydrogen offers tantalizing benefits as a flexible, zero-emitting fuel, but its physical characteristics also present some technical challenges. Hydrogen has the highest energy content by weight of common fuels – three times higher than gasoline – but is less dense than other fuels. It is highly transportable and can be stored for long periods of time once pressurized, though new technologies need to be developed to store and transport it more efficiently (NREL 2023).

When hydrogen is burned, the only significant by-product is water vapor. Thus, hydrogen has the potential to help reduce greenhouse gas emissions dramatically when used as a feedstock in chemical and industrial processes, in the transportation sector, and in the delivery of clean energy (U.S. DOE 2023a).

4.2.1 The Production of Hydrogen

Hydrogen is an energy carrier, like electricity, not an energy source, that exists abundantly in its pure form. Hydrogen itself must be produced by isolating it from other elements in compounds that contain it using feedstocks such as water, fossil fuels, or biomass.

There are multiple methods currently available for hydrogen production from a variety of domestic sources. Many of these processes are energy-intensive and rely on fossil fuels – or electricity generated from fossil fuels – as an energy source, with greenhouse gases emitted during production. Along with several emerging production methods for hydrogen being investigated, common mature processes include:

- *Natural gas reforming.* This method uses thermal processes to produce hydrogen and carbon dioxide from the methane in natural gas. Steam methane reforming (SMR), requiring high-temperature steam, is the most common method. Natural gas reforming in large, centralized facilities currently accounts for 95% of hydrogen production in the United States (U.S. DOE 2023b). Carbon emissions are produced as a direct result of the production process, as well as in the upstream production of the natural gas and steam required.

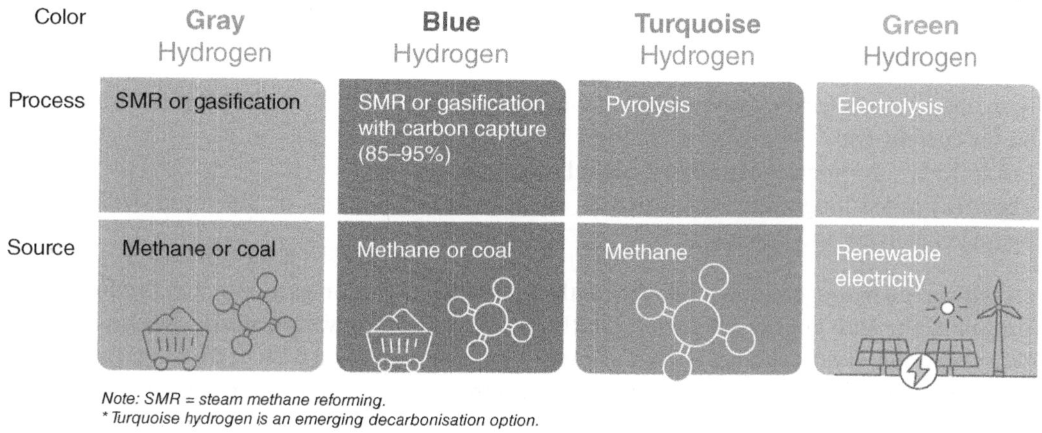

Note: SMR = steam methane reforming.
** Turquoise hydrogen is an emerging decarbonisation option.*

Figure 4.1 Selected shades of hydrogen (IRENA 2020).

- *Gasification.* This method involves the conversion of coal or biomass into a mixture of mainly hydrogen, carbon monoxide, and carbon dioxide. In addition to direct carbon emissions, there are emissions associated with feedstock processing and transport, especially where coal is used.
- *Electrolysis.* This method involves the splitting of water into hydrogen and oxygen using electricity. No greenhouse gases are emitted during the process. However, there are upstream emissions if the electricity used is generated from fossil fuels, which most currently is.

A commonly used but loosely defined system for categorizing hydrogen by its energy source and production process is by color code (Figure 4.1).

- *Gray hydrogen* makes up the largest category of hydrogen currently available. It refers to hydrogen produced using fossil fuels where the resulting greenhouse gas emissions are released into the atmosphere. Gray hydrogen is most commonly produced via SMR or coal gasification.
- *Blue hydrogen* is produced using similar methods as gray hydrogen, but the resulting carbon dioxide emissions are mitigated through carbon capture, utilization, and storage (CCUS). Blue hydrogen is sometimes referred to as "low carbon" hydrogen for this reason.
- *Green hydrogen* refers to hydrogen produced through electrolysis using energy solely from renewable sources that emit no greenhouse gases. It is the only true "clean" hydrogen. Green hydrogen currently accounts for only about 1% of global hydrogen production (IRENA 2023).

Additional and emerging "colors" of hydrogen include:

- *Turquoise hydrogen* made by splitting methane into hydrogen and solid carbon.
- *Yellow hydrogen* produced by electrolysis using grid electricity from a variety of sources.
- *Pink hydrogen* produced via electrolysis using nuclear power.
- *Black or brown hydrogen* made with coal or lignite through the process of gasification.

4.2.2 Clean, Green Zero-Carbon Hydrogen

Clean, green hydrogen provides the most environmental benefit because it releases no climate-warming emissions. There are no direct emissions from the production process or upstream emissions from the electricity produced using renewable energy. The widespread production of clean hydrogen is an ambitious goal, with some challenges to overcome before a hydrogen energy future can be fully realized including:

- Zero-emission electricity from renewable sources has a high relative cost and limited availability on the current grid
- Electrolyzer units require significant capital investment
- There is limited long-route infrastructure, such as fueling stations or pipelines, dedicated to hydrogen delivery and transport

Overall, green hydrogen production currently costs three to five times as much as production of hydrogen in the blue or gray categories. (Hiltbrand et al. 2021) However, steady decreases in the cost of renewable energy sources like wind and solar power – coupled with decreasing costs of the electrolyzer equipment – are expected to drive down the cost of green hydrogen and make it more widely available and attractive to produce (IRENA 2020).

Another factor likely to increase green hydrogen production is the potential for it to be integrated with intermittent power generation from renewable resources. Renewable energy sources can be variable, i.e., solar power on a cloudy day, resulting in periods of either low or excess electricity generation. Rather than pausing electricity generation when there is excess supply, the excess could be shifted to the production of hydrogen to make the most efficient use of the available resources (U.S. DOE 2023c). Likewise, the hydrogen can be stored for use at times when power generation is low. This synergy offers a powerful incentive that could boost the production of both green hydrogen and renewable energy while stabilizing the energy grid.

4.2.3 Low-Carbon Blue Hydrogen

As the pursuit of cost-effective green hydrogen continues, there is likely to be substantial attention paid to blue hydrogen as a "bridge" fuel in the interim. The technology and facilities to produce blue hydrogen may be achievable sooner because existing infrastructure could be retrofitted with CCUS capabilities.

However, over-investment in blue-hydrogen production – processes that still rely on fossil fuels – could hinder emissions reductions and delay the transition to a sustainable energy system. For blue hydrogen to have any significant impact on lowering emissions, CCUS must be proven technically effective and economically feasible. Estimates suggest that about 85–95% of carbon emissions would be captured and stored or reused. CCUS above 95% would be costly (IRENA 2020).

In addition, the fact that blue hydrogen production utilizes natural gas carries the inherent risk of upstream methane leaks where the gas is extracted. Methane is itself a dangerous greenhouse gas, with 84 times the global warming potential of carbon dioxide over a 20-year period (U.S. EPA 2023). Bringing its own environmental risks and costs, blue hydrogen production must be evaluated carefully before being adopted as a short-term compromise in the quest for net-zero energy.

4.2.4 Fuel Cells

Hydrogen has been employed in fuel cell technology for decades. A notable early example was in providing on-board electricity and water supplies on the 1969 Apollo 11 mission (NASM 2023). Using a chemical catalyst, a fuel cell (see also Chapter 6) separates hydrogen (or other fuel's) molecules into protons and electrons. The electrons follow a pathway to an external circuit, generating a flow of electricity. When pure hydrogen is used as the fuel, the only by-products are water and heat.

The amount of power generated by a fuel cell varies depending on its size and other factors, and single fuel cells can be combined in series into a "stack" to increase power output. Hydrogen fuel cells can be used flexibly and in a wide variety of applications, including:

- To supply electricity to the power grid
- As backup or emergency power in buildings
- To power fuel cell electric vehicles
- As large portable supplies of electricity for places that are not connected to the power grid, and as portable sources of electricity for small applications, such as laptops and cell phones (U.S. EIA 2022)

4.2.5 Hydrogen's Potential Uses in Decarbonization

The United States currently produces approximately 10 million metric tons (MMT) of hydrogen each year, mainly for use as a feedstock for crude oil refining, in the production of ammonia and other chemicals, and in steel and cement production (U.S. DOE 2023a). But hydrogen's versatility as an "enabling" technology means it could be a critical part of a wide array of industrial, transportation, and utility applications as these sectors shift toward decarbonization. Such applications include:

- *Energy storage and grid stability*. As noted above, hydrogen can be used flexibly along with intermittent renewable energy sources to meet end-use energy demands and to enable long-term energy storage.
- *Power generation and blending with natural gas*. There is the potential for hydrogen to be blended with natural gas to take advantage of existing delivery and distribution infrastructure. In addition, hydrogen can be used in power generation through gas turbines and fuel cell technology (U.S. DOE 2020a).
- *Industrial and chemical processes*. Hydrogen serves as a feedstock in a range of chemical and industrial processes. It is used to manufacture ammonia for fertilizers, in the production of methanol for secondary chemical applications, in steelmaking and cement production, and in petrochemical and refinery processing. There is significant potential for clean or low-carbon hydrogen to be used in place of current hydrogen feedstocks that are produced from fossil fuels.
- *Transportation*. Hydrogen fuel cell vehicles (FCEVs), including heavy-duty trucks, buses, and commercial vehicles, have been in use for some time and the market for them is growing. There is increasing interest in fuel cell-enabled off-road functions where electrification is less feasible, such as in rail transportation and maritime applications. In addition, hydrogen and hydrogen-derived synthetic fuels are being investigated for use in aviation (IEA 2022).

4.2.6 Challenges

Several challenges remain in bringing clean hydrogen into more widespread use are described below:

- *Cost and competitiveness*. The high cost of renewable energy sources relative to fossil fuels and of production equipment are major barriers to the adoption of clean hydrogen. Both producers and end-users will be reluctant to enter the market when lower-cost alternatives to hydrogen exist. These costs are expected to decrease, however, as electrolyzer manufacturing increases and the cost of renewable energy steadily decreases.

- *Infrastructure for storage and transportation.* Hydrogen's low density requires pressurization for it to be easily stored and transported. Very little infrastructure dedicated to hydrogen is currently in place, and it will need to be built before hydrogen can be deployed in any significant way. There is an ongoing investigation into whether hydrogen blended with natural gas can use existing pipelines and equipment, but questions remain about the long-term feasibility of this option due to the likelihood of leaks and pipe embrittlement over time (IEA 2022).
- *Inefficiencies.* Energy loss occurs throughout the clean hydrogen supply chain, from electrolysis to transport to use in fuel cells, thus requiring incrementally more renewable energy to produce it. Capacity for renewable energy production will need to increase to accommodate both direct end-use demand and the production of clean hydrogen (IRENA 2020).

4.3 Hydrogen Energy Policy Priorities

As technology advances and opportunities are emerging in various sectors, this is a critical time for responsible and forward-thinking policies and regulations to take shape regarding hydrogen implementation. The current boom in investment and research lends itself to a "free for all" environment, where, without adequate oversight and guidelines, efforts and focus could be misplaced onto technologies and uses that are ineffective or are merely replacing current polluting streams with new ones. A foremost priority for policymakers is to ensure that the dispatch of new hydrogen technologies emphasizes climate and environmental goals. In order to guarantee that these priorities are adopted, the following recommendations are explained in further detail in the next several subsections:

- Keep the focus on climate goals and deploy hydrogen strategically
- Prioritize equity and public health
- Consider long-term efficiency and cost
- Adopt rigorous standards and definitions for clean hydrogen

4.3.1 Keep the Focus on Climate Goals and Deploy Hydrogen Strategically

When considering where to invest money and resources across a range of emerging technologies, policymakers must maintain a focus on the overarching goal of lowering climate-warming greenhouse gas emissions, and not on expediting hydrogen technology for its own sake. It is important to recognize that hydrogen is one of several options for decarbonization.

Hydrogen will not serve as an effective replacement for fossil fuels in the near term in most sectors, and even in the long term, other available solutions will also be required alongside hydrogen. Energy efficiency is already one of the most cost-effective and attainable strategies for reducing energy use and, therefore, greenhouse gas emissions. Direct electrification is also a shorter-term and more viable alternative to hydrogen for many applications, and its potential for decarbonization is growing rapidly as electricity generation from renewables steadily increases.

The deployment of green hydrogen should be focused on hard-to-decarbonize applications where practical alternatives are not available. Sectors such as ammonia production and refining already utilize hydrogen in significant quantities, but prioritizing their shift to clean hydrogen would provide direct benefits quickly and more cost-effectively than introducing a green hydrogen stream where no infrastructure already exists. These high-value applications offer the most policy support and overall return on investment. Figure 4.2 illustrates where green hydrogen can complement existing methods for decarbonization.

	Renewables	Direct electrification	Energy efficiency	Green hydrogen
Heating	• Solar water heaters, direct geothermal use, biomass (low-grade heating)	• Heat pumps	• Retrofit of buildings • Technological advancement	• High-grade heating
Industry	• Solar drying, biomass (productive uses)	• Electric industrial application (e.g., arc furnaces)	• Use of best available technologies	• Steelmaking refineries • Chemical industry
Land transport	• Biofuels	• Battery electric vehicles	• Performance standards • Travel avoidance • Engine design	• FCEVs
Shipping	• Biofuels • Wind energy	• Short-distance shipping	• Ship design • Operation optimisation • Travel avoidance	• Green ammonia • Methanol
Aviation	• Biojet fuels	• Short-distance aviation	• Plane design • Travel avoidance	• Hydrogen and synthetic fuels for aviation

Figure 4.2 Hydrogen is a complement to alternative ways to decarbonize end uses (IRENA 2020).

4.3.2 Prioritize Equity and Public Health

A focus on public health and equity works in tandem with the goals of lowering climate-harming greenhouse gas emissions. The effects of climate change are already disproportionately impacting populations that experience systemic environmental and economic inequities. The costs and benefits of the current U.S. energy system are inequitably distributed, with communities of color and low-income communities having lower access to technology and capital to benefit from clean energy deployment.

Fence-line communities can face long-term negative health impacts from some extractive and industrial-scale power and manufacturing facilities.

Policies that aim to accelerate the adoption of clean hydrogen must at the same time address historical and systemic inequities of the current energy system. Government investments must emphasize the interests and participation of marginalized and excluded communities, enabling access to the economic, environmental, and health benefits of the zero-emissions energy system. New or retrofitted production, storage, and distribution facilities must be designed to minimize the environmental and public health impacts to nearby populations to ensure that the hydrogen supply chain is not simply replacing the negative impacts of fossil fuel systems with new and different threats.

4.3.3 Consider Long-Term Efficiency and Cost

While green hydrogen solutions are not readily available for hard-to-decarbonize sectors, such as aviation, continued investments in research and development are critical to the eventual transition away from fossil fuels (IRENA 2020). However, with an abundance of funding available, there is a risk that existing power suppliers will push for investment in hydrogen-based applications that fit into their existing infrastructure but that may not significantly decrease climate-warming emissions. Further, hydrogen production itself incurs energy losses, multiplying the inefficiencies of these misplaced investments where better alternatives may be available (Liebreich 2020). Thus, policymakers must be wary of investing in hydrogen technologies that could raise the long-term costs of transitioning away from fossil fuels and delay the effort.

4.3.4 Adopt Rigorous Standards and Definitions for Clean Hydrogen

Hydrogen can only be a tool for reaching net-zero emissions if the hydrogen itself is produced from certifiably renewable resources. While the widely used term "green hydrogen" is generally accepted to refer to hydrogen produced with 100% renewable resources, there is no industry-wide standard or accounting system for the lifecycle greenhouse gas emissions of hydrogen. There are evolving "Guarantee of Origin" schemes in use in countries and localities around the world, though they vary widely in how they define and measure clean hydrogen.

Policymakers should take critical steps to establish a similar scheme in the United States to ensure end-users have reliable and transparent information about the sourcing of hydrogen. A scheme should be based on clean hydrogen with high emissions thresholds to consistently promote the lowest-emissions sources (United Nations 2021). It should include reporting on emissions from production to transport to end-use, with clear comparisons available for the lifecycle emissions of alternative production pathways (including fossil fuel-based production) and transparency into the systems and processes involved.

4.4 U.S. Federal Energy Policies and Regulatory Frameworks

4.4.1 Hydrogen in Historical U.S. Energy Policy

Policies related to hydrogen use in the United States have been shaped historically by federally funded research and incentives. There have not been broad federal strategies to shape hydrogen production or use, but rather only smaller-scale programs. Early federal initiatives funded research focused mainly on hydrogen use in the transportation sector. Spurred largely by rising oil prices and a dependence on imported supplies, the government supported the development of electric and hybrid vehicle technologies, including research into fuel cells. Subsequent legislation throughout the 1990s and early 2000s provided continued frameworks and funding for the U.S. DOE's Hydrogen Program and research into hydrogen FCEVs (U.S. DOE 2020b). Table 4.1 provides a summary of key U.S. hydrogen-related legislation and policies through 2005.

The Energy Policy Act of 2005 was passed with a range of clean energy goals, and specifically addresses hydrogen and fuel cell technology as part of a "sustainable national energy economy" (U.S. Congress 2005). It served to continue funding for ongoing research and foster coordination with private industry. Beyond transportation, the law also sought to expand research into hydrogen's potential for use in the industrial, commercial, and utility sectors (Connolly 2006).

Federal policy around hydrogen was largely unchanged for more than a decade, though funding for research programs and initiatives continued. Between 2000 and 2020, DOE invested more than $4 billion in hydrogen-related projects including "hydrogen production from diverse domestic sources, hydrogen delivery and storage, and conversion technologies including fuel cells and

Table 4.1 Key hydrogen-related legislation and policies through 2005.

Legislation	Purpose
Electric and Hybrid Vehicle Research, Development, and Demonstration Act of 1976	To "promote basic and applied research . . . conduct demonstration programs, ascertain consumer needs and desires on electric and hybrid vehicles, and ascertain long-term changes in road design, urban planning, utility rate structures, and tax policies needed to manufacture and use such vehicles" (U.S. Congress 1976).
Spark M. Matsunaga Hydrogen, Research, Development and Demonstration Act of 1990	Calls for a "comprehensive five-year management plan for a research and development program designed to identify and address areas of research critical to the realization of a domestic hydrogen fuel production capability within the shortest time practicable." Also establishes a "Hydrogen Technical Advisory Panel . . . to implement a program in the interest of accelerating broader application of hydrogen technologies (U.S. Congress 1990).
Energy Policy Act of 1992	Calls for "oil and gas demand reduction and substitution," including concentration on fuel cells for transportation and renewable hydrogen energy systems and defines hydrogen as an "alternative fuel" (U.S. Congress 1992).
Hydrogen Future Act, 1996	To "conduct a research, development, and demonstration program leading to the production, storage, transport, and use of hydrogen for industrial, residential, transportation, and utility applications" (U.S. Congress 1996).
Energy Policy Act of 2005, including Title VIII (Spark M. Matsunaga Hydrogen Act of 2005)	To "enable and promote comprehensive development, demonstration, and commercialization of hydrogen and fuel cell technology in partnership with industry; . . . build a mature hydrogen economy that creates fuel diversity in the massive transportation sector of the United States; . . . to eliminate most emissions from the transportation sector, and . . . to create, strengthen, and protect a sustainable national energy economy" (U.S. Congress 2005).

turbines" (U.S. DOE 2020b). Even so, investments had remained far below what would be required to transform the energy landscape into one that relied heavily on hydrogen (Talus and Martin 2022).

4.4.2 Significant Federal Policies and Initiatives Since 2016

There has been renewed interest in hydrogen in the late 2010s and early 2020s brought on by several factors. In 2015, countries around the world committed to reducing climate-warming greenhouse gas emissions under the historic Paris Agreement. The global COVID-19 pandemic, supply chain disruptions, and geopolitical instability have led to rapid fluctuations in gas prices and fears about energy security. In the United States, several policy initiatives and new legislative efforts have been undertaken that invigorate the federal government's investment in hydrogen technologies as a secure, domestic, zero-emitting energy enabler.

4.4.2.1 H2@Scale, 2016

A steady rise in federal funding for hydrogen research followed the 2016 launch of the DOE's H2@Scale program. The initiative is intended to facilitate collaborative research projects among national laboratories, academic institutions, and industry to investigate and demonstrate how hydrogen can fit into applications across many parts of the interconnected energy system. The focus on partnerships

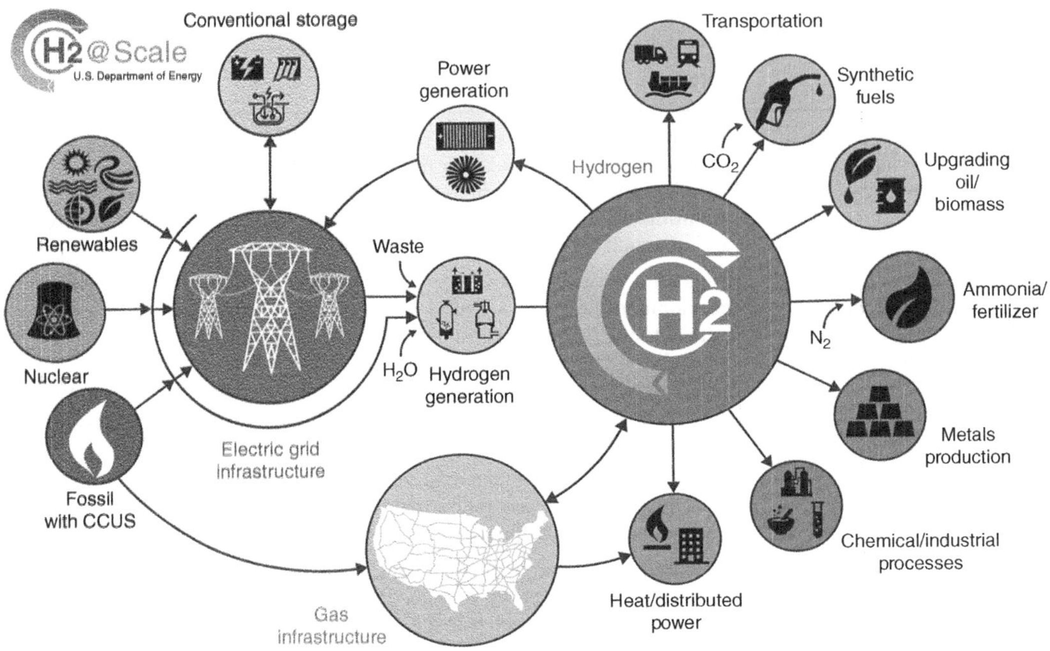

Figure 4.3 H2@Scale Concept (U.S. DOoE 2023d).

with the private sector is notable, aimed at moving research projects more easily into the demonstration phase within existing facilities and systems (U.S. DOE 2023d). Figure 4.3 shows the H2@Scale concept, illustrating how hydrogen can enable decarbonization across a range of sectors.

4.4.2.2 Hydrogen Program Plan, 2020

The DOE's Hydrogen Program Plan lays out a comprehensive, high-level view of research and initiatives across multiple industrial sectors, involving numerous offices within the DOE and in coordination with other federal agencies. The plan builds on existing strategies and demonstrates a renewed emphasis on the government's role in pushing proven hydrogen technologies to market by addressing key challenges and market barriers including (U.S. DOE 2020b):

- Technical challenges of cost, durability, reliability, and performance
- The lack of hydrogen infrastructure
- Lifecycle cost and environmental impact
- The need for harmonized codes and standards, and best practices for safety
- The need for a robust supply chain and workforce

4.4.2.3 Hydrogen Shot, 2021

The DOE Energy Earthshots Initiative was launched in 2021 to advance clean energy breakthroughs that will facilitate the nation's goal of net zero-carbon emissions by 2050. The first of these, the Hydrogen Energy Earthshot – or Hydrogen Shot – is aimed at reducing the cost of clean hydrogen by 80% to \$1/kg by 2030 (Hydrogen Shot, 2023). Aligned with the goals outlined in the agency's Hydrogen Program Plan and engaging an array of stakeholders, the Hydrogen Shot is again indicative of the government's objective to remove technical and market barriers to hydrogen technology's widescale implementation (U.S. DOE 2023e).

4.4.2.4 Bipartisan Infrastructure Law, 2021

The Infrastructure Investment and Jobs Act (H.R.3684, commonly referred to as the Bipartisan Infrastructure Law, or "BIL"), passed in 2021, was driven by a number of policy goals, with the broad aims of increasing U.S. competitiveness while making much-needed improvements to the nation's physical infrastructure. The law contains several provisions that address climate vulnerability, clean energy, and an upgraded power grid and includes investments of $9.5 billion into research and deployment of various initiatives focused on clean hydrogen.

Key components of the BIL include (U.S. DOE 2022a):

- It designates $8 billion for the DOE to develop Regional Clean Hydrogen Hubs. Locating hubs in various geographic regions allows for the use of diverse feedstocks in clean hydrogen production and makes clean hydrogen available for a wider variety of end uses. As such, hubs can generate their own circular economies and attract other segments of the supply chain to build nearby. In addition, regional hubs allow for wider, more equitable access to the economic, employment, and educational benefits of those economies.
- It provides $1 billion for a Clean Hydrogen Electrolysis Program to support research and deployment of more efficient and cost-effective electrolyzers and related technologies, with the goal of reducing the cost of clean hydrogen production.
- It allots $500 million for Clean Hydrogen Manufacturing and Recycling activities to support domestic manufacturing of clean hydrogen equipment, and to improve efficiency and cost-effectiveness of recycling and resource recovery from clean hydrogen technologies. This effort is intended to break down supply chain barriers and encourage participation of domestic manufacturers.
- It provides funding for a variety of alternative fuels, including hydrogen, through the following programs:
  - DOE National Alternative Fuels Corridors
  - DOT Port Infrastructure Development Program
  - DOT Emerging Alternative Fuel Vehicle (AFV) Study
  - DOT Low or Zero Emission Ferry Program
  - EPA Clean School Bus Program
- It calls for the DOE to produce a National Clean Hydrogen Strategy and Roadmap to be updated and refined every 3 years. Importantly, the Roadmap emphasizes how energy justice and equity fit into the larger hydrogen strategy. It calls out opportunities where the equitable removal of market barriers will necessarily address access and well-being of overburdened and underserved communities, and involve stakeholders from environmental justice groups, Tribal Nations, and labor unions.

4.4.2.5 Inflation Reduction Act, 2022

The Inflation Reduction Act is widely viewed as the U.S. federal government's most significant action to date to address climate change. It calls for investments of $370 billion across the economy to spur clean energy innovation and adoption. In conjunction with provisions from the BIL, it is estimated that the Inflation Reduction Act will reduce U.S. greenhouse gas emissions by 40% below 2005 levels by 2030 (U.S. DOE 2022b).

The law introduces a range of policies and incentives to boost a U.S. market for clean hydrogen specifically, and includes two significant tax credits (The White House 2023):

- The new Clean Hydrogen Production Tax Credit (45 V) for qualified U.S. clean hydrogen production facilities. By incentivizing domestic production, the credit is intended to make clean hydrogen more cost-competitive and thereby expand its deployment. The size of the credit depends on lifecycle greenhouse gas emission reductions and is applicable to the production of both blue and green hydrogen.

- An increased Carbon Sequestration Tax Credit (45Q). Originally created in the Energy Improvement and Extension Act of 2008, the Inflation Reduction Act increases the 45Q credit in the hopes of increasing the use of CCUS. It adds enhanced incentives for direct air capture of emissions and lowers the capture threshold requirement for some production facilities.

4.4.3 Current Federal Regulation of Hydrogen

In addition to the emergence of a broad national strategy for hydrogen, it is critical to consider the regulatory framework that will govern the hydrogen supply infrastructure as it expands. There are many federal entities with regulating authority across the production, transportation and distribution, storage, and end usage sectors of the nation's existing energy systems and for hydrogen in its current uses. There are also international, state, and local jurisdictions enforcing regulations from the International Fire Code to local building codes (Baird et al. 2021).

Hydrogen is regulated for its current usage in relatively small markets and as a by-product in industrial processes. It is addressed in discrete areas throughout the Code of Federal Regulations (CFR) and United States Code (USC), but there is not, as yet comprehensive regulation of the hydrogen industry. As more hydrogen-based processes are adopted in energy, transportation, and industrial sectors – and larger quantities of hydrogen are being produced, stored, and transported – agencies across the federal government will be tasked with filling in gaps in existing standards or engaging in the development of new ones. Further, programs that incentivize zero-emission hydrogen technologies may be subject to federal oversight that requires new and specific rules to be put into place (Baird et al. 2021). Figure 4.4 shows current regulatory bodies across various sectors of the hydrogen supply chain.

These various regulating authorities may need to revisit how hydrogen fits into their respective frameworks to keep pace with expanding adoption of hydrogen energy systems. For example:

- Agencies such as the Pipeline and Hazardous Materials Safety Administration (PHMSA) and Federal Energy Regulatory Commission (FERC) have regulatory authority over aspects of the existing energy infrastructure as it pertains to natural gas, such as natural gas pipelines and import/export terminals. While some regulations as written can be applied to hydrogen or other flammable gases in those same applications, others that are specific to natural gas would need to be modified.
- The Federal Railroad Administration (FRA) oversees standards that apply to locomotive safety designs, which would already cover hydrogen indirectly as an alternative fuel. It also regulates electrical systems generators and protection from hazardous gases from exhaust and batteries, but the standards are specific to combustion and battery venting and do not address hydrogen fuel cells (49 CFR Part 229).

Table 4.2 summarizes select regulations in the CFR and USC that address key sectors of the hydrogen supply chain, identifying which bodies have regulatory authority and whether a regulation in its current form applies specifically to hydrogen.

As new laws and policies go into effect, regulatory interpretation may be key to shaping their ultimate implementation and impact. For example, the Inflation Reduction Act includes only vague guidance defining the measurement of lifecycle greenhouse gas emissions in order to apply for the 45V tax credit. The language leaves room for the U.S. Department of the Treasury to issue standards that potentially differ from previous standards established by the EPA, and which do not apply

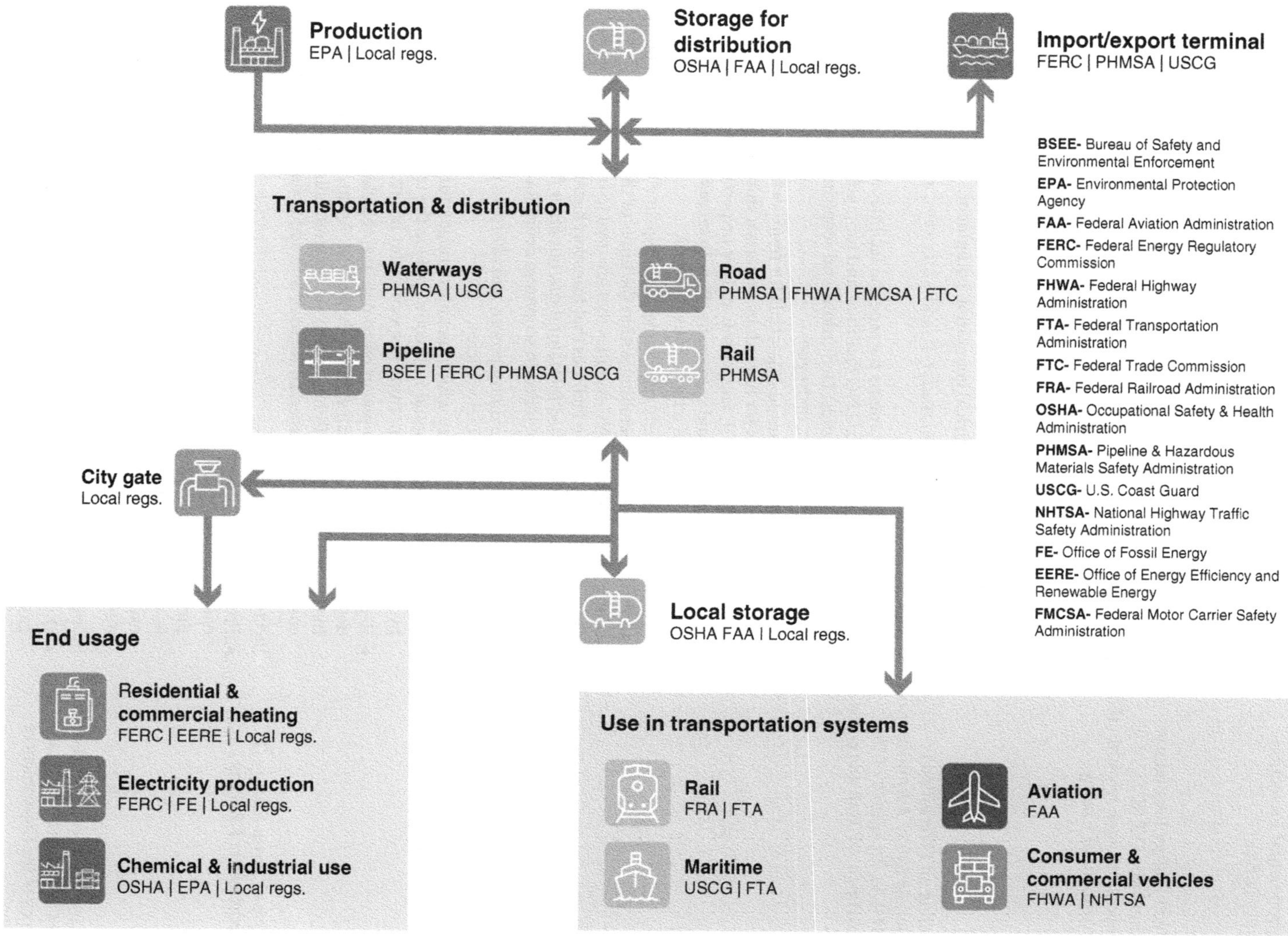

Figure 4.4 Hydrogen regulatory map (*Source:* Baird et al. (2021)/U.S. Department of Energy/Public Domain).

Table 4.2 Summary of select regulation and oversight of hydrogen systems.

Regulating Authority	Summary of Regulations and Applicability to Hydrogen
Environmental Protection Agency (EPA)	Hydrogen production and chemical/industrial use • Defines emissions thresholds for hydrogen production facilities. • Requires reporting of greenhouse gas emission due to combustion or use of products in a process, including production of gaseous hydrogen.
Federal Energy Regulatory Commission (FERC), independent agency within the Department of Energy	Transportation by pipeline, electricity production, import/export terminals, and residential and commercial heating • Regulates the siting and construction of facilities used for import or export of natural gas and the natural gas pipeline system. Provides regulation of energy sales and distribution of natural gas, specifically, but does not address hydrogen in this context. • Sets requirements for electricity production in a small power production or cogeneration facility. Hydrogen fuel cells are included in the definition of electrical generation equipment.
Pipeline and Hazardous Materials Safety Administration (PHMSA), an agency of the Department of Transportation	Transportation by pipeline, road, rail, and waterways; and import/export terminals • Prescribes safety requirements for pipeline facilities and the transportation of gas, for LNG facilities, and for pipelines that transport hazardous liquids. Definitions and scope include flammable gases such as hydrogen. • Lists and classifies hazardous materials for transportation and prescribes requirements for hazardous materials shipping and transporting by road, rail, and vessel, and for inspecting and maintaining packages and containers. Requirements cover the handling of hydrogen as a hazardous material. • Prescribes minimum safety requirements for pipeline facilities and the transportation of gas (including within the limits of the outer continental shelf), for LNG facilities that are used to transport gas via pipeline, and for pipeline facilities that transport hazardous liquids. Requirements are specified for natural gas, but other flammable gases, such as hydrogen, are included in scope and definition.
Federal Highway Administration (FHWA), an agency of the Department of Transportation	Transportation by road, auxiliary power, and alternative power supply, use in consumer/commercial vehicles • Regulates highway safety which includes bridges, tunnels, and other associated elements. Should indirectly apply to the transportation of hydrogen even though nothing is specified for hazardous materials. • Regulates additional equipment on commercial vehicles to ensure it does not reduce the overall safety of the vehicle. Again, hydrogen transport is not specifically addressed, but requirements are intended to ensure the safety of commercial vehicles with additional equipment.
Office of Fossil Energy (FE), Department of Energy	Electricity production • Prohibits any new baseload powerplant without the ability to use coal or another alternative fuel as a primary energy source and may prohibit existing power plants from using petroleum or natural gas as a primary energy source. Specified alternative fuels do not explicitly include hydrogen but note that fuels obtained from alternative fuel sources would be included.
Occupational Safety and Health Administration (OSHA), an agency of the Department of Labor	Storage and chemical and industrial use • Dictates the safety of the structural components and operations of gaseous and liquid hydrogen storage and delivery

Source: Adapted from Baird et al. (2021).

specifically to hydrogen. Stricter standards would bolster hydrogen production at the cleanest end of the spectrum, but fewer facilities would qualify to take advantage of the credit in the short term (Krupnick and Bergman 2022). Thus, regulators have a significant influence on the scope of current and future hydrogen energy policy, whether interpreting new laws or revising standards that apply to existing fuels and systems that hydrogen may ultimately complement or replace.

4.5 The Role of the States

States can develop their own sets of laws and regulations similar to those of the federal government. The Resource Conservation and Recovery Act (RCRA), for example, like most federal environmental legislation, encourages states to develop and run their own solid and hazardous waste programs as an alternative to EPA management. Thus, in a given state, the solid and hazardous waste regulatory program may be run by the EPA or by a state agency. For a state to have jurisdiction over its solid and hazardous waste program, it must receive approval from the EPA by showing that its program is at least as stringent as the EPA program.

States that are authorized to operate RCRA (or other regulatory) programs oversee the solid and hazardous waste tracking system in their state, operate the permitting system for waste facilities, and act as the enforcement arm in cases where an individual or a company practices illegal solid and hazardous waste management. If needed, the EPA steps in to assist the states in enforcing the law. The EPA can also act directly to enforce RCRA or other laws in states that do not yet have authorized programs. The EPA and the states currently act jointly to implement and enforce the regulations.

State agencies do not operate in a vacuum. Indeed, most trace their very existence to the passage of major federal laws. While federal environmental legislation, such as the Clean Air Act (CAA), for example, established national ambient air quality standards (NAAQS), it went on to direct that these standards be met by state implementation plans (SIPs). The National Permit Discharge Elimination System (NPDES), set up by the Clean Water Act, even anticipated a gradual handing off of control to states to administer their permitting systems once they evidenced adequate authority. Thus, a mix of federal and state statutory laws evolved along with the appropriate bureaucracies to implement them.

One major problem of having over 50 separate state environmental agencies is perhaps best illustrated by an initiative of the National Governor's Association (NGA 1997) to solve it. In 1996 the NGA recognized that ". . .the environmental technology industry has long been heralded as a key to enhanced environmental quality and accelerated economic development. . ." and that ". . . [t]his diverse industry develops technologies to prevent pollution, monitor, and control pollutant emissions, detect and measure contamination, and clean up environmental pollution. . ." It went on to point out, however, that ". . .developers and vendors of new environmental technologies are frustrated, in their efforts to penetrate the enormous U.S. market. . ." and that ". . .[m]any, observers place the blame at the door of the states, where a single technology may be *subject to 50 different procedures to gain a permit or approval. . .*". (emphasis added). Another problem is competition for authority in particular energy situations (i.e., whether federal law controls or state law prevails). Solutions are governed by the often-unclear doctrine of *preemption*. Under the doctrine of preemption, which is based on the Supremacy Clause of the U.S. Constitution, federal law preempts state law, even when the laws conflict. Thus, a federal court may require a state to stop certain behavior it believes interferes with, or conflicts with, federal law.

One can blame this patchwork quilt of laws, regulations, and standards on the notion of federalism on which the U.S. system of governance was built. Federalism is a sharing of power between the states and the national (federal) government. In 1787, the new federal government only exercised, limited, or enumerated powers granted to it by the Constitution, such as making treaties and printing money, The Tenth Amendment to the Constitution clarified that all other powers belonged to the states, i.e., "The powers not delegated to the United States by the Constitution, nor prohibited by it to the states, are reserved to the states respectively, or the people."

4.6 Global Hydrogen Energy Policies and Priorities

Global interest in hydrogen's capabilities has mirrored renewed efforts in the United States, driven largely by the climate crisis and in pursuit of energy security amidst geopolitical instability. Emissions reduction goals were established under the Paris Agreement developed at the United Nations Climate Change Conference (COP21) in 2015. Further, governments are committing to net-zero goals established as part of COP27 in 2022 (United Nations 2023).

4.6.1 Summary of Hydrogen Strategies in Key Global Markets

As of 2022, 26 countries had established national strategies focused on using hydrogen in their energy systems (IEA 2022). Research and policy activities of key nations and regions outside of the United States include the following (IRENA 2022):

- *China* does not have a national strategy but released its first hydrogen roadmap in 2016. It is the world's largest producer of hydrogen, mostly coal-based, but has undertaken more than 30 green hydrogen projects since 2019. It has a fleet of 8,400 FCEVs and leads the world in usage of fuel cell trucks and buses.
- *The European Union* established a hydrogen strategy in 2020 focused on renewable hydrogen and becoming a leader in clean hydrogen. It includes the goal of installing 40 gW of electrolyzers by 2030.
- *Japan* was the first nation to finalize a national hydrogen strategy in 2017 and is moving to incorporate hydrogen into all sectors of its economy. The government has invested heavily in hydrogen infrastructure, fuel cells, and fueling stations.
- *India* issued its national strategy in 2021, aiming to become a global hub for green hydrogen production and to achieve energy independence by 2047.
- *The Republic of Korea* released a hydrogen roadmap in 2019 with plans to use hydrogen to power 10% of its localities by 2030 and 30% by 2040. It emphasizes hydrogen as a key toward economic growth and aims to become a leader in production of large-scale stationary fuel cells and fuel cell electric vehicles.

4.6.2 Policy Priorities to Accelerate a Global Market for Clean Hydrogen

While different nations' economic circumstances, energy needs, and domestic capabilities vary widely, there are common priorities that can foster the development and growth of a global, interconnected hydrogen energy market.

4.6.2.1 Establishing Hydrogen Strategies

Some nations have moved forward with specific policies to implement their national hydrogen strategies, but there is not yet enough global movement to create demand and open viable markets. National hydrogen strategies are important indicators of a nation's commitment to hydrogen. They also signal support for private industry to help establish those markets and will play an important role in market development across the hydrogen supply chain (IRENA 2020).

In countries that have invested in hydrogen production, interest has diversified, shifting away from just the transportation sector, and committing to hydrogen strategies that include specific support for green hydrogen. This diversification is important to ensuring that hard-to-decarbonize sectors receive adequate support, where a hydrogen transition provides the highest environmental payoff (IRENA 2020).

4.6.2.2 Developing and Adopting International Codes and Regulations

As a global hydrogen energy market is established, standardization among nations will foster international projects and trade, promote knowledge-sharing, and enable more effective emissions reductions. Adopting common or complementary regulations and Guarantee of Origin schemes will ensure consistency and allow for the transparent comparison of lifecycle greenhouse gas emissions for clean hydrogen and common alternatives (IRENA 2020).

4.6.2.3 Leveraging Domestic Resources and Export Opportunities

A global energy system around clean hydrogen will be accelerated when nations take advantage of their individual resources and capacities and leverage those on the international market. By promoting domestic renewable resources and adapting infrastructure to the new hydrogen supply chain, nations can create opportunities for export and increase competitiveness in a hydrogen economy (IRENA 2020).

4.7 Summary

Hydrogen has enormous potential as an abundant, non-extractive energy resource that emits no climate-warming greenhouse gas pollution. Environmental challenges and opportunities exist on both the supply and demand sides of the emerging hydrogen economy. Hydrogen must be produced by isolating it from other elements in compounds that contain it including feedstocks such as water, fossil fuels, or biomass, using methods that may result in greenhouse gas emissions. A foremost priority for policymakers is to ensure that the dispatch of new hydrogen technologies emphasizes climate and environmental goals. The deployment of green hydrogen should be focused on hard-to-decarbonize applications where practicable alternatives are not available.

In the United States, a number of policy initiatives and new legislative efforts have been undertaken since 2016 that invigorate the federal governments' investment in hydrogen technologies. Regulators have significant influence on the scope of current and future hydrogen energy policy, whether interpreting new law or revising standards that apply to existing fuels and systems that hydrogen may ultimately complement or replace. While different nations' economic circumstances, energy needs, and domestic capabilities vary widely, there are common priorities that can foster the development and growth of a global, interconnected hydrogen energy market.

References

Baird, A.R., Ehrhart, B.D., Glover, A.M., and LaFleur, C.B. (2021). *Federal Oversight of Hydrogen Systems*. U.S. Department of Energy, Office of Energy Efficiency and Renewable Energy (EERE), Fuel Cell Technologies Program. https://doi.org/10.2172/1773235.

Connolly, M.J. (2006). *Developments in hydrogen production, technology and use under the energy policy act of 2005*. Mondaq. https://www.mondaq.com/unitedstates/chemicals/36954/developments-in-hydrogen-production-technology-and-use-under-the-energy-policy-act-of-2005 (accessed 4 February 2023).

Hiltbrand, G., Herndon, W., O'Rear, E.G., and Larsen, J. (2021). *Clean Hydrogen: A Versatile Tool for Decarbonization*. Rhodium Group. https://rhg.com/research/clean-hydrogen-decarbonization (accessed 15 February 2023).

International Energy Agency (2022). *Global Hydrogen Review 2022*. IEA. https://iea.blob.core.windows.net/assets/c5bc75b1-9e4d-460d-9056-6e8e626a11c4/GlobalHydrogenReview2022.pdf (accessed 3 February 2023).

International Renewable Energy Agency (2020). *Green Hydrogen: A Guide to Policy Making*. Abu Dhabi: IRENA. https://www.irena.org/-/media/Files/IRENA/Agency/Publication/2020/Nov/IRENA_Green_hydrogen_policy_2020.pdf (accessed 15 January 2023).

International Renewable Energy Agency (2022). *Geopolitics of the Energy Transformation: The Hydrogen Factor*. Abu Dhabi: IRENA. https://www.irena.org/-/media/Files/IRENA/Agency/Publication/2022/Jan/IRENA_Geopolitics_Hydrogen_2022.pdf?rev=1cfe49eee979409686f101ce24ffd71a (accessed 10 February 2023).

International Renewable Energy Agency (2023). *Hydrogen: Overview*. Abu Dhabi: IRENA. https://www.irena.org/Energy-Transition/Technology/Hydrogen (accessed 3 February 2023).

Krupnick, A. and Bergman, A. (2022). *Resource for the future. Incentives for clean hydrogen production in the inflation reduction act. Report 22–13*. https://media.rff.org/documents/Report_22-13_UNfLJLS.pdf.

Liebreich, M. (2020). *Separating hype from hydrogen – part two: the demand side*. BloombergNEF. https://about.bnef.com/blog/liebreich-separating-hype-from-hydrogen-part-two-the-demand-side (accessed 24 January 2023).

National Air & Space Museum, Smithsonian (2023). Fuel cell, Apollo. https://airandspace.si.edu/collection-objects/fuel-cell-apollo/nasm_A19730934000 (accessed 5 February 2023).

National Governor's Association (1997). *Interstate Cooperation to Speed Multistate Acceptance of Environmental Technologies*. Washington, DC: Natural Resources Policy Studies Division.

National Renewable Energy Laboratory (2023). *Hydrogen basics*. https://www.nrel.gov/research/eds-hydrogen.html (accessed 16 January 2023).

Talus, K. and Martin, M. (2022). A guide to hydrogen legislation in the USA: a renewed effort. *Journal of World Energy Law and Business* 15(6): 449–461. https://doi.org/10.1093/jwelb/jwac029.

The White House (2021). *The long-term strategy of the united states: pathways to net-zero greenhouse gas emissions by 2050*. https://www.whitehouse.gov/wp-content/uploads/2021/10/US-Long-Term-Strategy.pdf (accessed 14 February 2023).

The White House (2023). *Building a Clean energy economy: a guidebook to the inflation reduction act's investments in clean energy and climate action*. Version 2. https://www.whitehouse.gov/wp-content/uploads/2022/12/Inflation-Reduction-Act-Guidebook.pdf (accessed 6 August 2023).

U.S. Congress (1976). *Electric vehicle research, development, and demonstration act, 1976*. https://www.congress.gov/bill/94th-congress/house-bill/8800 (accessed 5 August 2023).

U.S. Congress (1990). *Spark M. Matsunaga hydrogen research, development, and demonstration act, 1990*. https://www.congress.gov/bill/101st-congress/senate-bill/639 (accessed 5 August 2023).

U.S. Congress (1992). *Energy policy act of 1992*. https://www.congress.gov/bill/102nd-congress/house-bill/776 (accessed 5 August 2023).

U.S. Congress (1996). *Hydrogen future act of 1996*. https://www.congress.gov/bill/104th-congress/house-bill/4138 (accessed 5 August 2023).

U.S. Congress (2005). *Energy policy act of 2005*. https://www.congress.gov/bill/109th-congress/house-bill/6 (accessed 5 August 2023).

U.S. Department of Energy (2020a). *Hydrogen strategy: enabling a low-carbon economy*. Office of Fossil Energy. https://www.energy.gov/sites/prod/files/2020/07/f76/USDOE_FE_Hydrogen_Strategy_July2020.pdf (accessed 28 January 2023).

U.S. Department of Energy (2020b). *Hydrogen program plan.* https://www.hydrogen.energy.gov/pdfs/hydrogen-program-plan-2020.pdf (accessed 6 February 2023).

U.S. Department of Energy (2022a). *DOE national clean hydrogen strategy and roadmap.* https://www.hydrogen.energy.gov/pdfs/clean-hydrogen-strategy-roadmap.pdf (accessed 22 January 2023).

U.S. Department of Energy (2022b). *The inflation reduction act drives significant emissions reductions and positions america to reach our climate goals.* https://www.energy.gov/sites/default/files/2022-08/8.18%20InflationReductionAct_Factsheet_Final.pdf (accessed 25 January 2023).

U.S. Department of Energy (2023a). *Hydrogen production.* Office of Energy Efficiency and Renewable Energy (EERE), Hydrogen and Fuel Cell Technologies Office. https://www.energy.gov/eere/fuelcells/hydrogen-production (accessed 16 January 2023).

U.S. Department of Energy (2023b). *Hydrogen production: natural gas reforming.* Office of Energy Efficiency and Renewable Energy (EERE), Hydrogen and Fuel Cell Technologies Office. https://www.energy.gov/eere/fuelcells/hydrogen-production-natural-gas-reforming (accessed 25 January 2023).

U.S. Department of Energy (2023c). *Hydrogen production: electrolysis.* Office of Energy Efficiency and Renewable Energy (EERE), Hydrogen and Fuel Cell Technologies Office. https://www.energy.gov/eere/fuelcells/hydrogen-production-electrolysis (accessed 22 January 2023).

U.S. Department of Energy (2023d). *H2@Scale: enabling affordable, reliable, clean, and secure energy across sectors.* Office of Energy Efficiency and Renewable Energy (EERE). https://www.energy.gov/eere/fuelcells/articles/h2scale-handout (accessed 5 February 2023).

U.S. Department of Energy (2023e). *Hydrogen shot.* Office of Energy Efficiency and Renewable Energy (EERE), Hydrogen and Fuel Cell Technologies Office. https://www.energy.gov/eere/fuelcells/hydrogen-shot (accessed 17 January 2023).

U.S. Energy Information Administration (2022). *Hydrogen explained. use of hydrogen.* https://www.eia.gov/energyexplained/hydrogen/use-of-hydrogen.php (accessed 18 January 2023).

U.S. Environmental Protection Agency (2023). *Understanding global warming potentials.* https://www.epa.gov/ghgemissions/understanding-global-warming-potentials (accessed 27 January 2023).

United Nations (2021). *Guiding principles for climate-aligned hydrogen deployment.* United Nations Framework Convention on Climate Change, Global Climate Action, Marrakech Partnership. https://racetozero.unfccc.int/wp-content/uploads/2021/10/Hydrogen-Guiding-Principles_vFinal.pdf (accessed 19 January 2023).

United Nations (2023). *The Paris agreement.* United Nationsl Climate Action. https://www.un.org/en/climatechange/paris-agreement (accessed 5 February 2023).

5

Thermodynamic Considerations

5.1 Introduction

Thermodynamics is the study of energy change from one state to another. The predictions that can be made using thermodynamic equations are essential for understanding fuel cells, a topic to be addressed in the next chapter. Thermodynamics was once defined as "the science that deals with the inter-transformation of heat and work." The fundamental principles of thermodynamics are contained in the first, second, and third laws of thermodynamics. These principles have been defined as "pure" or "theoretical" thermodynamics. These laws were developed and extensively tested in the latter half of the 19th century and are essentially based on experience. The third law was developed later in the 20th century.

Practically all thermodynamics, in the ordinary meaning of the term, is "applied thermodynamics" in that it is essentially the application of these three laws, coupled with certain facts and principles of mathematics, physics, and chemistry, to problems in engineering and science. The fundamental laws are of such generality that it is not surprising that these laws find application in other disciplines, including physics, chemistry, plus environmental, chemical and mechanical engineering.

The first law of thermodynamics is a conservation law for energy transformations. Regardless of the types of energy involved in the process – thermal, mechanical, electrical, elastic, magnetic, etc. The change in energy of a steady-state system is equal to the difference between energy input and energy output. The first law also allows free "convertibility" from one form of energy to another, as long as the overall quantity of energy is conserved. Thus, this law does not restrict the conversion of work into heat, or its counterpart – the conversion of heat into work.

Because work is 100% convertible to heat, whereas the reverse situation is *not* true, work is a more valuable form of energy than heat. This leads to an important second-law consideration – That is, energy has *quality* as well as *quantity*. Although it is not as obvious, it can also be shown through second-law principles and arguments that heat has quality in terms of its temperature. The higher the temperature at which heat transfer occurs, the greater the potential for energy transformation into work. Thus, thermal energy stored at higher temperatures is generally more useful to society than that available at lower temperatures. While there is an immense quantity of energy stored in the oceans and the Earth's core, for example, its present availability to society for performing useful tasks is essentially nonexistent. Theodore, Ricci, and VanVliet (2009) provide an additional qualitative review of the second law.

Hydrogen Energy: Principles and Applications, First Edition. Vincent J. DelGatto, Louis Theodore, R. Ryan Dupont, and Matthew C. Ogwu.

The choice of topics to be reviewed in this chapter was initially an area of debate, and after some deliberation, it was decided to introduce nine areas that many have included in this broad engineering subject. Each is discussed in the Sections detailed below:

5.1 Introduction
5.2 Energy Fundamentals and Principles
5.3 The First Law of Thermodynamics
5.4 Enthalpy Effects
5.5 Second Law Calculations
5.6 Phase Equilibrium
5.7 Stoichiometry
5.8 Chemical Reaction Equilibrium
5.9 Conservation Laws
5.10 Ideal Gas Law
References

The reader should note that there is an overlap between the topics of the first law of thermodynamics and the conservation laws as they are both related to the conservation law for energy. Also, it should be noted that the bulk of the material in this chapter has been drawn from Theodore and Reynolds (1991).

5.2 Energy Fundamentals and Principles

Energy can be defined as the ability to produce change or do work. Energy can produce light, heat, motion, and sound. In effect, it produces growth and powers technology.

There are basically two categories of energy – potential and kinetic. Some examples of these types of energy are detailed below

5.2.1 Potential Energy

This is stored energy and the energy of position or gravitational potential energy. There are several forms of potential energy, including:

1) *Chemical energy*. Energy stored in the bonds of atoms and molecules. It is the energy that holds these particles together. Foods we eat, biomass, petroleum, natural gas, and propane are examples of materials that contain stored chemical energy.
2) *Elastic energy*. Energy stored in objects by the application of a force. Compressed springs and stretched rubber bands are examples of elastic energy.
3) *Nuclear energy*. Energy stored in the nucleus of an atom – the energy that binds the nucleus together.
4) *Gravitational potential energy*. Energy of location or place.

5.2.2 Kinetic Energy

This is the energy of motion, for example, waves, electrons, atoms, molecules, substances, and objects. There are several forms of kinetic energy, including:

1) *Electrical energy*. Energy related to the movement of electrons is called electricity. Lightning is an example of electrical energy.

2) *Radiant energy or electromagnetic energy*. Energy travels in transverse waves. Radiant energy includes visible light, X-rays, gamma rays, and radio waves. Solar energy is an example of radiant energy.
3) *Thermal energy*. Energy, which is often described as heat, is the internal energy in substances, i.e., the vibration and movement of atoms and molecules within substances.
4) *Motion energy or mechanical energy*. Energy related to the movement of objects and substances from one place to another. Wind is an example of motion energy.
5) *Sound energy*. Energy related to the movement of longitudinal (compression/rarefaction) waves through substances.

5.2.3 Energy Fundamentals

Seven energy fundamentals are provided below:

1) Energy is a physical quantity that follows precise natural laws.
2) Physical processes on Earth are the result of energy flow through the Earth's system.
3) Biological processes depend on energy flow through the Earth's system.
4) Various sources of energy can be used to power human activities, and often this energy must be transferred from source to destination.
5) Energy decisions are influenced by economic, political, environmental, and social factors.
6) The amount of energy used by human society depends on many factors.
7) The quality of life of individuals and societies is affected by energy choices.

Principles associated with each of the fundamentals follow.

5.2.4 Energy Principles

Fundamental Statement 1 – Energy is a physical quantity that follows precise natural laws.

1.1 Energy is a quantity that is transferred from system to system.
1.2 The energy of a system or object that results in its temperature is called thermal energy.
1.3 Energy is neither created nor destroyed.
1.4 Energy available to do useful work decreases as it is transferred from system to system.
1.5 Energy comes in different forms and can be divided into categories.
1.6 Chemical and nuclear reactions involve the transfer and transformation of energy.
1.7 Many different units are used to quantify energy.
1.8 Power is a measure of energy transfer rate.

Fundamental Statement 2 – Physical processes on Earth are the result of energy flow through the Earth's system.

2.1 Earth is constantly changing as energy flows through the system.
2.2 Sunlight, gravitational potential, decay of radioactive isotopes, and rotation of the Earth are the major sources of energy driving physical processes on Earth.
2.3 Earth's weather and climate are mostly driven by energy from the Sun.
2.4 Water plays a major role in the storage and transfer of energy in the Earth system.
2.5 Movement of matter between reservoirs is driven by Earth's internal and external sources of energy.

2.6 Greenhouse gases affect energy flow through the Earth's system.
2.7 The effects of changes in Earth's energy system are often not immediately apparent.

Fundamental Statement 3 – Biological processes depend on energy flow through the Earth's system.

3.1 The Sun is the major source of energy for organisms and the ecosystems of which they are a part.
3.2 Food is a biofuel used by organisms to acquire energy for internal living processes.
3.3 Energy available to do useful work decreases as it is transferred from organism to organism.
3.4 Energy flows through food webs in one direction, from producers to consumers and decomposers.
3.5 Ecosystems are affected by changes in the availability of energy and matter.
3.6 Humans are part of Earth's ecosystems and influence energy flow through these systems.

Fundamental Statement 4 – Various sources of energy can be used to power human activities, and often this energy must be transferred from source to destination

4.1 Humans transfer and transform energy from the environment into forms useful for human endeavors.
4.2 Human use of energy is subject to limits and constraints.
4.3 Fossil fuels and biofuels are organic matter that contains energy captured from sunlight.
4.4 Humans transport energy from place to place.
4.5 Humans generate electricity in multiple ways.
4.6 Humans intentionally store energy for later use in a number of different ways.
4.7 Different sources of energy and the different ways energy can be transferred, transported, and stored each have different benefits and drawbacks.

Fundamental Statement 5 – Energy decisions are influenced by economic, political, environmental, and social factors.

5.1 Decisions concerning the use of energy resources are made at many levels.
5.2 Energy infrastructure has inertia.
5.3 Energy decisions can be made using a systems-based approach.
5.4 Energy decisions are influenced by economic factors.
5.5 Energy decisions are influenced by political factors.
5.6 Energy decisions are influenced by environmental factors.
5.7 Energy decisions are influenced by social factors.

Fundamental Statements 6 – The amount of energy used by human society depends on many factors.

6.1 Conservation of energy has two very different meanings – the physical law and the reduction of energy consumption.
6.2 One way to manage energy resources is through conservation.
6.3 Human demand for energy is increasing.
6.4 Earth has limited energy resources.

6.5 Social and technological innovation affects the amount of energy used by human society.
6.6 Behavior and design affect the amount of energy used by human society.
6.7 Products and services carry with them embedded energy.
6.8 The amount of energy used can be calculated and monitored.

Fundamental Statement 7 – The quality of life of individuals and societies is affected by energy choices.

7.1 Economic security is impacted by energy choices.
7.2 National security is impacted by energy choices.
7.3 Environmental quality is impacted by energy choices.
7.4 Increasing demand for and limited supplies of fossil fuels affect the quality of life.
7.5 Access to energy resources affects the quality of life.
7.6 Some populations are more vulnerable to the impacts of energy choices than others.

5.3 The First Law of Thermodynamics

For many environmental processes, the energy requirement represents a major item of the cost of operation, and one cannot arrive at a proper systems analysis and/or economic evaluation without performing an energy balance as well as a material balance. Just as practicing engineers rely on the law of conservation of mass for a material balance, they depend on the law of the conservation of energy for energy balance calculations.

These energy balance considerations are based on thermodynamics, the branch of science founded on laws of experience that deal with both energy and its conversion, as well as the transfer of energy in terms of heat and work as a system passes from one equilibrium state to another. Joule's experiments cleared the way for the enunciation of the first law of thermodynamics; namely, when a closed system is taken through a cyclic process, the work done on the surroundings equals the heat absorbed from the surroundings. This law for batch processes can be represented as:

$$\Delta E = Q + W \tag{5.1}$$

where potential, kinetic, and other energy effects have been neglected and E (often denoted as U) is the internal energy of the system, ΔE is the change in the internal energy of the system, Q is energy in the form of heat transferred across the system boundaries, and W is energy in the form of work transferred across system boundaries. In accordance with a recent change in convention, both Q and W are treated as positive terms if added to the system.

For practical purposes, the total work term, W, in the first law may be regarded as the sum of shaft work, W_s, and flow work, W_f

$$W = W_s + W_f \tag{5.2}$$

where W_s is work done on the fluid by some moving solid part within the system such as the vanes of a centrifugal pump. Note that in Equation 5.2, all other forms of work such as electrical, surface tension, and so on are neglected. The first law of thermodynamics for steady-state flow processes is then:

$$\Delta H = Q + W_s \tag{5.3}$$

where H is the enthalpy of the system and ΔH is the change in the system's enthalpy.

The internal energy and enthalpy in Equations 5.1 and 5.2, as well as other equations in this section may be on a mass basis, on a mole basis, or represent the total internal energy and enthalpy of the entire system. They may also be written on a time-rate basis if these equations are dimensionally consistent. For the sake of clarity, uppercase letters (e.g., H, E) represent properties on a mole basis, while lowercase letters (e.g., h, e) represent properties on a mass basis. Properties for the entire system will rarely be used and therefore require no special symbols.

Perhaps the most important thermodynamic function the engineer works with enthalpy. This is a term that requires additional discussion. Enthalpy is defined by the equation

$$H = E + PV \tag{5.4}$$

where P is once again the pressure of the system and V is the volume of the system. The terms E and H are state or point functions. By fixing a certain number of variables upon which the function depends, the numerical value of the function is automatically fixed; that is, it is single-valued. For example, fixing the temperature and pressure of a one-component single-phase system immediately specifies its enthalpy and internal energy.

5.4 Enthalpy Effects

There are many different types of enthalpy effects. These include:

1) Sensible (temperature)
2) Latent (phase)
3) Dilution (with water), e.g., HCl with H_2O
4) Solution (nonaqueous), e.g., HC1 with a solvent other than H_2O
5) Reaction (chemical)

This section is only concerned with Effects 1, 2, and 5. Details on Effects 3 through 4 are available in the literature (Theodore and Reynolds 1991; Smith, Van Ness, and Abbott 2001; Theodore, Ricci, and VanVliet 2009)

5.4.1 Sensible Enthalpy Effects of Heating

Sensible enthalpy effects are associated with temperature. There are methods that can be employed to calculate these changes. These methods include the use of:

1) enthalpy values
2) average heat capacity values
3) heat capacity as a function of temperature

If enthalpy values are available, the enthalpy change is given by

$$\Delta h = h_2 - h_1; \text{ mass basis} \tag{5.5}$$

$$\Delta H = H_2 - H_1; \text{ mole basis} \tag{5.6}$$

If average molar heat capacity data are available,

$$\Delta H = \bar{C}_p \Delta T \tag{5.7}$$

where $\bar{C}_p$ = average molar value of C_p in the temperature range ΔT. Average molar heat capacity data are provided in the literature (Theodore and Reynolds 1991; Pitzer 1995; Smith, Van Ness, and Abbott 2001; Perry and Green 2008; Theodore, Ricci, and VanVliet 2009).

A more rigorous approach to enthalpy calculations can be provided if the heat capacity variation with temperature is available. If the heat capacity is a function of the temperature, the enthalpy change is written in differential form:

$$dH = C_p dT \tag{5.8}$$

If the temperature variation of the heat capacity is given by

$$C_p = \alpha + \beta T + \gamma T^2 \tag{5.9}$$

Equation 5.8 may be integrated directly between some reference or standard temperature (T_o) and the final temperature (T_1) employing Equation 5.9:

$$\Delta H = H_1 - H_o \tag{5.10}$$

$$\Delta H = \alpha(T_1 - T_o) + (\beta/2)(T_1^2 - T_o^2) + (\gamma/3)(T_1^3 - T_o^3) \tag{5.11}$$

Equation 5.8 may also be integrated if the heat capacity is a function of temperature of the form:

$$C_p = a + bT + cT^{-2} \tag{5.12}$$

The enthalpy change is then given by

$$\Delta H = a(T_1 - T_o) + (b/2)(T_1^2 - T_o^2) + c(T_1^{-1} - T_o^{-1}) \tag{5.13}$$

Tabulated values of α, β, γ, and a, b, c for a host of compounds (including some chlorinated organics) are available in the literature (Theodore and Reynolds 1991).

5.4.2 Latent Enthalpy Changes

It has been observed that there is absorption of heat at constant temperature and pressure that accompanies the transition or equilibrium phase change from solid to liquid and liquid to gas. In terms of molecular theory, this latent enthalpy represents the energy required to overcome intermolecular forces of attraction and to permit molecules to pass from a more highly restrained condensed phase to a more mobile phase. This relationship indicates that the latent heat is the sum of a change in internal energy and energy of expansion (or contraction).

Common types of transition and the terminology applied to them are:

Type of transition:	Designation of energy change:
solid → liquid	enthalpy of fusion
liquid → gas	enthalpy of vaporization
solid → vapor	enthalpy of sublimation
solid → solid	enthalpy of transition

The last case represents a change from one crystalline modification to another. The enthalpy change involved is small compared with those accompanying the other three types and rarely finds application in energy management studies.

Since there is a relatively small difference in volume between a solid and the liquid to which it melts, the heat of fusion represents mainly an increase in internal energy. If the molar heat of fusion is not known, it may be approximated from the absolute temperature of fusion, T_f, using the following equation:

$$\Delta H_f / T_f = 2-3 \text{ for solid elements; } 5-7 \text{ for inorganic solids; } 9-11 \text{ for organic solids} \quad (5.14)$$

The ratio may be expressed in any consistent molar units, such as cal/(gmol-K) or Btu/(lbmol-°R).

Since there is a very substantial increase in volume in passing from the liquid to the vapor state, usually from a few hundred to over a thousand-fold, the heat of vaporization is large and important in many energy applications involving water. Information on properties of water is available in steam tables in the literature (Theodore and Reynolds 1991; Theodore, Ricci, and VanVliet 2009).

5.4.3 Chemical Reaction Enthalpy Effects

The standard enthalpy (heat) of reduction can be calculated from the standard enthalpy of formation data. To simplify the presentation that follows, examine the authors' favorite equation:

$$aA + bB = cC + dD \quad (5.15)$$

If the above reaction is assumed to occur at a standard (or reference) state, the standard enthalpy of reaction, ΔH°, is given by:

$$\Delta H^{\circ} = c\left(\Delta H_f^{\circ}\right)_C + d\left(\Delta H_f^{\circ}\right)_D - a\left(\Delta H_r^{\circ}\right)_A - b\left(\Delta H_f^{\circ}\right)_B \quad (5.16)$$

where $(\Delta H_f^{\circ})_i$ = standard enthalpy of formation of species i.

Thus, the (standard) enthalpy of a reaction is obtained by taking the difference between the (standard) enthalpy of formation of products and reactants multiplied by their respective stoichiometric coefficients. If the (standard) enthalpy of reaction or formation is negative (exothermic), as is the case with most combustion reactions, then energy is liberated due to the chemical reaction. Energy is absorbed and ΔH° is positive endothermic reaction.

Tables of enthalpies of formation and reaction are available in the literature (particularly thermodynamics text/reference books) for a wide variety of compounds (Theodore, Ricci, and VanVliet 2009). It is important to note that these are valueless unless the stoichiometric equation and the state of the reactants and products are included.

Theodore, Ricci, and VanVliet (2009) provide equations to describe the effect of temperature on the enthalpy of reaction. For heat capacity data in α, β, γ form:

$$\Delta H_T^{\circ} = \Delta H_{298}^{\circ} + \Delta\alpha\left(T-298\right) + \frac{1}{2}\Delta\beta\left(T^2-298^2\right) + \frac{1}{3}\Delta\gamma\left(T^3-298^3\right) \quad (5.17)$$

For the reaction presented in Equation 5.15:

$$\Delta\alpha = c\alpha_c + d\alpha_D - a\alpha_A - b\alpha_B \quad (5.18)$$

$$\Delta\beta = c\beta_c + d\beta_D - a\beta_A - b\beta_B \tag{5.19}$$

$$\Delta\gamma = c\gamma_c + d\gamma_D - a\gamma_A - b\gamma_B \tag{5.20}$$

For heat capacity in *a*, *b*, *c* form,

$$\Delta H_T^\circ = \Delta H_{298}^\circ + \Delta a(T - 298) + \frac{1}{2}\Delta b\left(T^2 - 298^2\right) + \Delta c\left(T^{-1} - 298^{-1}\right) \tag{5.21}$$

5.5 Second Law Calculations

The law of conservation of energy has already been defined as the first law of thermodynamics. Its application allows calculations of energy relationships associated with a wide variety of processes. The "limiting" law is called the second law of thermodynamics. Applications involve calculation of the maximum power output from a power plant and equilibrium yields of chemical reactions. In principle, this law states that water cannot flow uphill and heat cannot flow from a cold to a hot body of its own accord. Other defining statements for this law that have appeared in the literature are provided below (Theodore and Reynolds 1991):

1) Any process, the sole net result of which is the transfer of heat from a lower temperature level to a higher one, is impossible.
2) No apparatus, equipment, or process can operate in such a way that its only effect (on system and surroundings) is to convert heat absorbed completely into work.
3) It is impossible to convert the heat taken into a system completely into work in a cyclical process.

The second law also serves to define another important thermodynamic function called entropy. It is normally designated as *S*. The change in *S* for a reversible adiabatic process is always zero:

$$\Delta S = 0 \tag{5.22}$$

For liquids and solids, the entropy change for a system undergoing an absolute temperature change from T_1 to T_2 is given by:

$$\Delta S = C_p \ln\left(T_2/T_1\right);\ C_p = \text{constant} \tag{5.23}$$

The entropy change of an ideal gas undergoing a physical change of state from P_1 to P_2 at a constant temperature T is given by:

$$\Delta S_T = R\ln\left(P_1 / P_2\right);\ \text{Btu/lbmol} - {}^\circ\text{R} \tag{5.24}$$

The entropy change of 1 mol of an ideal gas undergoing a physical change of state from absolute temperatures T_1 to T_2 at a constant pressure is given by:

$$\Delta S_p = C_p \ln\left(T_2/T_1\right);\ C_p\left(gas\right) = \text{constant} \tag{5.25}$$

Correspondingly, the entropy change for an ideal gas undergoing a physical change from (P_1, T_1) to (P_2, T_2) is

$$\Delta S = R\ln\left(P_1/P_2\right) + C_p \ln\left(T_2/T_1\right) \tag{5.26}$$

Some fundamental facts relative to the entropy concept are discussed below. The entropy change of a system may be positive (+), negative (−), or zero (0); the entropy change of the surroundings during this process may likewise be positive, negative, or zero. However, note that the total entropy change, ΔS must be equal to or greater than zero:

$$\Delta S \geq 0 \tag{5.27}$$

The equality sign applies if the change occurs *reversibly* and *adiabatically*.

The third law of thermodynamics is concerned with the absolute values of entropy. By definition, the entropy of all pure crystalline materials at absolute zero temperature is exactly zero. Note, however, that chemical engineers and applied scientists are usually concerned with changes in thermodynamic properties, including entropy, rather than their absolute values.

5.6 Phase Equilibrium

Relationships governing the equilibrium distribution of a substance between two phases, particularly gas and liquid phases, are the principal subject matter of phase-equilibrium thermodynamics. These relationships form the basis of calculational procedures that are employed in the design and prediction of the performance of several unit operation equipment and processes involving mass transfer.

The most important equilibrium phase relationship is that between a liquid and a vapor. Raoult's and Henry's Laws theoretically describe liquid–vapor behavior and, under certain conditions, are applicable in practice. Raoult's Law is sometimes useful for mixtures of components of similar structures. It states that the partial pressure of any component in the vapor phase is equal to the product of the vapor pressure of the pure component and the mole fraction of that component in the liquid:

$$p_i = p_i' x_i \tag{5.28}$$

where p_i = partial pressure of component i in the vapor, p_i' = vapor pressure of pure component i at the same temperature, and x_i = mole fraction of component i in the liquid (Theodore and Ricci 2010; Theodore 2014).

This expression may be applied to all components. If the gas phase is ideal, this equation becomes:

$$y_i = \left(p_i'/P\right) x_i \tag{5.29}$$

where y_i = mole fraction of component i in the vapor, and P = total system pressure.

Unfortunately, relatively few mixtures follow Raoult's law, but it can be applied to estimating the vapor pressure of individual compounds above a hydrocarbon mixture, i.e., benzene above gasoline, in underground storage tanks and soil and groundwater remediation applications.

Henry's law is a more empirical relation used for representing data for many environmental engineering systems where the liquid phase is water. Henry's law is expressed as:

$$p_i = H_i x_i \tag{5.30}$$

where H_i = Henry's law constant for component i (in units of pressure). If the gas behaves ideally, the above equation may be written as:

$$y_i = K_H x_i \tag{5.31}$$

where K_H = dimensionless Henry's law constant (Theodore and Ricci 2010; Theodore 2014).

In some environmental engineering applications, mixtures of condensable vapors and non-condensable gases must be handled. A common example is water vapor and air; a mixture of organic vapors and air is another such example that often appears in air pollution applications. Condensers can be used to control organic emissions to the atmosphere by lowering the temperature of the gaseous stream, although an increase in pressure will produce the same result. The calculation for this is often accomplished using the phase equilibrium constant, K. This constant has also been referred to in the chemical industry as a phase componential split factor since it provides the ratio of the mole fractions of a component in the two equilibrium phases. The defining equation is

$$K_i = y_i / x_i \tag{5.32}$$

where K_i = phase equilibrium constant for component i (dimensionless).

As a first approximation, K_i is generally treated as a function only of the temperature and pressure. For ideal gas conditions, K_i may be approximated by

$$K_i = p_i' / P \tag{5.33}$$

where p_i' is the vapor pressure of component i, and P is the total system pressure.

Many of the phase equilibrium calculations involve hydrocarbons. Fortunately, most hydrocarbons approach ideal gas behavior over a fairly wide range of temperatures and pressures. Values for K_i for a large number of hydrocarbons are provided in two DePriester nomographs, which are available in the literature. These two nomographs were originally developed by DePriester in 1953 (DePriester 1953).

DePriester charts are a valuable source of vapor–liquid equilibrium data for many hydrocarbons that approach ideal behavior. However, it should be noted that the DePriester charts are based on experimental data. The fact that these compounds approach ideal behavior allows the data to be presented in a simple form, i.e., as a function solely of temperature and pressure.

5.7 Stoichiometry

Stoichiometry is that part of chemistry that deals with macroscopic weight relationships in chemical reactions and the calculations based on them. When chemicals react, they do so according to a strict proportion. When oxygen and hydrogen are combined to form water, the ratio of the amount of oxygen to the amount of hydrogen consumed is always 7.94 by mass and 0.500 by moles. The term *stoichiometry* refers to this phenomenon, which is sometimes called the chemical law of combining weights. The reaction equation for the combining of hydrogen and oxygen is:

$$2H_2 + O_2 \rightarrow 2H_2O \tag{5.34}$$

In chemical reactions, atoms are neither generated nor consumed, merely rearranged with different bonding partners. The manipulation of the coefficients of a reaction equation so that the number of atoms of each element on the left is equal to that on the right is referred to as balancing the equation. Once an equation is balanced, the small whole-number molar ratio that must exist between any two components of the reaction can be determined simply by observation: these are known as stoichiometric ratios. There are three such ratios (not counting the reciprocals) in the previous reaction (equation 5.3.4). These are the following:

2 mol H_2 consumed/mol O_2 consumed
1 mol H_2O generated/mol H_2 consumed
2 mol H_2O generated/mol O_2 consumed

The unit mol represents either the gmol or the lbmol. Using molecular weights, these stoichiometric ratios (which are molar ratios) may easily be converted to mass ratios. For example, the first ratio may be converted to a mass ratio by using the molecular weights of H_2 (2.016) and O_2 (31.999) as follows:

(2 gmol H_2 consumed) (2.016 g/gmol) = 4.032 g H_2 consumed
(1 gmol O_2 consumed) (31.999 g/gmol) = 31.999 g O_2 consumed

The mass ratio between the hydrogen and oxygen consumed is therefore
4.032/31.999 = 0.126 g H_2 consumed/g O_2 consumed

Molar and mass ratios are used in material balances to find the amounts or flow rates of components involved in chemical reactions.

Multiplying a balanced reaction equation through a constant does nothing to alter its meaning. The reaction shown is often written

$$H_2 + ½O_2 \rightarrow H_2O \tag{5.35}$$

There are times, however, when care must be exercised because the solution to the problem depends on the way the reaction equation is written. This is the case with chemical equilibrium problems and problems involving thermochemical reaction equations.

5.8 Chemical Reaction Equilibrium

With regard to chemical reactions, two important questions are of concern to the engineer: (1) how far will the reaction go and (2) how fast will the reaction go? Chemical thermodynamics provides the answer to the first question; however, it tells nothing about the second. Reaction rates fall within the domain of chemical reaction kinetics. To illustrate the difference and importance of both questions in engineering analysis of chemical reactions, consider the following process: Substance A, which costs 1 cent/ton, can be converted to Substance B, which is worth 1 million dollars/mg, by the reaction A → B. Chemical thermodynamics will provide information on the maximum amount of Substance B that can be formed. If 99.99% of Substance A can be converted to Substance B, the reaction would then appear to be highly economically feasible, from a thermodynamic point of view. However, a kinetic analysis might indicate that the reaction is so slow that, for all practical purposes, its rate is vanishingly small. For example, it might take 10^6 years to obtain 10^{-6}% conversion of Substance A. The reaction is then economically unfeasible. Thus, it can be seen that both equilibrium and kinetic effects must be considered in an overall engineering analysis of a chemical reaction (Theodore 2012).

A rigorous, detailed presentation of this equilibrium topic is beyond the scope of this chapter and this text. However, some discussion is presented to provide at least a qualitative introduction to chemical reaction equilibrium. If a chemical reaction is carried out in which reactants go-to products, the products will be formed at a rate governed (in part) by the concentration of the reactants and conditions such as temperature and pressure. Eventually, as the reactants form products and the products react to form reactants, the net rate of reaction must equal zero. At this point, equilibrium will have been achieved.

Chemical reaction equilibrium calculations are structured around a thermodynamic term referred to as free energy, G. This so-called Gibbs free energy is a thermodynamic property that represents the maximum useful work (excluding PV work associated with volume changes of the system) that a system can do on the surroundings when the process occurs reversibly at constant temperature and pressure. Note that free energy has the same units as enthalpy and may be used on a mole or total mass basis.

Consider the equilibrium reaction presented earlier as Equation 5.15:

$$aA + bB = cC + dD \tag{5.15}$$

For this reaction

$$\Delta G^\circ = c\left(\Delta G_f^\circ\right)_C + d\left(\Delta G_f^\circ\right)_D - a\left(\Delta G_r^\circ\right)_A - b\left(\Delta G_f^\circ\right)_B \tag{5.36}$$

The standard free energy of reaction ΔG° may thus be calculated from standard free energy of formation data in a manner similar to that for the standard enthalpy of reaction. The following equation is used to calculate the chemical reaction equilibrium constant K at a temperature T:

$$\Delta G_T^\circ = -RT\ln\left(K\right) \tag{5.37}$$

The Gibb's free energy, ΔG, for a chemical reaction, provides information related to the work capabilities of the process. For ideal conditions, ΔG represents the *maximum* amount of the work under so-called perfect reversible conditions (Theodore and Reynolds 1991; Pitzer 1995; Smith, Van Ness, and Abbott 2001; Theodore, Ricci, and VanVliet 2009).

For example, consider the combustion reaction of hydrogen with oxygen to produce water in the liquid state. For this reaction at standard conditions

$$H_2 + ½\,O_2 \rightarrow H_2O;\ \Delta H^\circ = 286\,\text{kJ/gmol},\ \Delta G 237\,\text{kJ/gmol} \tag{5.38}$$

Thus, the energy "lost" due to inefficiencies appears as heat, Q, which is simply the difference between ΔH° and ΔG°

$$Q = \Delta H^\circ - \Delta G^\circ = 286 - 237 = 49\,\text{kJ/gmol} \tag{5.39}$$

If the process involves a fuel cell (see next chapter), 237 kJ/gmol would represent the *electricity* produced.

5.9 Conservation Laws

This section examines and reviews some of the fundamental principles that engineers and applied scientists employ in performing design calculations and predicting the performances of energy equipment. Topics include the conservation laws for both mass and energy. The ideal gas law receives treatment in the next section. These basic principles will assist the reader in acquiring a better understanding of some of the material that is later in this book.

5.9.1 Conservation of Mass

The conservation law for mass can be applied to any process or system. The general form of this law is given in Equation 5.40.

$$\text{Mass in} - \text{Mass out} + \text{Mass generated} = \text{Mass accumulated} \tag{5.40}$$

Equation 5.40 may be applied to the total mass, or a particular species involved in a particular system either on a mole or mass basis. The conservation law for mass can be applied to steady-state or unsteady-state processes and batch or continuous systems. To isolate a system for study, it is separated from its surroundings by constructing a control boundary or envelope around it. This control boundary may be real (e.g., the walls of a vessel) or imaginary. Mass crossing the boundary and entering the system is part of the mass in terms of Equation 5.40, whereas leaving the system is part of the mass-out term. Equation 5.40 may be written for any compound whose quantity is not changed by chemical reaction, or for any chemical element, regardless of whether it has participated in a chemical reaction or not. It may be written for one piece of equipment, around several pieces of equipment, or around an entire process. It may be used to calculate an unknown quantity directly, to check the validity of experimental data, or to express one or more of the independent relationships among the unknown quantities in a particular problem (Reynolds 1990).

A steady-state process is one in which there is no change in conditions (temperature, pressure, etc.) or rates of flow with time at any given point in the system. Under steady-state conditions, the accumulation term in Equation 5.40 is zero. If there is no chemical reaction, the generation term is also zero. All other processes are classified as unsteady-state. In a batch process, the container holds the product or products. In a continuous process, reactants are continuously fed to a piece of equipment or to several pieces in series, and products are continuously removed from one or more points. A continuous process may or may not be operating at steady-state. As indicated previously, Equation 5.40 may be applied to the total mass of each stream (referred to as an overall or total material balance) or to the individual components of the streams (referred to as a componential or component material balance). Often, the primary task in preparing a material balance is to develop the quantitative relationships among the inflow and outflow streams (Reynolds 1990).

The conservation law for mass also finds application in performing resource utilization and pollution prevention audits. A *waste minimization opportunity assessment* is a planned procedure to identify ways to reduce or eliminate waste. The procedure consists first of a careful review of a plant's operation and waste streams and the selection of specific areas to assess. After a specific waste stream or area is established as the focus of the resource utilization and pollution prevention audit, several options that can potentially minimize materials use and waste production are developed and screened. The selected options are then evaluated for technical and economic feasibility. Finally, the most promising options are selected for implementation (Dupont, Ganesan, and Theodore 2017).

5.9.2 Conservation of Energy

A system may possess energy as a result of its temperature, velocity, position, molecular structure, surface, and so on. The energies corresponding to these conditions are internal, kinetic, potential, chemical, surface, and so on. Energy, like mass and momentum, is conserved. Application of the conservation law for energy gives rise to the first law of thermodynamics. This law, in steady-state form for batch processes, is presented here once again (potential, kinetic, and other energy effects are not considered).

$$\Delta E = Q + W \tag{5.1}$$

For a flow process,

$$\Delta H = Q + W_s \tag{5.3}$$

Most industrial facilities operate in a steady-state flow mode. If no significant mechanical or shaft work is added or withdrawn from the system, Equation 5.3 reduces to

$$Q = \Delta H \tag{5.41}$$

If a unit or system is operated adiabatically, where $Q = 0$, Equation 5.41 then becomes

$$\Delta H = 0 \tag{5.42}$$

Although the topics of material and energy balances are covered separately in this section, it should be emphasized that this segregation does not occur in reality; one almost always works with both energy and material balances simultaneously.

The terms E and H are state or point functions. By fixing a certain number of variables on which the function depends, the numerical value of the function is automatically fixed; that is, it is single-valued. For example, fixing the temperature and pressure of a one-component, single-phase system immediately specifies the enthalpy and internal energy of the system. The change in enthalpy as the system undergoes a change in state from (T_1, P_1) to (T_2, P_2) is given by

$$\Delta H = H_2 - H_1 \tag{5.43}$$

The correlations needed to calculate the values of enthalpy were presented earlier; rather, the reader is directed to the literature for additional information or online, for example, at UC Davis ChemWiki, http://chemwiki.ucdavis.edu/Reference.

5.10 Ideal Gas Law

The *ideal gas law* was derived from experiments in which the effects of pressure and temperature on gaseous volumes were measured over a moderate range of temperatures and pressures. As a general rule, this law works best when the molecules of the gas are far apart, that is, when the pressure is low, and the temperature is high. Under these conditions, the gas is said to behave ideally. For engineering calculations, the ideal gas law is almost always assumed to be valid since it generally works well for the temperature and pressure ranges used in most engineering and energy applications.

The two precursors of the ideal gas law are Boyle's law and Charles's law. Boyle found that the volume of a given mass of gas is inversely proportional to the absolute pressure if the temperature is kept constant; that is,

$$P_1V_1 = P_2V_2 \tag{5.44}$$

where V_1 and V_2 are the volumes of gas at absolute pressures P_1 and P_2, respectively. Charles found that the volume of a given mass of gas varies directly with the absolute temperature at constant pressure:

$$V_1/T_1 = V_2/T_2 \tag{5.45}$$

where V_1 and V_2 are the volumes of gas at absolute temperatures T_1 and T_2, respectively. Boyle's and Charles's laws may be combined into a single equation in which neither temperature nor pressure need be held constant:

$$P_1V_1/T_1 = P_2V_2/T_2 \tag{5.46}$$

This equation indicates that for a given mass of a specific gas, PV/T has a constant value. Since, at the same temperature and pressure, volume, and mass must be directly proportional, this statement may be extended to

$$PV/nT = R \tag{5.47}$$

where n is the number of moles and R is the universal gas constant. Equation 5.47 is called the *Ideal Gas Law*. Numerically, the value of R depends on the units of P, V, T, and n. Tabulated values of R are provided in Table 5.1.

Table 5.1 Values of R in various units.

R	Units of P	Units of V	Units of n	Units of T
10.73	psia	ft^3	lbmol	°R
0.7302	atm	ft^3	lbmol	°R
21.85	in Hg	ft^3	lbmol	°R
555.0	mm Hg	ft^3	lbmol	°R
297.0	in H_2O	ft^3	lbmol	°R
0.7398	bar	ft^3	lbmol	°R
1,545.0	psfa	ft^3	lbmol	°R
24.75	ft H_2O	ft^3	lbmol	°R
1.9872	—	—	lbmol	°R
0.0007805	—	—	lbmol	°R
0.0005819	—	—	lbmol	°R
500.7	—	—	lbmol	°R
1.314	atm	ft^3	lbmol	K
998.9	mm Hg	ft^3	lbmol	K
19.32	psia	ft^3	lbmol	K
62.361	mm Hg	L	gmol	K
0.08205	atm	L	gmol	K
0.08314	bar	L	gmol	K
8,314	Pa	L	gmol	K
8.314	Pa	m^3	gmol	K
82.057	atm	cm^3	gmol	K
1.9872	—	—	gmol	K
8.3144	—	—	gmol	K

References

DePriester, C. (1953). Light-hydrocarbon vapor-liquid distribution coefficients. *Chemical Engineering Progress Symposium Series* 49(7): 1–43.

Dupont, R.R., Ganesan, K., and Theodore, L. (2017). *Pollution Prevention: Sustainability, Industrial Ecology, and Green Engineering*, 2nd Edition. Boca Raton, FL: CRC Press/Taylor & Frances Group.

Perry, R. and Green, D. ed. (2008). *Perry's Chemical Engineers' Handbook*, 8th Edition. New York: McGraw-Hill.

Pitzer, K. (1995). *Thermodynamics*, 3rd Edition. New York: McGraw-Hill.

Reynolds, J. (1990). *Material and Energy Balances. A Theodore Tutorial*. East Williston, NY: Theodore Tutorials. Originally published by the USEPA/APTI, RTP, NC.

Smith, J., Van Ness, H., and Abbott, M. (2001). *Chemical Engineering Thermodynamics*, 6th Edition. New York: McGraw-Hill.

Theodore, L. (2012). *Chemical Reactor Analysis and Applications for the Practicing Engineer*. Hoboken, NJ: John Wiley & Sons.

Theodore, L. (2014). *Chemical Engineering: The Essential Reference*. New York: McGraw-Hill.

Theodore, L. and Reynolds, J. (1991). *Thermodynamics, A Theodore Tutorial*. East Williston, NY: Theodore Tutorials. Originally published by the USEPA/APTI, RTP, NC.

Theodore, L. and Ricci, F. (2010). *Mass Transfer Operatives for the Practicing Engineer*. Hoboken, NJ: John Wiley & Sons.

Theodore, L., Ricci, F., and VanVliet, T. (2009). *Thermodynamics for the Practicing Engineer*. Hoboken, NJ: John Wiley & Sons.

6

Fuel Cells

6.1 Introduction

Interest in hydrogen and fuel cells has increased dramatically in recent years. The main reason is that hydrogen may be an answer to the two major challenges facing the future global economy: climate change and the security of energy supplies. Both these challenges call for the development of new, highly efficient technologies that are either carbon neutral or low emitting technologies. High-temperature fuel cells, such as molten carbonate (MCFC) and solid oxide fuel cells (SOFC), are promising for the conversion of a fuel's chemical energy into electricity, attaining significantly higher efficiencies compared to similarly sized energy conversion devices such as gas turbines and internal combustion engines. Furthermore, these kinds of fuel cells produce high-temperature "waste" heat that can be used for cogeneration, which may be used to produce hydrogen for other uses (Leal and Brower 2005). Thus, the high efficiency and lower pollutant emission features of fuel cells compared to other technologies make them an attractive technology for energy generation.

When hydrogen is consumed in a fuel cell, it produces only water. Water electrolysis is the reverse of the fuel cell reaction, i.e., water dissociates into hydrogen and oxygen. Hydrogen can be produced from a variety of domestic resources, such as natural gas, nuclear power, biomass, and renewable power like solar and wind. These qualities make it an attractive fuel option for transportation and electricity generation applications. It can be used in cars, in houses, for portable power, and in many more applications.

Fuel cells are an electrochemical device in which the energy of a chemical reaction is converted directly into electricity. Unlike an electric cell or battery (to be discussed later), a fuel cell does not become depleted nor requires recharging; it operates as long as the fuel and an oxidizer are supplied continuously from outside the cell.

Part of the material appearing in this chapter was adopted from the literature (NETL 2004). Section titles for this chapter on fuel cells are presented below:

6.1 Introduction
6.2 Electricity Units
6.3 Fuel Cell Overview
6.4 Unit Cells
6.5 Critical Functions of Cell Components
6.6 Fuel Cell Stacking
6.7 Fuel Cell Systems

Hydrogen Energy: Principles and Applications, First Edition. Vincent J. DelGatto, Louis Theodore, R. Ryan Dupont, and Matthew C. Ogwu.

6.2 Electrical Units

Electrical units are used to express quantitative measurements of all types of electrostatic and electromagnetic phenomena and of the electrical characteristics of components of electrical circuits. Basic electrical units are part of the centimeter-gram-second (CGS) system but because, in most cases, these units are either too large or too small for convenient measurement, a number of practical units have been adopted for use in applied engineering.

The elemental unit of electricity is the absolute charge on a single electron or proton. The symbol for this unit is *e*. The CGS unit of electrical charge is the electrostatic unit (esu) which is defined as the quantity of electricity that when concentrated at a point in a vacuum will repel a like charge 1 cm away with a force of 1 dyne. The esu equals the aggregate charge carried by 2,082,000,000 electrons or protons.

The unit of electrical current in common use is the ampere, which is defined as 0.1 abamperes. The practical unit of electrical quantity is the coulomb, the amount of electricity passing a given point in a circuit in 1 s when a current of 1 A is flowing. The volt is the practical unit of potential difference. It is equal to 100 million abvolts and can be defined as the potential difference existing between two points when 1 J (10 million ergs) of work is required to move 1 C of electricity from one of the points to the other. The unit of electrical work is the watt. It represents the generation or use of electrical energy at the rate of 1 J/s. The kilowatt is equal to 1,000 W.

All components in electrical circuits exhibit one or more of the characteristics of resistance, capacitance, and inductance. The commonly used unit of resistance is the ohm, which is the resistance of a conductor in which a potential difference of 1 V causes a current flow of 1 A. The capacitance of a condenser is measured in farads. A condenser of 1 farad capacitance will exhibit a change in potential difference of 1 V between its plates when 1 C of electricity is transferred from one plate to the other. The Henry (H) is the unit of inductance. A coil has a self-inductance of 1 H when a change in current of 1 A/s produces a counter voltage of 1/V. In a transformer or in any two magnetically coupled circuits, a mutual induction of 1 H is that inductance that will induce a voltage of 1 V in the secondary when there is a change of 1 A/s in the primary.

6.3 Fuel Cell Overview

Fuel cells are electrochemical devices that convert chemical energy in fuels into electrical energy directly, promising power generation with high efficiency and low environmental impact. Because the intermediate steps of producing heat and mechanical work typical of most conventional power generation methods are avoided, fuel cells are not limited by thermodynamic limitations of heat engines such as the Carnot efficiency. In addition, because combustion is avoided, fuel cells produce power with minimal pollutants. However, unlike batteries, the reductant and oxidant in fuel cells must be continuously replenished to allow continuous operation. Fuel cells bear significant resemblance to electrolyzers. In fact, some fuel cells operate in reverse as electrolyzers, yielding a reversible fuel cell that can be used for energy storage.

Though fuel cells could, in principle, process a wide variety of fuels and oxidants, of most interest today are those fuel cells that use common fuels (or their derivatives) or hydrogen as a reductant, and ambient air as the oxidant. Most fuel cell power systems comprise several components:

1) *Unit cells*, in which the electrochemical reactions take place.
2) *Stacks*, in which individual cells are modularly combined by electrically connecting the cells to form units with the desired output capacity.
3) *Balance of plant* which comprises components that provide feed stream conditioning (including a fuel processor if needed), thermal management, and electric power conditioning among other ancillary and interface functions.

In the following, an overview of fuel cell technology is given according to each of these categories, followed by a brief review of key potential applications of fuel cells.

6.4 Unit Cells

6.4.1 Basic Structure

Unit cells form the core of a fuel cell. These devices convert the chemical energy contained in a fuel electrochemically into electrical energy. The basic physical structure, or building block, of a fuel cell, consists of an electrolyte layer in contact with an anode and a cathode on either side. A schematic representation of a unit cell with the reactant/product gases and the ion conduction flow directions through the cell is shown in Figure 6.1.

In a typical fuel cell, fuel is fed continuously to the anode (negative electrode), and an oxidant (often oxygen from the air) is fed continuously to the cathode (positive electrode). The electrochemical reactions take place at the electrodes to produce an electric current through the electrolyte while driving a complementary electric current that performs work on the load. Although a fuel cell is similar to a typical battery in many ways, it differs in several respects. A battery is an energy storage device in which all the energy available is stored within the battery itself. The battery will cease to produce electrical energy when the chemical reactants are consumed (i.e., when the battery is discharged). A fuel cell, on the other hand, is an energy conversion device to which fuel and oxidants are supplied continuously. In principle, the fuel cell produces power for as long as fuel is supplied.

Fuel cells are classified according to the choice of electrolyte and fuel, which in turn determine the electrode reactions and the type of ions that carry the current across the electrolyte. In theory, any substance capable of chemical oxidation that can be supplied continuously (as a fluid) can be burned galvanically as fuel at the anode of a fuel cell. Similarly, the oxidant can be any fluid that can be reduced at a sufficient rate. Though the direct use of conventional fuels in fuel cells would be desirable, most fuel cells under development today use gaseous hydrogen, or a synthesis gas rich in hydrogen, as a fuel. Hydrogen has a high reactivity for anode reactions and can be produced chemically from a wide range of fossil and renewable fuels, as well as via electrolysis. For similar practical reasons, the most common oxidant is gaseous oxygen, which is readily available from the air.

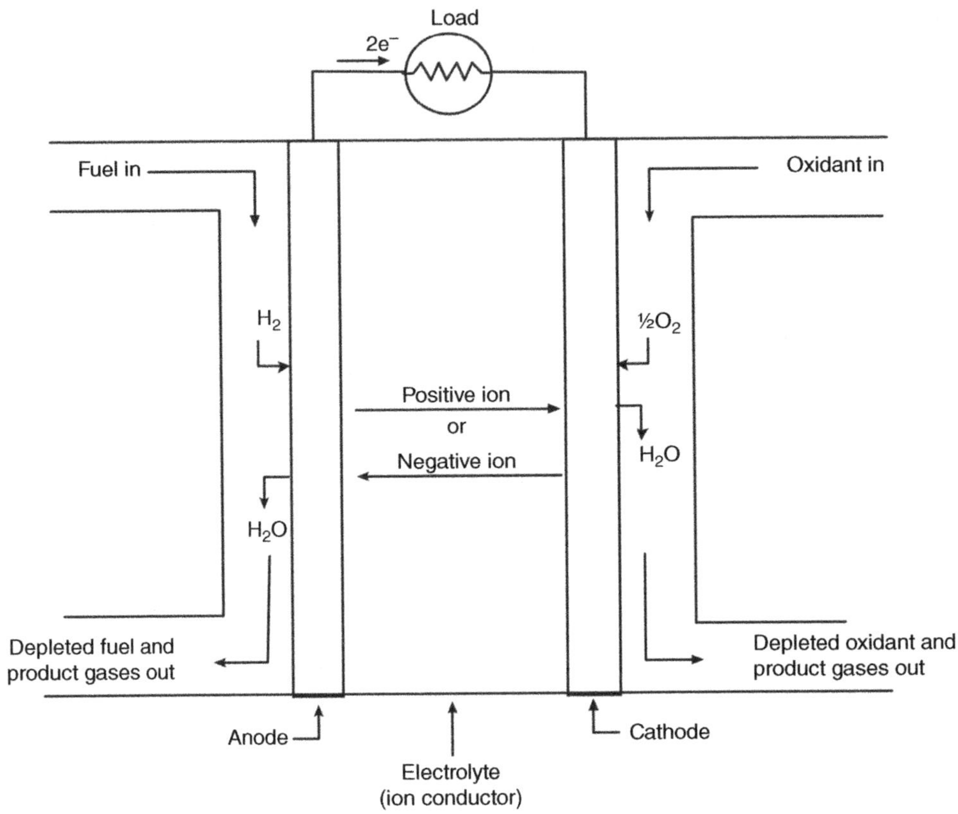

Figure 6.1 Schematic of an Individual Fuel Cell (*Source:* NETL (2004)/U.S. Department of Energy/Public Domain).

6.4.2 Internal Fuel Cell Process Details

To understand how electric energy is generated in a hydrogen-oxygen fuel cell, consider the fuel cell as a chemical reaction. The anode chemically changes the hydrogen *atoms* to positive hydrogen *electrons*, i.e.:

$$\text{Anode: } H_2 \rightarrow 2H^+ + 2e^- \tag{6.1}$$

The electrons move to the external current which the ions move to the cathode. At the cathode, the ions combine with oxygen and the electrons to form water and liberate energy, i.e.:

$$\text{Cathode: } 2H^+ + \tfrac{1}{2}O_2 + 2e^- \rightarrow H_2O \tag{6.2}$$

Equations 6.1 and 6.2 provide the overall reaction, i.e.:

$$H_2 + \tfrac{1}{2}O_2 \xrightarrow{298\,K} H_2O(l) \tag{6.3}$$

The energy liberated is given by the standard enthalpy change for this reaction.

$$H_2^{0\,kcal} + \tfrac{1}{2}O_2^{0\,kcal} \rightarrow H_2O - 68.4\,kcal \tag{6.4}$$

so that so that (see also Section 5.8)

$$\Delta H^\circ = -68.4 - 0 - 0 = -68.4\,\text{kcal}\left(-286\,\text{kJ}\right) \tag{6.5}$$

The standard free energy change for the reaction is similarly calculated

$$\Delta F^\circ = -56.6 - 0 - 0 = -56.6\,\text{kcal}\left(-237\,\text{kJ}\right) \tag{6.6}$$

Note that a decrease in the free energy indicates that the hydrogen and oxygen will react.

Some have arbitrarily defined efficiency, E, of the conversion of chemical energy to electrical energy as

$$H = \frac{\Delta F^\circ}{\Delta H^\circ} \tag{6.7}$$

Applying this equation to a hydrogen-oxygen fuel cell gives

$$E = 56.6/68.4 = 0.827 = 82.7\% \tag{6.8}$$

The efficiency is close to 95% of the calculation based on water in the vapor state. The reader should note that energy losses in the hydrogen-oxygen fuel cell are approximately 25% so the actual efficiency should be in the 55–60% range. This operating efficiency figure compares favorably to the 35% efficiency attained in a typical fossil fuel combustion system.

With reference to the above calculations, it should be noted that the reader was introduced to the Gibb's Free energy, ΔF°, in the last chapter. This term represents the work that can be recovered from a reaction where the compound – liquid water – is formed from its elements – all in a standard (or reference) state. Many have defined the efficiency of a fuel cell by dividing the electricity produced by the standard enthalpy of formation, ΔH°, with liquid water again formed from its elements. In this case, the total energy released, ΔH°, is comprised of both electrical energy, ΔG°, and thermal energy.

6.5 Critical Functions of Cell Components

A critical portion of most unit cells is often referred to as the three-phase interface. These mostly microscopic regions, in which the actual electrochemical reactions take place, are found where either electrode meets the electrolyte. For a site or area to be active, it must be exposed to the reactant, be in electrical contact with the electrode, be in ionic contact with the electrolyte, and contain sufficient electro-catalyst for the reaction to proceed at the desired rate. The density of these regions and the nature of these interfaces play a critical role in the electrochemical performance of both liquid and solid electrolyte fuel cells. In liquid electrolyte fuel cells, the reactant gases diffuse through a thin-electrolyte film that wets portions of the porous electrode and reacts electrochemically on their respective electrode surface. If the porous electrode contains an excessive amount of electrolyte, the electrode may "flood" and restrict the transport of gaseous species in the electrolyte phase to the reaction sites. The consequence is a reduction in the electrochemical performance of the porous electrode. Thus, a delicate balance must be maintained among the electrode, electrolyte, and gaseous phases in the porous electrode structure. Alternatively, in solid electrolyte fuel cells, the challenge is to engineer a large number of catalyst sites into the interface that is electrically and ionically connected to the electrode and the electrolyte, respectively, and that are efficiently exposed to the reactant gases. In most successful solid electrolyte fuel cells, a high-performance interface requires the use of an electrode that, in the zone near the catalyst, has mixed conductivity (i.e., it conducts both electrons and ions).

Over the past 20 years, the unit cell performance of at least some of the fuel cell technologies has been dramatically improved. These developments resulted from improvements in the three-phase boundary, reducing the thickness of the electrolyte, and developing improved electrode and electrolyte materials which broaden the temperature range over which the cells can be operated.

In addition to facilitating electrochemical reactions, each of the unit cell components has other critical functions. The electrolyte not only transports dissolved reactants to the electrode but also conducts ionic charge between the electrodes, thereby completing the cell electric circuit as illustrated in Figure 6.1. It also provides a physical barrier to prevent the fuel and oxidant gas streams from directly mixing.

The functions of porous electrodes in fuel cells, in addition to providing a surface for electrochemical reactions to take place, are to:

1) Conduct electrons away from or into the three-phase interface once they are formed (so an electrode must be made of materials that have good electrical conductance) and provide current collection and connection with either other cells or the load.
2) Ensure that reactant gases are equally distributed over the cell.
3) Ensure that reaction products are efficiently led away to the bulk gas phase.

As a consequence, the electrodes are typically porous and made of an electrically conductive material. At low temperatures, only a few relatively rare and expensive materials provide sufficient electro-catalytic activity, and so many catalysts are deposited in small quantities at the interface where they are needed. In high-temperature fuel cells, the electro-catalytic activity of the bulk electrode material is often sufficient.

Though a wide range of fuel cell geometries has been considered, most fuel cells under development now are either planar (rectangular or circular) or tubular (either single- or double-ended and cylindrical or flattened).

6.6 Fuel Cell Stacking

For most practical fuel cell applications, unit cells must be combined in a modular fashion into a cell stack to achieve the voltage and power output level required for the application. Generally, stacking involves connecting multiple unit cells in series via electrically conductive interconnects. Different stacking arrangements have been developed, which are described below in the next two sections.

6.6.1 Planar-Bipolar Stacking Fuel Cell (PBSFC)

The most common fuel cell stack design is the so-called planar-bipolar stack arrangement (Figure 6.2 depicts a PBSFC). Individual unit cells are electrically connected with interconnects. Because of the configuration of a flat plate cell, the interconnect becomes a separator plate with two functions:

1) To provide an electrical series connection between adjacent cells, specifically for flat plate cells, and
2) To provide a gas barrier that separates the fuel and oxidant of adjacent cells.

In many PBSFC designs, the interconnect also includes channels that distribute the gas flow over the cells. The PBSFC design is electrically simple and leads to short electronic current paths (which helps to minimize cell resistance).

PBSFCs can be further characterized according to the arrangement of the gas flow:

1) *Crossflow*. Air and fuel flow perpendicular to each other.
2) *Co-flow*. Air and fuel flow parallel and in the same direction. In the case of circular cells, this means the gases flow radially outward.

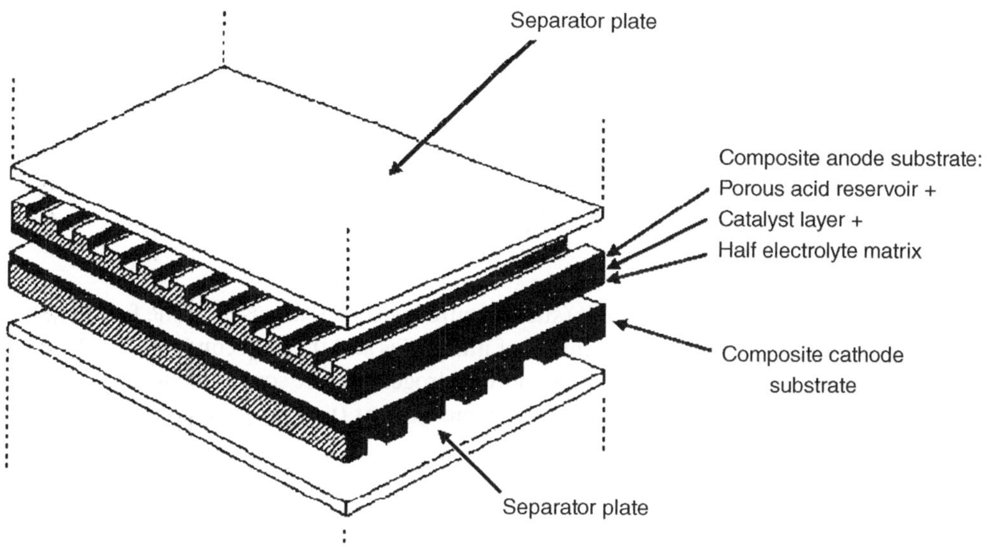

Figure 6.2 Expanded View of a Basic Fuel Cell Unit in a Fuel Cell Stack (*Source:* NETL (2004)/U.S. Department of Energy/Public Domain).

3) *Counter-flow*. Air and fuel flow parallel but in opposite directions. Again, in the case of circular cells, this means radial flow.
4) *Serpentine flow*. Air or fuel follows a zig-zag path.
5) *Spiral flow*. This applies to circular cells.

The choice of gas-flow arrangement depends on the type of fuel cell, the application, and other considerations. Finally, the manifold of gas streams to the cells in PBSFCs can be achieved in various ways:

1) *Internal* – the manifolds run through the unit cells.
2) *Integrated* – the manifolds do not penetrate the unit cells but are integrated into the interconnects.
3) *External* – the manifold is completely external to the cell, much like a wind box

6.6.2 Stacks with Tubular Cells

Especially for high-temperature fuel cells, stacks with tubular cells have been developed. Tubular cells have significant advantages in sealing and the structural integrity of the cells. However, they represent a special geometric challenge to the stack designer when it comes to achieving high-power density and short current paths. In one of the earliest tubular designs, current is conducted tangentially around the tube. Interconnects among the tubes are used to form rectangular arrays of tubes. Alternatively, the current can be conducted along the axis of the tube, in which case interconnection is done at the end of the tubes. To minimize the length of electronic conduction paths for individual cells, sequential series of connected cells is being developed. The cell arrays can be connected in series or parallel.

To avoid packing density limitations associated with cylindrical cells, some tubular stack designs use flattened tubes.

6.7 Fuel Cell Systems

In addition to the stack, practical fuel cell systems require several other sub-systems and components; the so-called balance of plant (BoP). Together with the stack, the BoP forms the fuel cell system. The precise arrangement of the BoP depends heavily on the fuel cell type, the fuel choice, and the application. In addition, specific operating conditions and requirements of individual cell and stack designs determine the characteristics of the BoP. Still, most fuel cell systems contain:

1) *Fuel preparation*. Except when pure fuels (such as pure hydrogen) are used, some fuel preparation is required, usually involving the removal of impurities and thermal conditioning. In addition, many fuel cells that use fuels other than pure hydrogen require some fuel processing, such as reforming, in which the fuel is reacted with some oxidant (usually steam or air) to form a hydrogen-rich anode feed mixture.
2) *Air supply*. In most practical fuel cell systems, this includes air compressors or blowers as well as air filters.
3) *Thermal management*. All fuel cell systems require careful management of the fuel cell stack temperature.
4) *Water management*. Water is needed in some parts of the fuel cell, while overall water is a reaction product. To avoid having to feed water in addition to fuel, and to ensure smooth operation, water management systems are required in most fuel cell systems.
5) *Electric power conditioning equipment*. Since fuel cell stacks provide a variable DC voltage output that is typically not directly usable for the load, electric power conditioning is typically required.

While perhaps not the focus of most development efforts, the BoP represents a significant fraction of the weight, volume, and cost of most fuel cell systems.

Figure 6.3 shows a simple rendition of a fuel cell power plant. Beginning with fuel processing, a conventional fuel (natural gas, other gaseous hydrocarbons, methanol, naphtha, or coal) is cleaned and then converted into a gas containing hydrogen. Energy conversion occurs when DC electricity is generated utilizing individual fuel cells combined in stacks or bundles. A varying number of cells or stacks can be matched to a particular power application. Finally, power conditioning converts the electric power from DC into regulated DC or AC for consumer use.

6.8 Fuel Cell Types

A variety of fuel cells are in different stages of development. The most common classification of fuel cells is by the type of electrolyte used in the cells and includes:

1) Polymer electrolyte fuel cells (PEFCs)
2) Alkaline fuel cells (AFCs)
3) Phosphoric acid fuel cells (PAFCs)
4) Molten carbonate fuel cells (MCFCs), and
5) Solid oxide fuel cells (SOFCs).

Broadly, the choice of electrolyte dictates the operating temperature range of the fuel cell. The operating temperature and useful life of a fuel cell dictate the physicochemical and thermomechanical

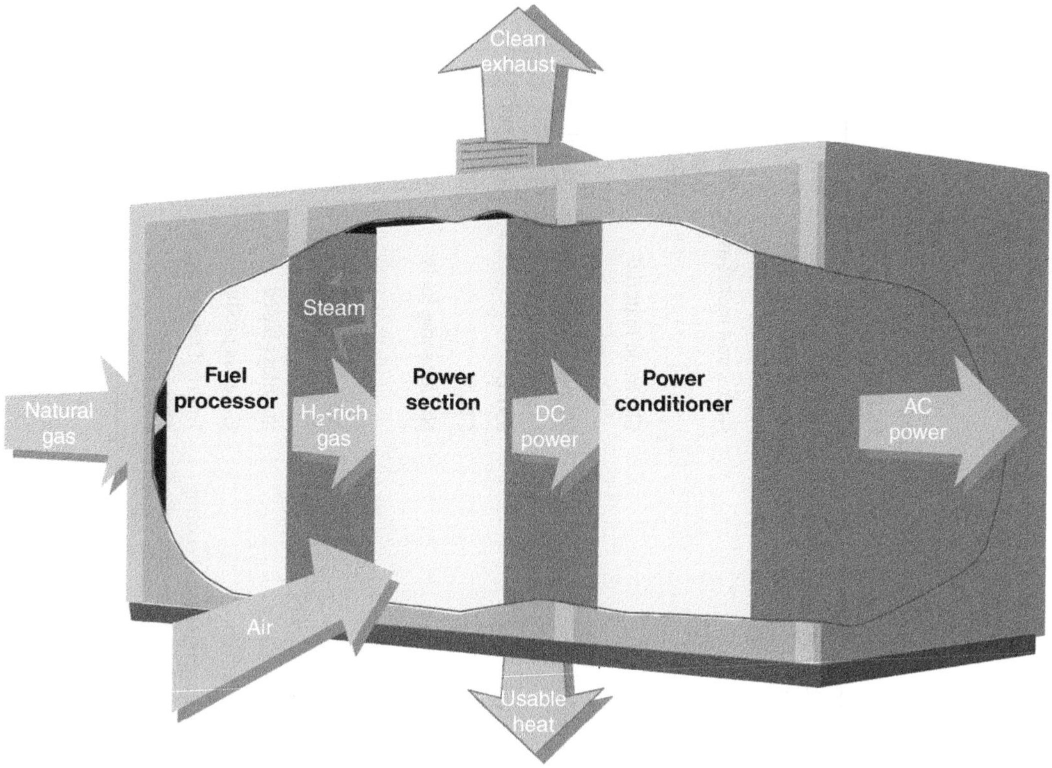

Figure 6.3 Fuel cell power plant major processes (*Source:* NETL (2004)/U.S. Department of Energy/Public Domain).

properties of materials used in the cell components (i.e., electrodes, electrolyte, interconnect, current collector). Aqueous electrolytes are limited to temperatures of about 200°C or lower because of their high vapor pressure and rapid degradation at higher temperatures.

Operating temperature also plays an important role in dictating the degree of fuel processing required. In low-temperature fuel cells, all the fuel must be converted to hydrogen prior to entering the fuel cell. In addition, the anode catalyst in low-temperature fuel cells (mainly platinum) is strongly poisoned by CO. In high-temperature fuel cells, CO and even CH_4 can be internally converted to hydrogen or even directly oxidized electrochemically. Table 6.1 provides an overview of the key characteristics of the main fuel cell types.

In parallel with the classification by electrolyte, some fuel cells are classified by the type of fuel used. Direct Alcohol Fuel Cells (DAFCs) or, more commonly, direct methanol fuel cells (DMFCs) use alcohol without reforming. Mostly, this refers to a PEFC-type fuel cell in which methanol or another alcohol is used directly, mainly for portable applications. In direct carbon fuel cells (DCFCs), solid carbon (presumably a fuel derived from coal, pet-coke, or biomass) is used directly in the anode, without an intermediate gasification step. Concepts with solid oxide, molten carbonate, and alkaline electrolytes are all under development. The thermodynamics of the reactions in a DCFC allows very high-efficiency conversion. Therefore, if the technology can be developed into practical systems, it could ultimately have a significant impact on coal-based power generation. A brief description of a few electrolyte cells of interest follows.

Table 6.1 Summary of major differences of the fuel cell types.

	PEFC	AFC	PAFC	MCFC	SOFC
Electrolyte	Hydrated polymeric ion exchange membranes	Mobilized or immobilized potassium hydroxide in asbestos matrix	Immobilized liquid phosphoric acid in SiC	Immobilized liquid molten carbonate in $LiAlO_2$	Perovskites (ceramics)
Electrodes	Carbon	Transition metals	Carbon	Nickel and nickel oxide	Perovskite and perovskite/metal cermet
Catalyst	Platinum	Platinum	Platinum	Electrode material	Electrode material
Interconnect	Carbon or metal	Metal	Graphite	Stainless steel or nickel	Nickel, ceramic, or steel
Operating temperature	40–80°C	65–220°C	205°C	650°C	600–1,000°C
Charge Carrier	H^+	OH^-	H^+	$CO_3^=$	$O^=$
External reformer for hydrocarbon fuels	Yes	Yes	Yes	No, for some fuels	No, for some fuels and cell designs
External shift conversion of CO to hydrogen	Yes, plus purification to remove trace CO	Yes, plus purification to remove CO and CO_2	Yes	No	No
Prime cell components	Carbon-based	Carbon-based	Graphite-based	Stainless-based	Ceramic
Product water management	Evaporative	Evaporative	Evaporative	Gaseous product	Gaseous product
Product heat management	Process gas + liquid cooling medium	Process gas + electrolyte circulation	Process gas + liquid cooling medium or steam generation	Internal reforming + process gas	Internal reforming + process gas

6.8.1 Polymer Electrolyte Fuel Cells

The electrolyte in this fuel cell is an ion exchange membrane (fluorinated sulfonic acid polymer or another similar polymer) that is an excellent proton conductor. The only liquid in this fuel cell is water, thus, corrosion problems are minimal. Typically, carbon electrodes with platinum electro-catalysts are used for both anode and cathode and with either carbon or metal interconnects.

Water management in the membrane is critical for efficient performance; the fuel cell must operate under conditions where the by-product water does not evaporate faster than it is produced because the membrane must remain hydrated. Because of the limitation on the operating temperature imposed by the polymer, usually less than 100°C, but more typically around 60–80°C, and because of problems with water balance, an H_2-rich gas with minimal or no CO (a poison at low temperature) is used. Higher catalyst loading (Pt in most cases) than that used in PAFCs is required for both the anode and cathode. Extensive fuel processing is required with other fuels, as the anode is easily poisoned by even trace levels of CO, sulfur species, and halogens.

PEFCs are being pursued in a wide variety of applications, especially for prime power for fuel cell vehicles (FCVs). As a consequence of the high interest in FCVs and hydrogen, the investment in PEFC over the past decade easily surpasses all other types of fuel cells combined. Although significant development of PEFC for stationary applications has taken place, many developers now focus on automotive and portable applications for this type of fuel cell.

6.8.1.1 Advantages

The PEFC has a solid electrolyte that provides excellent resistance to gas crossover. The PEFC's low operating temperature allows rapid start-up and, with the absence of corrosive cell constituents, the use of the exotic materials required in other fuel cell types, both in stack construction and in the BoP is not required. Test results have demonstrated that PEFCs are capable of high-power densities of over 2 kW/L and $2\,W/cm^2$. The PEFC lends itself, particularly to situations where pure hydrogen can be used as a fuel.

6.8.1.2 Disadvantages

The low and narrow operating temperature range makes thermal management difficult, especially at very high-power densities and makes it difficult to use the rejected heat for cogeneration or in bottoming cycles. Water management is another significant challenge in PEFC design, as engineers must balance ensuring sufficient hydration of the electrolyte against flooding the electrolyte. In addition, PEFCs are quite sensitive to poisoning by trace levels of contaminants including CO, sulfur species, and ammonia. To some extent, some of these disadvantages can be counteracted by lowering operating power density and increasing electrode catalyst loading, but both increase the cost of the system. If hydrocarbon fuels are used, the extensive fuel processing required negatively impacts system size, complexity, efficiency (typically in the mid-thirties), and system cost. Finally, the need for a hydrogen infrastructure to be developed for PEFCs poses a barrier to short-term commercialization.

6.8.2 Alkaline Fuel Cells (AFCs)

The electrolyte in this fuel cell is concentrated (85 wt%) KOH in fuel cells operated at high temperature (≈250°C), or less concentrated (35–50 wt%) KOH for lower temperature (<120°C) operation. The electrolyte is retained in a matrix (usually asbestos), and a wide range of electro-catalysts can be used (e.g., Ni, Ag, metal oxides, spinels, and noble metals). The fuel supply is limited to non-reactive constituents except for hydrogen. CO is a poison, and CO_2 will react with the KOH to form K_2CO_3, thus altering the electrolyte. Even a small amount of CO_2 in the air must be considered a potential poison for the alkaline cell. Generally, hydrogen is considered the preferred fuel for AFCs, although some DCFCs use alkaline electrolytes.

The AFC was one of the first modern fuel cells to be developed, beginning in 1960. The application at that time was to provide onboard electric power for the Apollo space vehicle. The AFC has enjoyed considerable success in space applications, but its terrestrial application has been challenged by its sensitivity to CO_2. Still, some developers in the United States and Europe pursue AFC for mobile and closed system (reversible fuel cell) applications.

6.8.2.1 Advantages

Desirable attributes of the AFC include its excellent performance on hydrogen (H_2) and oxygen (O) compared to other candidate fuel cells due to its active O_2 electrode kinetics and its flexibility to use a wide range of electro-catalysts.

6.8.2.2 Disadvantages

The sensitivity of the electrolyte to CO_2 requires the use of highly pure hydrogen as a fuel. As a consequence, the use of a reformer would require a highly effective CO and CO_2 removal system. In addition, if the ambient air is used as the oxidant, the CO_2 in the air must be removed. While this is technically not challenging, it has a significant impact on the size and cost of the system.

6.8.3 Phosphoric Acid Fuel Cells (PAFCs)

Phosphoric acid, concentrated to 100%, is used as the electrolyte in this fuel cell, which typically operates at 150–220°C. At lower temperatures, phosphoric acid is a poor ionic conductor, and CO poisoning of the Pt electro-catalyst in the anode becomes severe. The relative stability of concentrated phosphoric acid is high compared to other common acids; consequently, the PAFC is capable of operating at the high end of the acid temperature range (100–220°C). In addition, the use of concentrated acid (100%) minimizes the water vapor pressure so water management in the cell is not difficult. The matrix most commonly used to retain the acid is silicon carbide and the electro-catalyst in both the anode and cathode is Pt.

PAFCs are mostly developed for stationary applications. Both in the United States and Japan, hundreds of PAFC systems were produced, sold, and used in field tests and demonstrations. It is still one of the few fuel cell systems that are available for purchase. Development of PAFCs had slowed down in the past 10 years, in favor of PEFCs that were thought to have better cost potential. However, PAFC development continues.

6.8.3.1 Advantages

PAFCs are much less sensitive to CO than PEFCs and AFCs, PAFCs tolerate about 1% CO as a diluent. The operating temperature is still low enough to allow the use of common construction materials, at least in the BoP components. The operating temperature also provides considerable design flexibility for thermal management. PAFCs have demonstrated system efficiencies of 37–42% (based on the lower heating value [LHV] of natural gas fuel), which is higher than most PEFC systems could achieve (but lower than many of the SOFC and MCFC systems). In addition, the waste heat from PAFCs can be readily used in most commercial and industrial cogeneration applications and would technically allow the use of a bottoming cycle.

6.8.3.2 Disadvantages

Cathode-side oxygen reduction is slower than in AFCs and requires the use of a Pt catalyst. Although less complex than PEFCs, PAFCs still require extensive fuel processing, including typically a water gas shift reactor to achieve good performance. Finally, the highly corrosive nature of phosphoric acid requires the use of expensive materials in the stack (especially the graphite separator plates).

6.8.4 Molten Carbonate Fuel Cells (MCFCs)

The electrolyte in this fuel cell is usually a combination of alkali carbonates, which are retained in a ceramic matrix of $LiAlO_2$. The fuel cell operates at 600–700°C where the alkali carbonates form a highly conductive molten salt, with carbonate ions providing ionic conduction. At the high operating temperatures in MCFCs, Ni (anode) and nickel oxide (cathode) are adequate to promote the reaction. Noble metals are not required for operation, and many common hydrocarbon fuels can be reformed internally.

The focus of MCFC development has been larger stationery and marine applications, where the relatively large size and weight of MCFC and slow start-up time are not an issue. MCFCs are under development for use with a wide range of conventional and renewable fuels. MCFC-like technology is also considered for DCFCs. After the PAFCs, MCFCs have been demonstrated most extensively in stationary applications, with dozens of demonstration projects either underway or completed. While the number of MCFC developers and the investment level are reduced compared to a decade ago, development and demonstrations continue.

6.8.4.1 Advantages

The relatively high operating temperature of the MCFC (650°C) results in several benefits: no expensive electro-catalysts are needed as the nickel electrodes provide sufficient activity, and both CO and certain hydrocarbons are fuels for the MCFC, as they are converted to hydrogen within the stack (on special reformer plates) simplifying the BoP and improving system efficiency to the high forties to low fifties. In addition, the high-temperature waste heat allows the use of a bottoming cycle to further boost the system efficiency to the high fifties to low sixties.

6.8.4.2 Disadvantages

The main challenge for MCFC developers stems from the very corrosive and mobile electrolyte, which requires the use of nickel and high-grade stainless steel as the cell hardware (cheaper than graphite, but more expensive than ferritic steels). The higher temperatures promote material problems, impacting mechanical stability and stack life.

Also, a source of CO_2 is required at the cathode (usually recycled from anode exhaust) to form the carbonate ion, representing additional BoP components. High contact resistances and cathode resistance limit power densities to around 100–200 mW/cm^2 at practical operating voltages.

6.8.5 Solid Oxide Fuel Cells (SOFCs)

The electrolyte in this fuel cell is a solid, nonporous metal oxide, usually Y_2O_3-stabilized ZrO_2. The cell operates at 600–1,000°C where ionic conduction by oxygen ions takes place. Typically, the anode is Co—ZrO_2 or Ni—ZrO_2 cement, and the cathode is Sr-doped $LaMnO_3$.

Early on, the limited conductivity of solid electrolytes required cell operation at around 1,000°C, but more recently thin-electrolyte cells with improved cathodes have allowed a reduction in operating temperature to 650–850°C. Some developers are attempting to push SOFC operating temperatures even lower. Over the past decade, this has allowed the development of compact and high-performance SOFCs that utilized relatively low-cost construction materials.

Concerted stack development efforts, especially through the U.S. DOE's SECA program have considerably advanced the knowledge and development of thin-electrolyte planar SOFCs. As a consequence of the performance improvements, SOFCs are now considered for a wide range of applications, including stationary power generation, mobile power, auxiliary power for vehicles, and specialty applications.

6.8.5.1 Advantages

The SOFC is the fuel cell with the longest continuous development period, starting in the late 1950s, several years before the AFCs. Because the electrolyte is solid, the cell can be cast into various shapes, such as tubular, planar, or monolithic. The solid ceramic construction of the unit cell alleviates any corrosion problems in the cell. The solid electrolyte also allows precise engineering of the three-phase boundary and avoids electrolyte movement or flooding in the electrodes. The kinetics of the cell are relatively fast, and CO is a directly usable fuel as it is in the MCFCs. There is no requirement for CO_2 at the cathode as with the MCFCs. The materials used in SOFCs are modest in cost. Thin-electrolyte planar SOFC unit cells have been demonstrated to be cable of power densities close to those achieved with PEFCs. As with the MCFCs, the high operating temperature allows the use of most of the waste heat for cogeneration or bottoming cycles. Efficiencies ranging from around 40% (simple cycle small systems) to over 50% (hybrid systems) have been demonstrated, and the potential for 60+% efficiencies exists as it does for MCFCs.

6.8.5.2 Disadvantages

The high temperature of the SOFCs has its drawbacks. There are thermal expansion mismatches among materials and sealing between cells is difficult in the flat plate configurations. The high operating temperature places severe constraints on materials selection and results in difficult fabrication processes. Corrosion of metal stack components (such as the interconnects in some designs) is a challenge. These factors limit stack-level power density (though significantly higher than in PAFCs and MCFCs), and thermal cycling and stack life (though the latter is better than for MCFCs and PEFCs).

6.9 Fuel Cell Characteristics

The interest in terrestrial applications of fuel cells is driven primarily by their potential for high efficiency and very low environmental impact (virtually no acid gas or solid emissions). Efficiencies of present fuel cell plants are in the range of 30–55% based on the LHV of the fuel. Hybrid fuel cell/reheat gas turbine cycles that offer efficiencies greater than 70% of LHV, using demonstrated cell performance, have been proposed. PAFC units produce extremely low emissions even compared to the Los Angeles Basin (South Coast Air Quality Management District) requirements, the strictest requirements in the U.S. Measured emissions from the PAFC units are <1 ppm of NO_x, 4 ppm of CO, and < 1 ppm of reactive organic gases (non-methane) (NETL 2004). In addition, fuel cells operate at a constant temperature, and the heat from the electrochemical reaction is available for cogeneration applications. Table 6.2 summarizes the impact of the major constituents within fuel gases on the various fuel cells.

Another key feature of fuel cells is that their performance and cost are less dependent on the scale than other power technologies. Small fuel cell plants operate nearly as efficiently as large

Table 6.2 Summary of major fuel constituents' impact on PEFCs, AFCs, PAFCs, MCFCs, and SOFCs.

	PEFC	AFC	PAFC	MCFC	SOFC
H_2	Fuel	Fuel	Fuel	Fuel	Fuel
CO	Poison (reversible; 50 ppm/stack)	Poison	Poison (<0.5%)	Fuel	Fuel
CH_4	Diluent	Poison	Diluent	Diluent	Fuel
CO_2 and H_2O	Diluent	Poison	Diluent	Diluent	Diluent
S as H_2S and COS	No data	Poison	Poison (<50 ppm)	Poison (<0.5 ppm)	Poison (<1.0 ppm)

ones, with equally low emissions, and comparable costs. This opens up applications for fuel cells where conventional power technologies are not practical. In addition, fuel cell systems can be relatively quiet generators.

To date, the major impediments to fuel cell commercialization have been insufficient longevity and reliability, unacceptably high cost, and lack of familiarity of markets with fuel cells. For fuel cells that require special fuels (such as hydrogen), the lack of a fuel infrastructure also limits commercialization.

A baker's dozen of the characteristics that fuel cells and fuel cell plants offer are:

1) Direct energy conversion (no combustion)
2) No moving parts in the energy converter
3) Quiet
4) Demonstrated high availability of lower temperature units
5) Siting ability
6) Fuel flexibility
7) Demonstrated endurance/reliability of lower temperature units
8) Good performance at off-design load operation
9) Modular installations to match load and increase reliability
10) Remote operation
11) Unattended operation
12) Size flexibility
13) Rapid load-following capability

General negative features of fuel cells include:

1) Market entry cost high
2) Endurance/reliability of higher temperature units not demonstrated
3) Unfamiliar technologies to the power industry
4) Limited infrastructure currently

6.10 Overall Advantages/Disadvantages

The fuel cell types addressed here have significantly different operating regimes. As a result, their materials of construction, fabrication techniques, and system requirements differ. These distinctions result in individual advantages and disadvantages that govern the potential of the various cells to be used for different applications. Developers use the advantages of fuel cells to identify early applications and address research and development issues to expand applications.

The characteristics, advantages, and disadvantages summarized in the previous section form the basis for the selection of the candidate fuel cell types to respond to a variety of application needs. The major applications for fuel cells are as stationary electric power plants (including cogeneration units), of motive power for vehicles, and as onboard electric power for space vehicles or other closed environments.

6.11 Batteries

Electric vehicles (EVs) run on batteries that must be plugged in to recharge. What is a battery? It is an energy storage system. It is important to realize that they do not produce electricity – they store electricity produced elsewhere, e.g., from coal, uranium, natural gas-powered plants, diesel-fueled generators, etc. Therefore, an EV is *not* a zero-emission vehicle unless charged exclusively by a renewable power source. Since approximately 40% of the electricity generated in the United States is from coal-fired plants, it follows that 40% of the EVs on the road are coal-powered. The only question of concern is what produces the power since it does not come from the battery as the battery is only the storage device, like a gas tank in a car.

There are two types of batteries, rechargeable, and single-use. The most common single-use batteries are A, AA, AAA, C, D, 9V, and lantern types. Those dry-cell batteries use zinc, manganese, lithium, silver oxide, or zinc and carbon to store electricity chemically. Please note they all contain toxic, heavy metals.

Rechargeable batteries only differ in their internal materials, usually lithium-ion, nickel-metal oxide, or nickel-cadmium. The United States uses three billion batteries a year, and most are not recycled; they end up in landfills. California is the only state which requires all batteries to be recycled.

All batteries are self-discharging. That means even when not in use, they leak tiny amounts of energy. When a battery runs down and can no longer power a machine or light, it is dead. However, it continues to leak small amounts of electricity. As the chemicals inside it run out, pressure builds inside the battery's metal casing, and eventually, it cracks. The metals left inside are toxic and then ooze out. All batteries eventually rupture; it just takes rechargeable batteries longer to end up in a landfill.

In addition to dry-cell batteries, there are also wet cell ones used in automobiles, boats, and motorcycles. Ninety percent are recycled. Unfortunately, no procedure is in place on how to recycle single-use ones properly.

A typical EV battery weighs 1,000 lb, about the size of a travel trunk. It contains 25 lb of lithium, 60 lb of nickel, 44 lb of manganese, 30 lb of cobalt, 200 lb of copper, and 400 lb of aluminum, steel, and plastic. Inside are over 6,000 individual lithium-ion cells. All these (toxic) components come from mining. To manufacture each EV auto battery, 25,000 lb of brine for lithium, 30,000 lb of ore for cobalt, 5,000 lb of ore for nickel, and 25,000 lb of ore for copper must be processed. All told, nearly 500,000 lb of the Earth's crust is required for just one battery.

Sixty-eight percent of the world's cobalt, a significant part of a battery, comes from the Congo. Their mines have no pollution controls, and they employ children who die from handling this toxic material. Some claim that these factors should be included in the cost of driving an EV.

Hydrogen Fuel Cell Vehicles (FCVs) are similar to electric vehicles EVs in that they use an electric motor instead of an internal combustion engine to power the wheels. However, while EVs run on batteries, FCVs generate their electricity onboard. In a fuel cell, H_2 gas from the vehicle's fuel tank combines with O_2 from the air to generate electricity with only water and heat as byproducts of the process. Similar to EVs, FCVs are zero-emission vehicles when operating, i.e., they have no smog-related or greenhouse gas tailpipe emissions. However, emissions can be created in the process of producing and transporting hydrogen fuel. Fueling a hydrogen FCV is similar to refilling a gas tank. One simply attaches a nozzle from a designated hydrogen dispenser at a public station to fill up the tank. The refueling times are short, as little as 5 minutes, and some FCVS can get over 300 miles on one tank of hydrogen fuel.

6.12 Summary

Hydrogen is a versatile energy carrier that can be used to power nearly every end-use energy need. The fuel cell – an energy conversion device that can efficiently capture and use the power of hydrogen may be the key to making the widespread adoption of hydrogen happen. Some of the positive features of fuel cells are listed below (U.S. DOE 2023).

1) Stationary fuel cells can be used for backup power, power for remote locations, distributed power generation, and cogeneration (in which excess heat released during electricity generation is used for other applications).
2) Fuel cells can power almost any portable application that typically uses batteries, from hand-held devices to portable generators.
3) Fuel cells can also power transportation, including personal vehicles, trucks, buses, and marine vessels, as well as provide auxiliary power to traditional transportation technologies. Hydrogen can play a particularly important role in the future by replacing the imported petroleum we currently use in our cars and trucks.
4) Fuel cells directly convert the chemical energy in hydrogen to electricity, with pure water and potentially useful heat as the only byproducts.
5) Hydrogen-powered fuel cells are not only pollution-free but also can have two to three times the efficiency of traditional combustion technologies. A conventional combustion-based power plant typically generates electricity at efficiencies of 33–35%, while fuel cell systems can generate electricity at efficiencies up to 60% (and even higher with cogeneration).
6) Gasoline engines in a conventional car are less than 20% efficient in converting the chemical energy in gasoline into power that moves the vehicle, under normal driving conditions. Hydrogen FCVS, which use electric motors, are much more energy efficient and use 40–60% of the fuel's energy – corresponding to more than a 50% reduction in fuel consumption compared to a conventional vehicle with a gasoline internal combustion engine.
7) In addition, fuel cells operate quietly, have fewer moving parts, and are well-suited to a variety of applications.

Reducing cost and improving durability are the two most significant challenges to fuel cell commercialization. Fuel cell systems must be cost-competitive with, and perform as well or better than, traditional power technologies over the life of the system.

Present research is focused on identifying and developing new materials that will reduce the cost and extend the life of fuel cell stack components including membranes, catalysts, bipolar plates, and membrane-electrode assemblies. Low-cost, high-volume manufacturing processes will also help to make fuel cell systems cost-competitive with traditional technologies.

References

Leal, E. and Brower, J. (2005). Thermodynamic analysis of production of hydrogen using high-temperature fuel cells. *Proceedings of IMEC, ASME*, Orlando, FL. New York: ASME.

National Energy Technology Laboratory (2004). *Fuel Cell Handbook*, 7th Edition. *DOE/NETL-2004/1206*. https://doi.org/10.2172/834188.

U.S. Department of Energy (2023). Hydrogen program. *Fuel Cells*. Office of Energy Efficiency and Renewable Energy (EERE) https://www.hydrogen.energy.gov/fuel_cells.html (accessed 11 August 2023).

Part II

Select Hydrogen Energy Topics

Chapters in this part key on hydrogen and the role hydrogen is certain to play in the energy future. The first chapter of this part should be viewed as an overview of hydrogen energy considerations while the remaining chapters provide an expanded presentation of the topics highlighted in this introductory chapter.

Chapters

Hydrogen Energy: Principles and Applications, First Edition. Vincent J. DelGatto, Louis Theodore, R. Ryan Dupont, and Matthew C. Ogwu.

7

Hydrogen Energy Overview

7.1 Introduction

Hydrogen is poised to become a major future energy supply/carrier (not an energy source) but here is the challenge. If an alternative fuel to fossil fuel, then hydrogen should be:

1) Technically feasible from a process standpoint.
2) Economically attractive.
3) Durable.
4) Easily converted to another energy form.
5) Of minimal or manageable safety concerns.
6) Of minimal or manageable environmental concern.
7) Reliable from a process perspective.

The U.S. government and many others presently think hydrogen energy can someday satisfy the above criteria. The technical community may not be so sure. It is a given that hydrogen can be produced from a variety of resources, including both renewable and nonrenewable ones. The overall process is depicted in Figure 7.1. The reader should note that "conversion steps" are considered an integral part of "uses." Here is the thinking about hydrogen energy. Create it from hydrogen-bearing compounds with processes using water, biomass, fossil fuels, or waste materials. Then store that hydrogen for as long as needed, and then convert it into electricity or energy in the form of heat or other useful end products.

Hydrogen is the most abundant element in the universe; however, it is rarely found in its elemental form on Earth. Nonetheless, hydrogen is a unique energy carrier that can transform the present fossil fuel economy into a hydrogen economy. It may be the fuel of the future. Once hydrogen is produced, it can be used to store, move, and deliver low or zero-carbon energy.

Hydrogen is a versatile energy carrier but still not a primary energy source. Therefore, hydrogen is often referred to as a secondary energy source. It can be produced from multiple feedstocks and can be used across virtually any application (see Figure 7.1).

Hydrogen can be produced in large centralized production facilities or in smaller distributed production facilities and can be transported via truck, pipeline, tanker, or other means.

Hydrogen, as a versatile energy carrier and chemical feedstock, offers advantage that can potentially unite the nation's energy resources – renewables, nuclear, and fossil fuels – and enables innovation in energy production and end uses that can help decarbonize three of the most

Hydrogen Energy: Principles and Applications, First Edition. Vincent J. DelGatto, Louis Theodore, R. Ryan Dupont, and Matthew C. Ogwu.

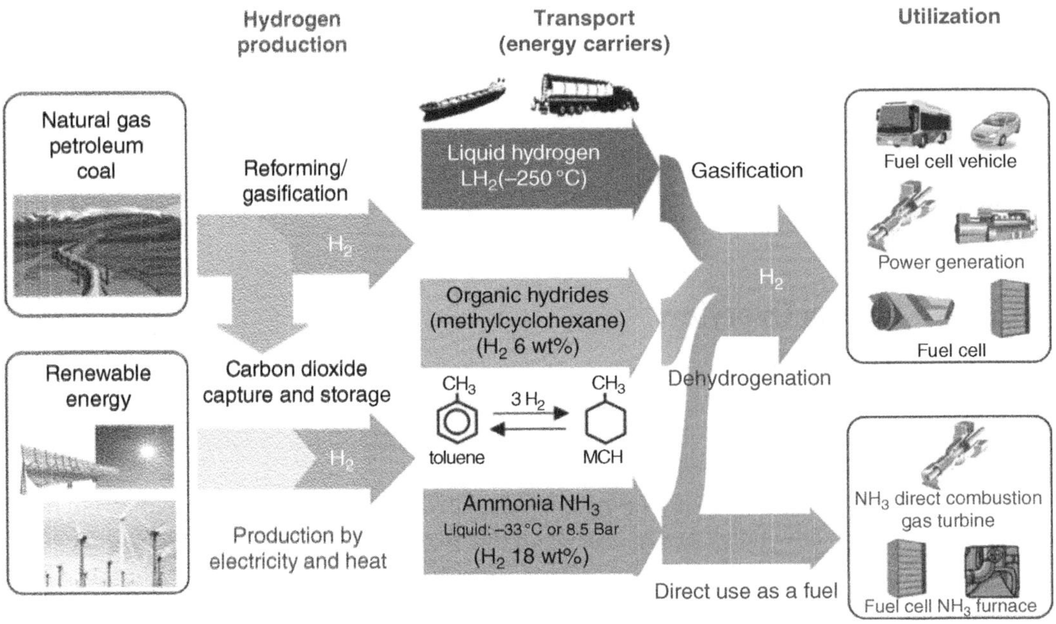

Figure 7.1 Overall hydrogen process options.

energy-intensive sectors of the economy: transportation, electricity generation, and manufacturing (U.S. DOE 2020).

Some key benefits of hydrogen are listed below:

1) Hydrogen can be produced from diverse domestic resources for use in multiple sectors, or for export.
2) Hydrogen has the highest energy content by weight of all known fuels – three times higher than gasoline – and is a critical feedstock for the entire chemicals industry, including for liquid fuels.
3) Hydrogen, along with fuel cells or combustion-based technologies, can enable zero or near zero-carbon emissions in transportation, stationary or remote power, and portable power applications, producing only water at low combustion temperatures.
4) Hydrogen can be used for gigawatt-hours of energy storage and as a "responsive load" on the grid to enable grid stability, increasing the utilization of power generators, including nuclear, coal, natural gas, and renewables.
5) Hydrogen can be used in a variety of domestic industries, such as the manufacturing of steel, cement, ammonia, and other chemicals.

Hydrogen and related technologies, such as fuel cells and turbines, have transitioned from highly specialized applications to commercially available products. Thousands of fuel cells are already in use in passenger and commercial vehicles, forklifts, and stationery and backup power units throughout the United States. The number of retail hydrogen fueling stations has grown in recent years. Recent advances in power generation have led to commercialization of large-frame turbines that can fire hydrogen/natural gas blends. The Department of Energy (DOE) has invested more than $4 billion in a number of hydrogen-related areas, including hydrogen production from diverse domestic sources, hydrogen delivery and storage, and conversion technologies including fuel cells and turbines (U.S. DOE 2020).

Technical challenges for hydrogen and related technologies are cost, durability, reliability, performance, and lack of hydrogen infrastructure. For widespread commercialization, hydrogen utilization technologies must enter larger markets and be able to compete with current technologies in terms of life-cycle cost, performance, durability, and environmental impact. This will first require developing and harmonizing codes and standards, fostering best practices for safety. The hydrogen fuel alternatives to the present fuels employed must be technically feasible, economically viable, easily converted to another energy form when combusted, be safe to use, and be potentially harmless to the environment. The current focus of the technical community is directed toward hydrogen as a replacement fuel source since it is a clean alternative that produces insignificant greenhouse gas emissions, although storage and transport remain major issues due to hydrogen's low density. The question thus remains: will the investment in a new hydrogen infrastructure be too costly? Unfortunately, the question is presently still being debated while the United States and many other countries are undertaking early-stage development.

The reader should note that a portion of the material presented in this chapter has been drawn in part from the literature (Skipka and Theodore 2014; U.S. DOE 2020; IRENA 2022; University of Michigan), and this chapter is intended to be introductory in nature. The remaining chapters in Part II of the text will expand on some of these key subject areas. For example, processing, storage, transportation and transmission, conversion, and uses. Environmental issues receive treatment in the Part III.

Section numbers and their accompanying titles of this chapter "Hydrogen Energy Overview" are listed below.

7.2 Early History

The Sun is basically a giant ball of hydrogen and helium gases. In a process called fusion, four hydrogen atoms combine to form one helium atom, releasing energy as radiation. This radiant energy is Earth's most abundant energy source. It provides light and heat and makes plants grow. It also causes the wind to blow and thc rain to fall, and it is stored as chemical energy in fossil fuels. Most of the present sources of energy originally came from the Sun.

In the 1847 novel, *The Mysterious Island*, Jules Verne amazingly envisioned the use of hydrogen as an energy carrier. "Yes, my friends, I believe that ordinary water will one day be employed as fuel, that hydrogen and oxygen, which constitute it, used singly or together, will furnish an inexhaustible source of heat and light." Today Verne's dream is being taken seriously by many practicing scientists and engineers. Hydrogen has the necessary properties and can fulfill the role of an energy carrier that can be derived from water (Skipka and Theodore 2014).

In recent years, hydrogen has been touted as the answer to many of society's energy problems. It is the perfect clean fuel. It has multiple applications through direct combustion and conversion to electricity in fuel cells. Fuel, cell technology in vehicles provides an opportunity to address and possibly solve petroleum and urban air pollution in the future.

Fuel cell technology is not new. In fact, it was invented in 1839 by Sir William Robert Grove. It has only been recently that this technology has become a potentially viable alternative to low-cost fossil fuels. Fuel cells have already found numerous niche applications in highly selected areas such as the NASA Space Program and as an electrical backup system to protect sensitive electronic equipment should the power grid temporarily go down. Fuel cells have also found use in mass transit. The reader may choose to review Part I, Chapter 6 for additional information on fuel cell technology.

7.3 Processing

Estimates of U.S. and world production of hydrogen for the Year 2019 indicate that the United States (9.04 million metric tons, MMT) produced approximately 14% of the world's total hydrogen output of 64.46 MMT (U.S. DOE 2019). In the United States, the oil refining industry is responsible for the majority of both hydrogen production and consumption, with the *usage* of hydrogen by refineries for the purpose of hydrocracking, ultra-low-sulfur diesel hydrotreating, fluid catalytic cracking, gas oil hydrotreating, and naphtha hydrotreating. Other demands for pure hydrogen include that for ammonia production, transportation, and power-to-gas applications. Hydrogen which is part of a mixture of gases can be used as fuel or feedstock for applications such as methanol production and iron reduction (U.S. DOE 2019). More discussion of hydrogen uses is provided in Section 7.6.

Pure hydrogen does not occur naturally. As noted in the Introduction, hydrogen is not a basic energy resource (except for fusion power in the Sun); it must be produced by using some other basic energy resource to generate hydrogen from water or other hydrogen-containing chemical compounds. Thus, the extraction step is, in effect, the processing step.

There are a host of methods that can be employed to process/produce hydrogen. Seven such methods are described here including some developmental projects.

1) Steam reforming (Theodore, Ricci, and VanVliet 2009; Theodore 2012). Industry produces the hydrogen it needs through a process called *steam methane reforming (SMR)*. High-temperature steam separates hydrogen from the carbon atoms in methane (CH_4). The hydrogen produced by this method is not used as a fuel but for industrial processes. This is the most cost-effective way to produce hydrogen today, but it uses fossil fuels both in the manufacturing process and as the heat source. The general reforming reaction is

$$C_nH_m + nH_2O \rightarrow nCO + \frac{(2n+m)H_2}{2} \tag{7.1}$$

The CO can then assist in the further conversion to H_2, as described in the next process.

2) Partial oxidation of hydrocarbon (Theodore 2012). This has been defined as gasification of heavy hydrocarbons to a syngas, with the CO all subsequently shifted to H_2. The process may be represented in equation form as:

$$C_nH_m, \text{S}, O_2, \text{and } H_2O \rightarrow CO, H_2, \text{and } H_2S \tag{7.2}$$

The reaction may be extended to produce additional hydrogen.

3) Electrolysis (Theodore 2012). Another way to make hydrogen is by electrolysis – splitting water into its basic elements: hydrogen and oxygen. Electrolysis involves passing an electric current through water to separate the atoms (2 H_2O + electricity = 2 H_2 + O_2). Hydrogen collects at the cathode and oxygen at the anode. Hydrogen produced by electrolysis is extremely pure, and electricity from renewable sources can power the process but is presently very expensive. Today, hydrogen produced from electrolysis is 10 times more costly than from natural gas, and three times more costly than gasoline per unit of energy generated. However, water is abundant and renewable, and technological advances in renewable electricity could make electrolysis a more attractive method to produce hydrogen in the future.
The overall electrolysis reaction may be described as follows. Electric energy is supplied to a fuel cell and hydrogen and oxygen are evolved at the two electrodes. The cathode supplies electrons and hydrogen is generated as follows:

$$2Na^+ + 2H_2O + 2e^- \rightarrow 2NaOH + H_2 \quad (7.3)$$

At the anode, oxygen is generated, and electrons are given up:

$$2OH^- \rightarrow O_2 + H_2 + 2e^- \quad (7.4)$$

Combining Equations 7.3 and 7.4 leads to the overall reaction shown in Equation 7.5:

$$2Na^+ + 2H_2O + 2OH^- \rightarrow 2NaOH + 2H_2 + O_2 \quad (7.5)$$

4) Coal-steam-oxygen. This process involves coal gasification to a syngas followed by conversion to hydrogen.
5) Steam-iron process. Steam converts a mixture of 5% Fe and 95% FeO to 80% Fe_3O_4, and 20% FeO_4.

$$Fe_3O_4 + 4CO \rightarrow 3Fe + 4CO_2 \quad (7.6)$$

$$H_2O + 3FeO \rightarrow H_2 + Fe_3O_4 \quad (7.7)$$

The reaction is then followed by:

$$Fe_3O_4 + CO \rightarrow 3Fe + CO_2 \quad (7.8)$$

6) Nuclear heat-steam-coal. The process using nuclear heat is essentially the same as the coal-steam-oxygen process, except that the heat is supplied by a loop from a nuclear reactor into the gasifier unit instead of oxygen feed to burn some of the coal.
7) Emerging technologies. There are also several experimental methods of producing hydrogen. *Photoelectrolysis* uses sunlight to split water molecules into its components. A semiconductor absorbs the energy from the Sun and acts as an electrode to separate the water molecules. In *biomass gasification*, wood chips and agricultural wastes are superheated until they turn into hydrogen and other gases. Biomass can also be used to provide the heat. Scientists have also discovered that some *algae* and *bacteria* produce hydrogen under certain conditions using lignin or sunlight as their energy source. Experiments are underway to find ways to induce these microbes to produce hydrogen efficiently. Domestic biomass and waste stream resources, with the potential for over a billion dry tons of feedstock annually, can also be leveraged for sustainable hydrogen production. Applicable categories of feedstocks include primary biomass

energy sources such as poplar, willow, and switchgrass, as well as biogas produced from anaerobic digestion of organic residues from sources such as landfills, agricultural waste, and municipal solid waste.

With reference to Method 3 above, Electrolysis, there are numerous types of fuel cells in existence today, the most common being the hydrogen fuel cell. As noted in Part I, Chapter 6, a fuel cell consists of two electrodes sandwiching an electrolyte. In the case of a hydrogen fuel cell, hydrogen passes over the negative electrode (i.e., the anode), and oxygen passes over the positive electrode (i.e., the cathode). When the hydrogen molecule comes in contact with a (platinum) catalyst, a hydrogen ion is formed. The proton passes through the electrolyte, called a proton exchange membrane, while the electron passes through an external circuit. The electron then returns to the cathode, where it is reunited with the proton in the presence of oxygen from outside air. This final reaction produces pure water and heat. This is simply the combustion reaction of hydrogen, but combustion in the sense of burning (as with oil) does not occur in the cell. The heat liberated is nearly one-quarter of a million J/gmol H_2O formed. Thus, this external circuit can convert this heat to power any type of electrical device.

7.4 Storage

Hydrogen has almost three times the energy content per unit mass of gasoline (the energy content of hydrogen is 33 kWh/kg, while that of gasoline is 12 kWh/kg), but the volumetric energy density of gaseous hydrogen is low, making it difficult to store in compact containers. To overcome this challenge, hydrogen is usually stored using *physical processes*, as a gas or cryogenic liquid; it can also be stored using material-based processes that incorporate hydrogen in other chemical compounds. The current portfolio of hydrogen storage options includes physical-based gaseous and liquid storage in tanks and reversible and nonreversible materials-based storage. Reversible materials are able to release or uptake hydrogen directly through changes in temperature or pressure, whereas nonreversible materials require additional chemical or physical processing to effect reuptake of hydrogen (also referred to as "regenerable" materials). Each of these two options is briefly discussed below and an expanded presentation is provided in Chapter 13.

7.4.1 Physical-Based Storage

For a range of transportation and other stationery and power generation applications, gaseous hydrogen is usually stored in pressurized tanks, which are typically constructed of all-metal or composite-overwrapped pressure vessels. Examples include carbon fiber composite-overwrapped tanks with metal or polymer liners for onboard storage of hydrogen at a pressure of 350 bar and 700 bar in commercial material-handling equipment and fuel cell electric vehicles, respectively. Larger all-metal or composite-overwrapped pressure tanks are used for bulk gaseous hydrogen storage at hydrogen refueling stations, and in various chemical and stationary power applications. Large-scale geologic storage within salt caverns, saline aquifers, depleted natural gas or oil reservoirs, and engineered hard rock reservoirs also offers opportunities for long-duration energy storage applications (U.S. DOE 2020).

7.4.2 Materials-Based Storage

As an alternative to gaseous or liquid storage, hydrogen can also be densely stored at low pressures in certain material compounds. Different categories of material-based storage include metal hydrides, adsorbents, and chemical hydrogen storage. Metal hydrides store hydrogen atoms by chemically bonding them to atoms in a compound structure. Examples include complex hydrides such as magnesium borohydride. Adsorbents, such as micro-porous super-activated carbons or metal-organic frameworks (MOFs), utilize weak bonding between molecular hydrogen and adsorbent surfaces, and typically require lower storage temperatures. Hydrogen storage via metal hydrides and adsorbents is considered reversible since hydrogen uptake and release can be controlled by changing the temperature and/or the pressure of the material. Many chemical hydrogen carriers, including materials such as methylcyclohexane, have the potential to store large quantities of hydrogen on a mass and volume basis. With these materials, however, thermal or catalytic chemical reactions are needed both to bind and release the hydrogen, and these processes can result in significant round-trip energy losses. Liquid chemical hydrogen carriers compatible with large-scale storage and transport include common compounds such as ammonia and methanol (U.S. DOE 2020).

7.5 Transportation and Transmission

Hydrogen can be transported in liquid or gaseous form. In gaseous form, hydrogen can be transmitted in much the same way that natural gas is today – by means of pipelines. The movement of gas by pipeline is one of the cheapest ways of moving energy, i.e., pipelines are clearly an attractive method of transportation of energy. Hydrogen can also be transported in liquid form, which it assumes at approximately −420°F and 1 atm. This liquid "system" would require continual cooling of the hydrogen. Liquid hydrogen has a relatively high energy per unit mass; thus, it is currently useful as a fuel for aircraft and rockets.

Hydrogen, as an energy carrier, may be stored in the gaseous form in an underground space. It may also be stored as a metal hydride. Aquifers and underground rock formations have been used for natural gas storage and can also be used for hydrogen, although the problems associated with the permeability of rocks are much more serious for hydrogen storage because of their low molecular weight. Finally, hydrogen has also been stored in steel cylinders at elevated pressures of approximately 125 atm.

As the production of electricity from hydrogen increases, so will the need for both energy storage and transportation. Hydrogen may be the perfect carrier for this energy. It can store the energy and distribute it to wherever it is needed. However, it is estimated that transmitting electricity long distances is four times more expensive than shipping hydrogen by pipeline.

To summarize, to support a wide range of applications, delivery infrastructure for hydrogen may incorporate multiple technology pathways capable of transporting hydrogen in various forms, including as a gas in pipelines and high-pressure tube trailers, as a liquid via tanker trucks, and using chemical hydrogen carriers. Different technologies for dispensing hydrogen may also be needed depending on how the hydrogen is transported, stored, and utilized. The technologies required to support these delivery pathways are at various stages of development, but they must all ultimately be both affordable and meet or exceed the level of safety, convenience, reliability, and energy efficiency expected from existing infrastructure for other fuels currently used in commerce. At present, there are four main methods of hydrogen delivery at scale: gaseous tube trailers, liquid tankers, pipelines (for gaseous hydrogen), and chemical hydrogen carriers (U.S. DOE 2020). Each of these pathways is described later in Chapter 12.

7.6 Uses

Hydrogen has the potential for use in diverse applications across multiple sectors, where it can provide substantial environmental and economic benefits, as well as improved energy security and resiliency. Large amounts of hydrogen can be used in many applications across multiple sectors, including transportation, power generation, and industrial and manufacturing sectors, which can enable economies of scale and support a robust domestic supply chain. Integrated energy systems, which can span sectors offer additional opportunities by using hydrogen as an energy carrier to improve the economics of existing and emerging electric power generation systems. In summary, hydrogen can have broad fuel and feedstock applications across many energy areas including: Transportation, Industrial Processes, Electric Power Generation and Integrated Hybrid Energy Systems.

In 2021, almost 47% of the global hydrogen production was from natural gas, 27% from coal, 22% from oil (as a by-product) and only around 4% came from electrolysis. Electricity had a global average renewable share of about 33% in 2021, which means that only about 1% of global hydrogen output is produced with renewable energy. Electrolytic hydrogen from dedicated production remained limited to demonstration projects adding up to a total capacity of 0.7 GW in 2021. In contrast, the 1.5°C Scenario would need 4–5 TW by 2050, requiring a faster rate of growth than that experienced by solar photovoltaic (PV) and wind to date.

Hydrogen used as an energy carrier currently remains limited to principally road vehicles. By June 2021 more than 40,000 fuel cell electric vehicles were in circulation around the world, with almost 90% of those in four countries: Korea, the United States, the People's Republic of China, and Japan. By the end of 2020, there were about 6,000 fuel cell electric buses (95% of those in China) and more than 3,100 fuel cell electric trucks. These totals are small fractions of the global vehicle fleet. Honda leased a fuel cell sedan called Clarity FCX from 2017 to 2021 at 16 dealerships in California. This fuel cell-powered Honda stopped production in 2021 reportedly due to the lack of hydrogen fueling stations.

The full value of hydrogen can be fully realized when it is further converted to derivatives. Hydrogen can be combined with carbon from CO_2 to produce virtually any hydrocarbon. It can be used to produce ammonia, which can be used as feedstock for fertilizers (most of the current use) or as fuel for new applications such as shipping.

Hydrogen can also be used to produce methanol, synthetic fuels, or even as a reducing agent to replace coal in iron production. Once it is converted to these commodities, the energy density is increased further, making long-distance transport and long-term storage cost-effective. Thus, the conversion to hydrogen derivatives effectively unlocks global renewable energy trade. For instance, liquid ammonia has almost eight times the energy density (MJ/m^3) of lithium-ion batteries and more than 20 times the gravimetric energy density (MJ/kg). See Figure 7.2 for comparative energy densities of various energy mediums.

The higher energy density of hydrogen-derived commodities effectively increases the distance that energy can be transported in a cost-effective way, connecting low-cost renewable energy regions with demand centers that have either limited renewable potential or costly renewable energy. Global energy trade through hydrogen derivatives would provide economic benefits as importing countries can tap into cheaper (than domestic) resources, improving the resilience of the system since there are alternative ways to satisfy final energy demand, and hence strengthening energy security.

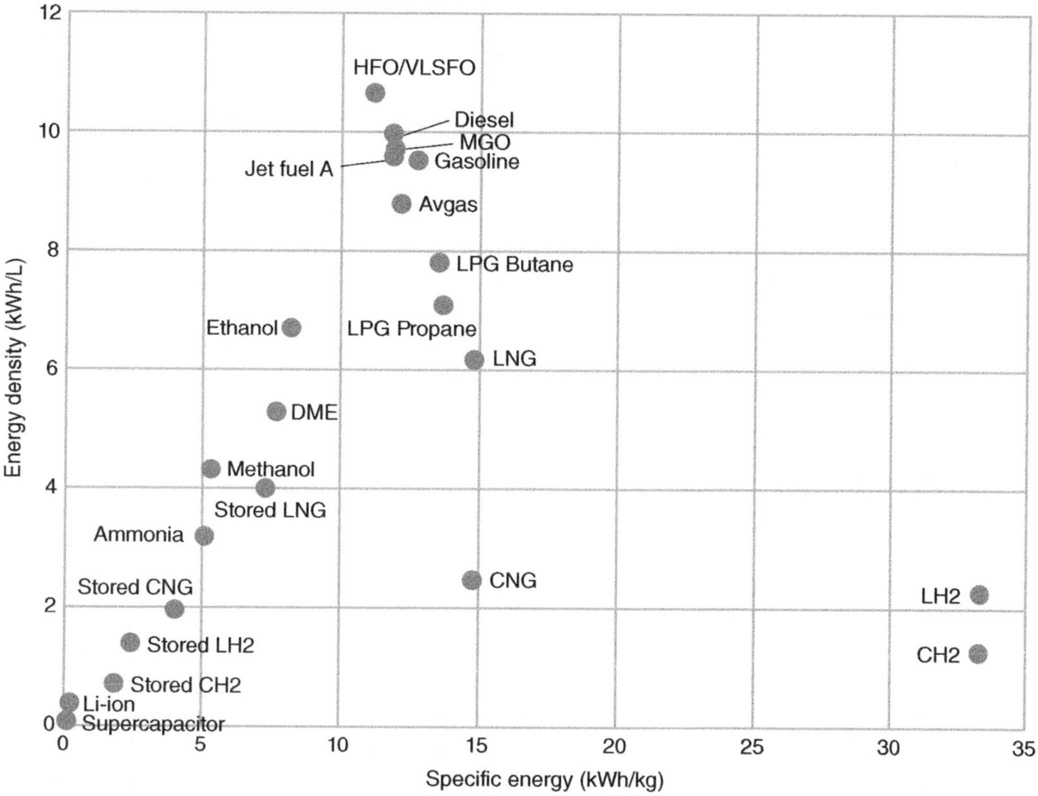

Notes: Avgas = aviation gasoline; CH2 = hydrogen compressed at 70 MPa; CNG = natural gas compressed at 25 MPa; DME = dimethyl ether; HFO/VLSFO = heavy fuel oil/very low sulphur fuel oil; LH2 = liquefied hydrogen; Li-ion = lithium-ion battery; LNG = liquefied natural gas; LPG = liquefied petroleum gas; Stored CNG = Type IV tank at 250 bar; Stored CH2 = best available CH2 tanks at 70 MPa; Stored LH2 = current small-scale LH2 on-board tanks; Stored LNG = small-scale storage at cryogenic conditions; MGO = maritime gasoil. Numbers are expressed on a lower heating value (LHV) basis. Weight of the storage equipment is included.

Figure 7.2 Energy density of various energy mediums compared to hydrogen and derivatives (*Source:* IRENA (2022), Global hydrogen trade to meet the 1.5 °C climate goal: Part II – Technology review of hydrogen carriers, International Renewable Energy Agency, Abu Dhabi).

7.6.1 Potential Role of Ammonia for Alternative Vehicle Fuel in a Hydrogen Economy

Hydrogen's low energy density is perhaps one of the greatest barriers to the implementation of hydrogen-fueled fuel cell vehicles. Several chemical and solid-state approaches have been examined, and are currently being investigated, that could lead to higher stored energy density. The possible role ammonia might play in implementation of a hydrogen economy as an alternative vehicle fuel. Ammonia has several desirable characteristics that suggest its use as a medium to store hydrogen. First, it can be liquefied under mild conditions. The vapor pressure of ammonia at room temperature is 9.2 bar (~121 psig). Its physical properties are like those of propane. This means that ammonia can be stored in a simple, inexpensive pressure vessel. Second, ammonia has a large weight fraction of hydrogen. Hydrogen constitutes 17.65% of the mass of ammonia.

When these two factors are combined, the result is a liquid that is simply contained, with a volumetric hydrogen density about 45% higher than that of liquid hydrogen. Ammonia can be decomposed (cracked) over a catalyst to produce the desired fuel – hydrogen (H_2) along with nitrogen (N_2) a nontoxic, nongreenhouse gas. In addition, ammonia might be an excellent transition fuel. It can be burned directly in an internal combustion engine (ICE) with no carbon emission, converted to electricity directly in an alkaline fuel cell, or cracked to provide hydrogen for nonalkaline fuel cells (FC). Central production from hydrocarbon fuels (e.g., natural gas) would provide opportunities for CO_2 capture; transportation and distribution would be simpler and cheaper than hydrogen delivery; and final use in an ICE or fuel cell would produce no carbon dioxide. Ammonia synthesis can be coupled with hydrogen production to increase efficiency or biomass. These feedstocks are generally gasified to form synthesis gas (CO and H_2), which can then be reacted with water and nitrogen to produce ammonia.

7.7 Environmental Issues

Hydrogen is an ecologically friendly fuel. It does not produce other chemical byproducts, except for a small amount of nitrogen oxides (at elevated temperatures). Hence, a key feature of hydrogen as a fuel is that it is relatively nonpolluting. Regarding safety from unwanted explosions, hydrogen fuel in automotive vehicles is probably as safe as gasoline. The technical community now realizes that the safe handling of hydrogen presents a challenge to work out the methods necessary to utilize hydrogen for commercial, residential, and transportation uses. However, the risks of ignition or spreading flames are actually less than those for gasoline. Since, in open air and well-ventilated places, leaks, or spills diffuse so rapidly (because hydrogen is the lightest of all elements), the basic risk-related requirements are appropriate ventilation, leak prevention, and elimination of ignition (Theodore and Dupont 2012; Dupont, Ganesan, and Theodore 2017; M.K. Theodore and L. Theodore 2021). The *Hindenburg* dirigible accident is a constant reminder of the need for safety precautions when using hydrogen.

Regarding fuel cells, the main reason that they have gained such notoriety in recent years is that they are environmentally friendly. This zero-emissions energy device has potential benefits to the environment by reducing pollution, including CO_2, the largest contributor to the greenhouse effect (M.K. Theodore and L. Theodore 2021). Widespread use of fuel cells in this country, especially in the transportation industry, could also reduce the reliance on fossil fuels. This benefit has tremendous implications. Imagine a hydrogen economy in which all domestic energy needs are produced in the United States. Imagine cell phones and laptops whose battery life is measured in months instead of days. Imagine vehicles operating silently and emitting harmless water vapor. Imagine individual homes generating their electricity and heat independently from the utility grid with an efficiency greater than any combustion device. It might sound too amazing now, but fuel cells and fuel cell technology have the potential to make a hydrogen economy a reality.

Technically sound codes and standards, effective safety- and quality-related technologies and processes, and widely disseminated safety information will all play a key role in realizing a meaningful hydrogen economy (Theodore and Dupont 2012). They are essential for deploying hydrogen-related technologies; ensuring quality, consistency, and interoperability; and providing regulatory bodies, manufacturers, system operators, and end users with the tools they need to ensure that emerging technologies are at least as reliable, safe, and high-performing as the incumbents. They also improve confidence in the commercial viability of the technologies among all stakeholders, which can further accelerate adoption and encourage investment in this new and sustainable energy source.

There is also a need to improve hydrogen sensing and contaminant detection, and development of quantitative risk assessment tools, and streamline permitting processes to help with siting hydrogen infrastructure and further reduce barriers to deployment. Risk mitigation strategies, best safety practices, and lessons learned must be identified and evaluated, and ongoing support will be needed to develop and sustain collaborative institutional processes for disseminating safety information about this technology.

Hydrogen codes and standards are in use today and critical in current industrial-scale hydrogen technologies (such as reforming, coal gasification, refineries, etc.), as are safety technologies such as sensors for monitoring and control at industrial plants and refineries. As new applications emerge – such as marine, rail, and heavy-duty vehicles – additional efforts in all aspects of safety, codes, and standards may be required to address needs specific to each application. This topic will be revisited in Chapter 17.

References

Dupont, R.R., Ganesan, K., and Theodore, L. (2017). *Pollution Prevention: Sustainability, Industrial Ecology, and Green Engineering*, 2nd Edition. Boca Raton, FL: CRC Press/Taylor & Frances Group.

IRENA (2022). International Renewable Energy Agency – Hydrogen Overview.

Skipka, K. and Theodore, L. (2014). *Energy Resources: Availability, Management, and Environmental Impacts*. Boca Raton, FL: CRC Press/Taylor & Francis Group.

Theodore, L. (2012). *Chemical Reactor Analysis and Applications for the Practicing Engineer*. Hoboken, NJ: John Wiley & Sons

Theodore, L. and Dupont, R.R. (2012). *Environmental Health and Hazard Risk Assessment: Principles and Calculations*. Boca Raton, FL: CRC Press/Taylor & Francis Group.

Theodore, L., Ricci, F., and VanVliet, T. (2009). *Thermodynamics for the Practicing Engineer*. Hoboken, NJ: John Wiley & Sons.

Theodore, M.K. and Theodore, L. (2021). *Introduction to Environmental Management*, 2nd Edition. Boca Raton, FL: CRC Press/Taylor & Francis Group.

U.S. Department of Energy (2019). Current hydrogen market size: domestic and global. DOE Hydrogen and Fuel Cells Program Record, Record #19002. https://www.hydrogen.energy.gov/pdfs/19002-hydrogen-market-domestic-global.pdf (accessed 3 September 2023).

U.S. Department of Energy (2020). Hydrogen program plan. Roles of Ammonia in a Hydrogen Economy A Study of Issues Related to the Use Ammonia for On-Board Vehicular Hydrogen Storage U.S. Department of Energy 2020.pdf (accessed 2 October 2024).

8

Government Hydrogen Programs

8.1 Introduction

A number of issues will be important to the development of a hydrogen energy economy in the United States, and the technical community needs to bear these in mind as this industry continues to develop. Two such issues include financial considerations and public education. These are briefly discussed below.

From a financial perspective, the availability of federal and state incentives, such as grants, loan guarantee programs, rebate programs, and tax incentives are crucial to the future development and adoption of hydrogen energy systems. In addition, accessibility to project financing will be an issue for new hydrogen projects, particularly those proposing the use of new technology or technology that has not yet been tested. Given the risks – both real and perceived – that hydrogen presents and the nature of large-scale deployment of hydrogen, insurance coverage considerations also will play an important role for the industry and its participants.

Another issue will be stakeholder engagement. Industry will need to invest in educating and engaging with the public and lawmakers to help ensure continued and expanded support, as well as promulgation of well-reasoned and transparent laws and regulations of the developing industry.

The U.S. federal government and many state governments already have recognized the potential that large-scale deployment of hydrogen energy holds. Support for continued and increased development exists in the form of not only grant programs but also tax incentives. These programs offer significant opportunities for the type of research and development (R&D) that is needed to make hydrogen cost-competitive and drive demand. Continuation and expansion of these programs will be a major factor in the development of a hydrogen energy economy.

Section numbers and their accompanying titles for this chapter concerned with government hydrogen energy programs are listed below.

Hydrogen Energy: Principles and Applications, First Edition. Vincent J. DelGatto, Louis Theodore, R. Ryan Dupont, and Matthew C. Ogwu.

The reader should note that the bulk of the material presented in this chapter was adapted from two key references in the literature that include: Skipka and Theodore (2014), and K&L Gates (2020).

8.2 Department of Energy Programs

The Department of Energy (DOE) is a cabinet-level branch of the U.S. government created by an act of Congress in 1977. The DOE is headed by the Secretary of Energy, who is appointed by the president with the advice and consent of the Senate. The DOE was established by Congress to deal with urgent national problems of dwindling supplies of coal, oil, and natural gas and the increasing dependence of the U.S. economy on foreign sources of fossil fuel, particularly oil. It is responsible for overseeing the management of the fuel and energy resources of the United States and for fostering more efficient technologies for the utilization of those resources. The DOE is also involved with exploring how best to use social media to accomplish its mission, engage the public in discussion, include people in the governing process, and collaborate internally and externally.

The DOE has its headquarters in Washington, DC, and several operations offices throughout the United States. It has assumed control of the facilities and personnel of five regional administrations for the transmission and marketing of electric power and strategic reserves of oil, shale, and uranium. Many research laboratories, including, for example, Brookhaven National Laboratory in Upton, New York, and the nuclear research facility at Oak Ridge, Tennessee, are also part of the DOE.

With regard to hydrogen, the federal government's main hydrogen R&D office is the Hydrogen and Fuel Cells Program which includes hydrogen production, delivery, infrastructure, storage, fuel cells, and multiple end-uses across transportation, industrial, and stationary power applications. The program also manages activities in technology validation, manufacturing, analysis, systems development and integration, safety, codes and standards, education, and workforce development.

Recent calls for grant proposals have dealt with advanced hydrogen fueling technologies for medium- and heavy-duty fuel cell vehicles and also proposals that address technical barriers to hydrogen blending in natural gas. In July 2020, DOE announced 18 awards for a total of $64 million for proposals in these priority areas.

Hydrogen technologies also would qualify for DOE Improved Energy Technology Loans through the DOE Loan Guarantee Program. Eligible projects for this program would reduce air pollution and greenhouse gas emissions and support early commercial use of advanced technologies, including biofuels and alternative fuel vehicles. The program is intended to accelerate commercial use of improved energy technologies. Eligible projects may include the deployment of fueling infrastructure, including associated hardware and software for alternative fuels (DOE 2023).

DOE has also announced its intention to invest up to $100 million over 5 years in two new DOE consortia to advance hydrogen and fuel cell technology R&D. These consortia are dependent on federal appropriations. One consortium would conduct R&D to achieve large-scale, affordable electrolyzers, which use electricity to split water into hydrogen and oxygen and can be powered by various energy sources, including natural gas, nuclear, and a host of renewables. This R&D will complement and support large industry deployment by enabling more durable, efficient, and low-cost electrolyzers. The other consortium will be involved with accelerating the development of fuel cells for heavy-duty vehicle applications, including long-haul trucks. This initiative will have a 5-year goal of proving the ability to manufacture a fully competitive heavy-duty fuel cell truck that can meet all of the durability, cost, and performance requirements of the trucking industry.

8.3 Other Federal Programs

In addition to the DOE programs, there are several other existing federal programs that support the production and deployment of hydrogen in the United States. Three such programs are summarized below:

1) The Airport Zero Emission Vehicle (ZEV) and Infrastructure Incentives is managed through the Federal Aviation Administration (FAA). Through this program, the FAA can award Airport Improvement Program (AIP) grants for the acquisition and operation of ZEVs at an airport. Grant funding must be used for airport-owned, on-road vehicles used exclusively for airport purposes (DOE 2023).
2) The Alternative Fuel and Advanced Vehicle Technology Research and Demonstration Bonds program allows qualified state, tribal, and local governments to issue Qualified Energy Conservation Bonds subsidized by the U.S. Department of the Treasury at competitive rates to fund capital expenditures on qualified energy conservation projects. Eligible activities include R&D projects related to nonfossil fuels, as well as advanced battery manufacturing technologies (DOE 2023).
3) The Low and Zero Emission Public Transportation Research, Demonstration, and Deployment Fund provides financial assistance to local, state, and federal government entities; public transportation providers; private and nonprofit organizations; and higher education institutions for research, demonstration, and deployment projects involving low and/or zero emission public transportation vehicles (DOE 2023).

Hydrogen technologies also qualify for various federal tax credits such as the Alternative Fuel Infrastructure Tax Credit that expired on 31 December 2020, and the ongoing Alternative Fuel Tax Exemption that will be discussed in more detail below.

8.4 State Programs

States have the ability to develop their own sets of laws and regulations similar to those of the federal government. Most federal environmental legislation encourages states to develop and run their own programs as an alternative to federal management as long as the state programs are as stringent as federal ones.

As many states look to reduce carbon emissions, promote renewable energy and clean transportation, and investigate seasonal energy storage, programs that help support hydrogen production and distribution have emerged. The list below provides a high-level overview of four state-level programs that show the range of programs states have begun to implement to reach their decarbonation goals (K&L Gates 2020).

One major problem of having 50 separate state agencies addressing energy issues is perhaps best illustrated by an initiative of the National Governor's Association (NGA) to solve problems with the widespread adoption of environmental control technologies (NGA 1997). The NGA pointed out, that ". . . developers and vendors of new . . . technologies are frustrated in their efforts to penetrate the enormous US market" and that "Many observers place the blame at the door of the states, where a single technology may be subject to 50 different procedures to gain a permit or approval." Another problem is a competition for authority in particular energy situations (i.e., whether federal law controls or state law prevails). Solutions are governed by the often unclear doctrine of preemption.

One can blame this patchwork quilt of laws, regulations, and standards on the notion of *federalism* on which the U.S. system of governance was built. Federalism is basically a sharing of' power between the states and national (federal) government. In 1787, the new federal government only exercised limited or enumerated powers granted to it by the Constitution, such as making treaties and printing money. The Tenth Amendment to the Constitution clarified that all other powers belonged to the states – "The powers not delegated to the United States by the Constitution, nor prohibited by it to the states, are reserved to the states respectively, or to the people" (Theodore 2022).

Additional updates on state laws and programs related to hydrogen energy and decarbonation efforts can be found through the DOE (2023) at https://afdc.energy.gov/fuels/laws/HY.

8.4.1 California

California has several regulations and incentives related to the production, distribution, and use of hydrogen. Most of these regulations and incentives are tied into broader clean transportation programs administered by the California Air Resources Board (CARB) or local air districts or utilities. The California Energy Commission (CEC) also makes grant funding available for the R&D of new hydrogen technologies and fueling stations.

The CARB and local air district programs include several rebate programs and mandates for ZEVs, which include hydrogen-powered vehicles and/or grant preferential treatment for hydrogen fueling stations (e.g., the Alternative Fuel Vehicle Parking Incentive Program). CARB also must periodically evaluate the need for publicly available hydrogen fueling stations and submit a report to the CEC. In turn, CEC can then make funding decisions for fueling infrastructure required to meet California's goal of 100 publicly available hydrogen fueling stations. Such stations are eligible to generate credits under California's Low Carbon Fuel Standard (LCFS).

California also requires its state agencies and utilities to broadly consider the energy landscape through the middle of the 21st century. Every 2 years, the CEC must submit an Integrated Energy Policy Report (IEPR). Utilities also must submit Integrated Resource Plans (IRPs) to the California Public Utilities Commission. In the past several years, the IEPR and IRPs have considered different types of energy storage technologies to integrate into California's renewables-heavy electricity sector. Hydrogen will begin to play a larger role in the IEPR and IRPs in the coming years. One high-profile example is the Los Angeles Department of Water and Power's plan to convert the Intermountain Power Plant in Delta, Utah, from coal to a blend of natural gas and green hydrogen sourced from nearby wind and solar generators and stored seasonally in giant salt caverns.

8.4.2 Oregon

Like California, Oregon's regulations and incentives for hydrogen tie closely to its clean transportation and ZEV programs. When hydrogen is used as a transportation fuel, it can also generate credits under Oregon's Clean Fuels Program, which is similar to California's Low Carbon Fuel Standard.

In September 2019 Oregon took a significant, step in greening its natural gas infrastructure with the passage of Senate Bill 98 (SB 98). SB 98 sets targets for utilities to deliver an increasing percentage of "renewable" natural gas to retail customers and allows utilities to include the higher cost of obtaining renewable natural gas in the rate base. Critically, SB98 includes renewable hydrogen in its definition of renewable natural gas. Renewable hydrogen is hydrogen produced from excess wind, solar, and hydropower and can be used by itself in the transportation sector and for industrial use or can be blended into the natural gas pipeline system and delivered to traditional gas-fired resources.

8.4.3 Washington

Like Oregon and California, Washington state has several hydrogen incentives embedded in its clean transportation initiatives. Most of these programs, including grant funding for fueling infrastructure, are administered by the Washington State Department of Transportation.

Washington state also adopted tax relief for key components of the hydrogen economy. For example, sales and use taxes do not apply to the sale of property used for hydrogen fueling infrastructure. Washington's 6.5% retail sales and state use tax does not apply to the sale or lease of new or used hydrogen-powered passenger vehicles, light-duty trucks, and medium-duty passenger vehicles.

Finally, in April 2019, Substitute Senate Bill 58 authorized Washington's public utilities to produce, distribute, and sell hydrogen produced from renewable resources such as wind, solar, and hydropower. The bill was seen as a win for environmentalists, as well, because utilities would use excess midday wind and solar power and reduce excess spill from hydropower facilities, which has the potential to harm fish. While utilities may include the higher cost of acquiring green hydrogen in their rate base, the Washington bill (unlike Oregon's SB 98) stops short of setting targets for the delivery of renewable natural gas or renewable hydrogen to retail customers.

8.4.4 South Carolina

Because it hosts several domestic and foreign automobile manufacturing facilities interested in positioning themselves for hydrogen fuel cell vehicles, South Carolina was an early adopter of hydrogen-friendly regulations and incentives. Several universities, federal labs, and state agencies founded the South Carolina Hydrogen and Fuel Cell alliance to advance the cause of hydrogen fuel cells in the state. In June 2010, South Carolina passed the Hydrogen and Fuel Cell Permitting Law that placed statewide permitting authority within the Office of the State Fire Marshal. The South Carolina Research Authority administered the South Carolina Hydrogen Infrastructure Development Fund until 2012, although that program appears to have expired. South Carolina also offers a sales tax exemption that helps to promote hydrogen production and fuel cell technologies.

8.5 Tax Incentives

Several U.S. federal income and excise tax credits encourage investment in hydrogen projects. These credits include:

1) The investment tax credit (ITC) available under Tax Code Section 48 for qualified fuel cell property that generates electricity.
2) A federal income tax credit for fuel cell motor vehicles placed in service during a given tax year under Code Section 30B.
3) A federal income tax credit for placing alternative fuel vehicle refueling property into service during a given tax year under Code Section 30C.
4) A federal excise tax credit on the sale or use of liquefied hydrogen under Code Section 64264(d).

These four tax credit programs are described in more detail below (K&L Gates 2020).

8.5.1 ITC for Fuel Cell Property

The ITC for qualified fuel cell property provides a percentage credit (subject to a per-kilowatt cap) against U.S. federal income tax for property placed in service before 1 January 2024. Property is qualified fuel cell property, if it is a "fuel cell power plant" with a nameplate capacity rating of at least 0.5 kW and has an electricity-only generation efficiency greater than 30%. A fuel cell power plant is defined as "an integrated system comprised of a fuel cell stack assembly and associated balance of plant components which converts a fuel into electricity using electrochemical means." Thus, an operational hydrogen fuel cell with a nameplate capacity of at least 0.5 kW generally qualifies for the ITC.

Financing structures involving fuel cells vary depending on the size of the fuel cell and financing facility. Taxpayers who have the need for both the electricity produced by hydrogen fuel cells (e.g., data center operators) and the ITC have the option to own the fuel cells outright but should plan for asset retirement and disposition. For all other taxpayers, sale-leaseback structures are particularly useful for smaller fuel cells, e.g., cells used to power warehouse equipment, and may also be used for larger, stationary fuel cells if the right counterparty can be obtained. Partnership flip structures and lease passthrough structures are also available.

8.5.2 New Qualified Fuel Cell Motor Vehicle Credit

New qualified fuel cell motor vehicles placed in service after 2010 generally qualify for a credit up to a per-vehicle maximum of $4,000–$40,000 depending upon the gross vehicle weight rating for the vehicle. The allowable credit amount will be increased by between $1,000 and $4,000 per vehicle if such a vehicle achieves a fuel efficiency of between 150% and 300% of a statutory baseline fuel efficiency standard. This credit was available only for motor vehicles placed in service through 31 December 2020.

A motor vehicle is a new qualified fuel ceil motor vehicle if:

1) It is powered by a hydrogen fuel cell.
2) It meets applicable clean air standards.
3) Its original use commences with the taxpayer claiming the credit.
4) It is acquired for use or lease by the taxpayer rather than for resale.
5) It was made by a manufacturer.

For this purpose, a motor vehicle is a vehicle that is capable of operating on public roads (or exclusively on rails) and has four wheels.

8.5.3 Alternative Fuel Vehicle Refueling Property Credit

A credit is also available for qualified alternative fuel vehicle refueling property that is placed in service no later than 31 December 2020. The credit is up to 30% of the cost of the property, subject to a cap of $30,000 for depreciable property (i.e., property used in a trade or business) and $1,000 for any other property placed in service by the taxpayer "at a location."

Qualified alternative fuel vehicle refueling property must be:

1) Depreciable
2) Owned by the taxpayer that put it to original use
3) Used for the storage or dispensing of, among other things, fuel the volume of which is composed of at least 85% hydrogen
4) Not be installed on property that is used as the principal residence of the taxpayer claiming the credit

To the extent that all or a portion of this credit is attributable to property for which a Code Section 38(b) credit is allowed (e.g., new markets tax credit property), the alternative fuel vehicle refueling property credit is not allowed.

8.5.4 Alternative Fuel Credit

The alternative fuel credit is available as an offset to the Code Section 4041 fuel excise lax. The credit is $0.50 per gallon equivalent of an alternative fuel (that includes liquefied hydrogen) produced in the United States and sold by a taxpayer for use as a fuel in a motor vehicle or motorboat, or for use as fuel in aviation or consumed by the taxpayer. This fuel must meet carbon recapture requirements which can be accomplished by the fuel being certified as having been derived from coal produced at a gasification facility that separates and sequesters not less than 75% of such facility's total carbon dioxide emissions. This credit was available for fuel produced through 2020.

Tax-exempt entities such as state and local governments that dispense qualified fuel from on-site refueling stations for use in vehicles should generally qualify for this credit but must register with the IRS. The credit must first be taken against the entity's alternative fuel tax liability, with any excess claimed as a direct payment from the IRS.

8.6 Project Financing

While hydrogen investing has been increasing, project financing for hydrogen has been very limited outside of select government-supported projects and small niche applications. In an article by Lafargue (2020) of Barclays, he suggested hydrogen could be a $2.5 trillion market by 2050. If the growth projections for new hydrogen applications over the next two decades are to become reality, then a massive expansion of hydrogen project financing options and expertise will be necessary.

As with any emerging industry or technology, leveraging existing project finance tools will be difficult until market and technological certainty has become more widely accepted. This is especially true for hydrogen because hydrogen production, particularly green and blue hydrogen, currently is more expensive than natural gas. Cost reductions will require scale, which will almost certainly require continued and potentially expanded government investment or subsidies similar to those discussed for the initial commercial-scale growth of the industry. In fact, the Hydrogen Council (2020) estimated that the hydrogen market requires more than $70 billion of investment over the next decade to become competitive with the lowest-cost low-carbon alternative. Some of this investment, including investment in the buildout of necessary supporting infrastructure, will be in the type of capital-intensive assets that are suited for project finance structures.

Project finance refers to financing with limited recourse from lenders directly to the owner/sponsors. Financing is based on the economic viability of the project itself and relies on project revenue to service financing payments. With capital-intensive projects, project financing typically requires some form of long-term revenue certainty.

In the near term, new hydrogen projects will carry market and regulatory uncertainty and will rely on technology that is new or not proven, at scale, which will add risk, making traditional project finance difficult, and certainly much more expensive. Government support that offsets project-level risk or provides subsidies to create higher returns likely will be necessary to support project financing.

Beyond economics, public perception and politics likely will define the development of financing markets for the hydrogen economy as investors and lenders are, in significant numbers, searching for access to the growing renewable energy economy.

8.7 Insurance Coverage

The energy industry is among the most dynamic in the world, as are the risks that it faces. Those risks have materialized into some of the most devastating and costly losses of any industry. Significant losses affecting the energy industry have occurred in the past (Theodore and Dupont 2012; K&L Gates 2020).

More specific to hydrogen, the diverse methods of its production carry high risks, such as the release of harmful or flammable gasses, including carbon monoxide, carbon dioxide, and methane. Additional traditional risks of all forms of hydrogen production, transportation, storage, and end-use include design flaws, human error, equipment failure, and natural disasters. As such, the development of opportunities in any segment of the energy industry, including the hydrogen market, carries with it the need for careful and sophisticated risk assessment and management (Theodore and Dupont 2012; K&L Gates 2020).

The transfer of risk through insurance is a critical part of any enterprise's risk management program, Moreover, insurance is an important corporate asset and is often a company's largest source of contingent capital. Companies participating in the hydrogen market must carefully construct their insurance programs to meet the considerable risks of their operations. Insurance programs for energy companies also will reflect the complexity of their operations. This will require the placement of an insurance program of sufficient breadth and likely will require spreading the risk across multiple insurers so as to provide sufficient amounts of coverage. Consideration also must be given to incorporating appropriate deductibles or self-insured retentions.

In addition to a sophisticated insurance approach, another important risk management tool for companies in the hydrogen market is the proactive and coordinated management of contract-driven relationships with contractors and suppliers. Vendor contracts routinely include indemnification obligations and insurance requirements. These contract provisions should be managed for consistency across vendors and for coordination with the company's own insurance program.

8.8 Stakeholder Engagement

If the United States begins to adopt and integrate hydrogen as an energy source into the national infrastructure, public stakeholder engagement could be one of the most critical elements to ensure success. A lack of familiarity and experience with hydrogen on the part of the general public could lead to distrust and open opposition. For most of the U.S. public, the mention of hydrogen conjures images of the Hindenburg Airship disaster in New Jersey in the 1930s. Moreover, opposition movements to infrastructure, i.e., Not In My Back Yard (NIMBY); Not On Planet Earth (NOPE); and Build Absolutely Nothing Anywhere Near Anybody (BANANA) – do not discriminate against the infrastructure they oppose, as many U.S. wind and solar power developers have experienced.

Much of the U.S. legal structure is built on the idea of public engagement in the democratic process, whether that is through special interest lobbying in Washington, D.C., regulatory proceedings before federal agencies, or civic engagement in local issues and elections. Frequently, defined processes afford opportunities for stakeholder engagement on safety and security issues, in particular.

Such processes include opportunities for written comments, public scoping meetings, technical conferences, and legal challenges and appeals to agencies' or executives' decisions. Industry participants that develop hydrogen infrastructure must:

1) Understand any existing regulatory framework for engaging hydrogen safety and security issues.
2) Establish a baseline understanding of hydrogen safety and security issues applicable to each particular project.
3) Identify key safety and security messages that are critical to convey to stakeholders.

Regulators and other governmental decision-makers should intentionally explain the regulatory processes, highlighting elements that factor into the decision-makers conclusions, including:

1) Coordination requirements between a project developer, state, and local officials with regard to safety, risk, and security issues.
2) Development of any emergency response plans that may be required, including any procedures required for notification of the public and identification of any evacuation routes in the event of an accident.
3) Mitigation strategies that a developer may be required to develop to supplement the standard safety and security practices employed by the developer, including any ongoing training programs and any customized public action plans.

There will be no way to eliminate public participation and any potential opposition to the development of the hydrogen economy in the United States. Each hydrogen infrastructure project will present new and different challenges requiring tailored engagement plans and outreach. As a result, the hydrogen industry must be deliberate and intentional about educating and engaging thoughtfully with impacted stakeholders and the general public.

References

Hydrogen Council (2020). *Path to Hydrogen Competitiveness: A Cost Perspective*. Washington, DC: Hydrogen Council. https://hydrogencouncil.com/wp-content/uploads/2020/01/Path-to-Hydrogen-Competitiveness_Full-Study-1.pdf.

K&L Gates (2020). *The H_2 Handbook. Legal, Regulatory, Policy, and Commercial Issues Impacting the Future of Hydrogen*. Washington, DC: K&L Gates LLP. https://www.klgates.com/The-Hydrogen-Handbook#:~:text=The%20H2%20Handbook%20provides,part%20of%20the%20energy%20mix.

Lafargue, J. (2020). *Hydrogen Energy, the Fuel of the Future*. Barclays Bank Plc. https://privatebank.barclays.com/news-and-insights/2020/october/hydrogen-energy (accessed 22 September 2023).

National Governor's Association (1997). *Interstate Cooperation to Speed Multistate Acceptance of Environmental Technologies*. Washington, DC: National Governor's Association, Natural Resources Policy Studies Division.

Skipka, K. and Theodore, L. (2014). *Energy Resources: Availability, Management, and Environmental Impacts*. Boca Raton, FL: CRC Press/Taylor & Francis Group.

Theodore, L. (2022). On the constitution. *The Theodore Newsletter*. August. www.theodorenewsletter.com.

Theodore, L. and Dupont, R.R. (2012). *Environmental Health and Hazard Risk Assessment: Principles and Calculations*. Boca Raton, FL: CRC Press/Taylor & Francis Group.

U.S. Department of Energy (2023). *Hydrogen Laws and Incentives*. Washington, DC: Alternative Fuels Data Center. https://afdc.energy.gov/fuels/laws/HY (accessed 2 October 2024).

9

Hydrogen Physical and Chemical Properties

*Onwukaeme Chibuzo Kenneth**

*School of Natural Science, Inha University, Incheon, South Korea

9.1 Introduction

This chapter examines the physical and chemical properties of hydrogen. Hydrogen is the simplest element in nature and possesses the capacity for clean energy production, making it a potential replacement for existing fossil fuel technology. With the active pursuit of net-zero energy goals for industrial and residential applications, this chapter explores the unique properties of hydrogen that qualify it as a viable candidate for a 21st century clean energy source.

Section numbers and their accompanying titles for this chapter concerned with the physical and chemical properties of hydrogen are listed below.

9.2 Physical and Chemical Properties of Matter

Both the physical and chemical properties of materials are important considerations in any hydrogen energy study. A substance may exhibit certain characteristics under one particular set of reaction or process conditions of temperature, pressure, and composition. However, these conditions are often changed when an attempt is made to implement an energy strategy. To promote a better understanding of such effects, many of which are covered in this section, a qualitative review of atomic theory is included. This is followed by material related to macroscopic physical and chemical properties, and properties of mixtures which are briefly reviewed in the next section.

Hydrogen Energy: Principles and Applications, First Edition. Vincent J. DelGatto, Louis Theodore, R. Ryan Dupont, and Matthew C. Ogwu.

Mass in the sciences is defined as the amount of matter that a body contains and as a measure of the inertial property of that body, i.e., of its resistance to change of motion. Mass is different from weight. Weight on Earth, although proportional to mass, varies with the position of a given mass relative to the Earth. Thus, equal masses at the same location in a gravitational field will have equal weights.

At the atomic level matter consists of atoms which are made of elementary particles that include electrons, protons, and neutrons, which retain their identity when chemical reactions take place. Their diameter is in the Angstrom range. An atom consists of one nucleus and one or more electrons. The nucleus is several orders of magnitude smaller than an atom and has a positive charge while electrons are negatively charged. Matter is electrically neutral; therefore, it contains equal amounts of positive and negative electrical charge. Chemical composition reveals the types and numbers of atoms, which in turn defines the physical properties of the substance. The constituent atoms possess distinctive properties that depend on factors such as the number of electrons in their valence shells, the mass of their nucleus, physical appearance, and response to external fluctuations such as variations in temperature and pressure. Hydrogen is composed of a single atom with one electron orbiting around a nucleus containing a single proton.

On a macroscopic level, matter can be divided roughly into three main categories – gases, liquids, and solids; these are often described as the three chief states of matter. The gaseous state is characterized by a marked sensitivity of the volume to changes in temperature and pressure; it tends to fill completely any available space. One of the main characteristics of a gas is that its molecules are not held together but move about freely in a volume significantly larger than the volume of the molecules themselves. The molecules of gas exert pressure on the walls of its container by striking the walls and rebounding from them. If the volume occupied by the gas is decreased, the molecules strike the walls more often and the pressure of the gas increases accordingly.

A liquid, similar to a gas, has no definite shape and takes the shape of the vessel in which it is contained. A liquid, such as liquid hydrogen, is in equilibrium with its own vapor. The pressure corresponding to the concentration of gas molecules in equilibrium with the pure liquid is called the *vapor pressure of the liquid* at a given temperature. The vapor pressure of every liquid increases with increasing temperature. An increase in temperature produces an increase in molecular activity, with increased molecular oscillation in gases and increased molecular vibration in both liquids and solids.

9.2.1 Physical Properties

Density. The *density* of a substance is the ratio of its mass to its volume and may be expressed in units of pounds per cubic foot (lb/ft^3) or kilograms per cubic meter (kg/m^3). For solids, density can be determined easily by placing a known mass of a substance in a liquid and measuring the displaced volume. The density of a liquid can be measured by weighing a known volume of the liquid in a graduated cylinder. For gases, such as hydrogen, the Ideal Gas Law can be used to calculate the density from the temperature, pressure, and molecular weight of the gas.

Viscosity. *Viscosity* is a property associated with a fluid's resistance to flow; more precisely, this property accounts for the energy losses resulting from shear stresses that occur between different portions of the fluid, moving at different velocities. The absolute viscosity μ has units of mass per length-time; the fundamental unit is the poise, which is defined as 1 g/cm-s. Viscosities are frequently given in centipoises (0.01 P), the abbreviation for which is cP. In English units, the absolute viscosity is expressed as pounds (mass) per foot-second (lb/ft-s). The absolute viscosity depends primarily on temperature and to a lesser degree on pressure.

The kinematic viscosity, v, the absolute viscosity divided by the density of the fluid, is useful in certain fluid flow applications. The units for this quantity are length squared per time (e.g., ft^2/s). Because fluid viscosity changes rapidly with temperature, a numerical value of viscosity has no significance unless the temperature is specified.

Heat capacity. The *heat capacity* of a substance is defined as the quantity of heat required to raise the temperature of a unit mass by 1°. The term *specific heat* is frequently used in place of *specific heat capacity*. This is not strictly correct because, traditionally, specific heat has been defined as the ratio of the heat capacity of a substance to the heat capacity of water. However, since the specific heat of water is approximately 1 cal/g-°C or 1 Btu/lb-°F, the term specific heat has come to imply heat capacity. For gases, the addition of heat to cause a 1° temperature rise may be accomplished either at constant pressure or at constant volume. Since the amounts of heat necessary are different for the two cases, subscripts are used to identify which heat capacity is being used – C_p for constant pressure, and C_v for constant volume. This distinction does not have to be made for liquids and solids, since there is little difference between the two. Values of heat capacity are available in the literature (e.g., Theodore and Reynolds 1992; Theodore et al. 2009) or online, for example, The Engineering Toolbox, www.EngineeringToolBox.com.

Thermal conductivity. Energy in the form of heat will flow from a region of high temperature to that of low temperature until thermal equilibrium is reached. This mode of heat transfer, in which vibrating molecules pass kinetic energy through a solid, is called conduction. Liquids and gases also transport heat in this fashion. *Thermal conductivity* provides a measure of how fast heat flows through a substance and is defined as the amount of heat that flows in a unit of time through a unit surface area of unit thickness as a result of a unit difference in temperature. Typical units for thermal conductivity are Btu/h-ft-°F.

Diffusivity. The *diffusion coefficient*, or *diffusivity*, D, is a measure of the rate of transfer of one substance through another by molecular diffusion. The diffusivity for a component A migrating through a solute B, for example, is given by the ratio of the flux J_A (mass transferred per unit time through a unit area) to its concentration gradient dC_A/dz. Diffusivity is a function of temperature, pressure, concentration, phase, and the number and nature of the other components within a mixture (Theodore and Ricci 2010).

Vapor pressure. *Vapor pressure* is an important property of liquids and, to a much lesser extent, of solids. If a liquid is allowed to evaporate in a confined space, the pressure of the vapor phase increases as the amount of vapor within the space increases. If there is sufficient liquid present, the pressure in the vapor space eventually comes to equal exactly the pressure exerted by the liquid at its own surface. At this point, a dynamic equilibrium exists in which vaporization and condensation take place at equal rates and the pressure in the vapor space remains constant. This gas pressure exerted at equilibrium is called the vapor pressure of the liquid. Solids, like liquids, also exert a vapor pressure. Evaporation of solids (sublimation) is noticeable only for the few solids characterized by materials with an appreciable vapor pressure.

Boiling point. The *boiling point* is the temperature at which the vapor pressure of a liquid is equal to its surrounding pressure. Since the vapor pressure remains constant during boiling, so does the liquid temperature. The higher the system pressure, the higher the temperature must be to induce boiling. The boiling point is specific to each individual substance and is a strong function of pressure.

Freezing point. The *freezing point* is the temperature at which a liquid and solid state can coexist in equilibrium at a given pressure. At this point, the rate at which the substance leaves the solid state equals the rate at which it leaves the liquid state.

9.2.2 Chemical Properties

Flammability (or explosion) limit. Flammability limits for a flammable gas define the concentration range of a gas-air mixture within which an ignition source can start a self-propagating reaction. The minimum and maximum fuel concentrations in air that will produce a self-sustaining reaction under given conditions are called the lower flammability limit (LFL) and the upper flammability limit (UFL), respectively. The abbreviations LEL and UEL, for lower and upper explosive limits, are sometimes used. The range of flammability limits becomes greater when the ignition energy is higher, moving the UFL to a higher concentration. The flammability limits also increase when the initial pressure and/or initial temperature at which the ignition source is activated increases. In addition, the increase in pressure and temperature will increase the rate at which a flame propagates through a gas. Tables of flammability limits of flammable gases are available in the literature, for example, Theodore et al. (1988) or Theodore and Dupont (2012), or online, for example, at The Engineering Toolbox, www.EngineeringToolBox.com.

Flash point. The *flash point* of a flammable liquid is defined as the temperature at which the vapor of the liquid is the same as the vapor pressure corresponding to the LFL concentration. The three major methods of measuring the flash point are:

1) The Cleveland open cup method.
2) The Penskey-Martens closed cup tester.
3) The tag closed cup method.

Autoignition temperature (AIT). The *autoignition temperature (AIT)*, or the maximum spontaneous ignition temperature, is defined as the minimum temperature at which combustion occurs in a bulk gas mixture when the temperature of a flammable gas-air mixture is raised in a uniformly heated apparatus. The AIT represents a threshold below which chemicals and combustibles can be handled safely. The AITs of selected substances are available in the literature (Theodore et al. 1988; Theodore and Dupont 2012) or online, for example, The Engineering Toolbox, http://www.engineeringtoolbox.com/fuels-ignition-temperatures-d_171.html.

Heat (enthalpy) of reaction. Many chemical reactions evolve or absorb heat. When applying energy balances in technical calculations the heat (enthalpy) of reaction is often indicated in mole units so that they can be directly applied to show its chemical change. Tables of standard enthalpies of formation combustion and reaction are available in the literature (Theodore and Reynolds 1992; Theodore et al. 2009) or online, for example, The Engineering Toolbox, www.EngineeringToolBox.com, for a wide variety of compounds. It is important to note that these are useless unless the stoichiometric equation and the state of reactants and products are included. Refer to Part I, Chapter 5 for more details on this chemical property.

Heat (enthalpy) of combustion. The heat (enthalpy) of combustion at temperature T is defined as the enthalpy change during the chemical reaction where 1 mol of material is burned in oxygen, where all reactants and products are at temperature T. This quantity finds extensive application in calculating enthalpy changes for incineration reactions and is often given in the literature for a standard state temperature of 60°F (16°C). Although much of the literature data on standard heats of reaction are given for 25°C, there is little sensible enthalpy change between these two temperatures and the two sets of data may be considered compatible. Thus, the standard heat enthalpy of combustion is obtained by taking the difference between the standard heat of formation, ΔH_f°, of the products and that of the reactants. If the standard heat of combustion is negative (exothermic), as is the case with most combustion/incineration reactions, energy is liberated as a result of the reaction. Energy is absorbed if the enthalpy of combustion is positive (endothermic). Extensive standard heat of formation and standard enthalpy of combustion data at 298 K are provided in the literature (Theodore and Reynolds 1992; Theodore et al. 2009) or online, for example, at UC Davis ChemWiki, http://chemwiki.ucdavis.edu/Reference. Refer to Part I, Chapter 5 for more details on this chemical property.

Gross heating value. The heating value of a fuel is one of its most important chemical properties. This value represents the amount of heat evolved in the complete reaction (usually combustion) of a given quantity of waste and/or fuel. The gross heating value or higher heating value (HHV) represents the enthalpy change or heat released when a compound is stoichiometrically combusted at 60°F, with the final (flue) products at 60°F and any water present in the liquid state. Stoichiometric combustion requires that no oxygen be present in the flue gas following combustion of the starting fuel. Refer to Part I, Chapter 5 for more details on this chemical property.

Net heating value. Due to the varying amounts of water that may be produced in a combustion reaction, a second method for expressing heating value is also in use, the *net heating value*. The net heating value is similar to the gross heating value except that it is calculated with the produced water in the vapor state. The net heating value is also known as the lower heating value (LHV). The LHV is equal to the HHV minus the heat of vaporization of the water in the final combustion gas. Refer to Part I, Chapter 5 for more details on this chemical property.

Theoretical adiabatic flame temperature (TAFT). Most combustion reactions do not operate at stoichiometric or 0% excess air. Incomplete combustion and high carbon monoxide levels would result under this condition. If all the heat liberated by the reaction goes into heating up the products of combustion (the flue gas), the temperature achieved is defined as the *flame temperature*. If the reaction process is conducted adiabatically, with no heat transfer loss to the surroundings, the final temperature achieved in the flue gas is defined as the *adiabatic flame temperature*. If the combustion process is conducted with theoretical or stoichiometric air (0% excess air), the resulting temperature is defined as the *theoretical flame temperature*. Theoretical or stoichiometric air is defined as the exact amount of air required to completely react with a compound to produce oxidized end products. Any air in excess of this stoichiometric amount is defined as excess air. If the reaction is conducted both adiabatically and with stoichiometric air, the resulting temperature is defined as the *theoretical adiabatic flame temperature* (TAFT). It represents the maximum temperature the products of combustion (flue) can achieve if the reaction is conducted both stoichiometrically and adiabatically (Theodore 2014). Refer to Part I, Chapter 5 for more details on this chemical property.

The Wobbe index. The *Wobbe index* is defined as the heating value divided by the square root of the specific gravity (relative to air) of the fuel. Two chemically different fuels that have the same Wobbe index will produce the same heat output in a burner designed to operate at a given pressure, making the Wobbe index a key parameter for evaluating fuel interchangeability. Pipelined natural gas in the United States has a specified range of Wobbe index to ensure that commercial and residential burners produce constant heat with fuel pressure regulation (Green and Southard 2019).

Flame speed. *Flame speed* is a measured property of each fuel and should be considered a descriptor of how "fast" a flame can propagate in a fuel. The actual flame speed in a given device depends on many factors (i.e., temperature, fuel/air ratio, turbulence level). It is commonly known that hydrogen has a very high flame speed compared to methane and factors unique to hydrogen combustion must be considered in devices that use premixed hydrogen flames.

9.3 Properties of Mixtures

pH. An important chemical property of an aqueous solution is its *pH* value, which indicates the acidity or basicity of the substance. In a neutral solution, i.e., pH = 7.0, such as pure water, the hydrogen (H^+) and hydroxyl (OH^-) ion concentrations are equal. At ordinary temperatures, this concentration is:

$$\left[H^+\right]=\left[OH^-\right]=10^{-7}\,\text{g-ion/L} \tag{9.1}$$

where $[H^+]$ = hydrogen ion concentration, and $[OH^-]$ = hydroxyl ion concentration. The notation g-ion is the unit for gram-ion, which represents Avogadro's number (6.02×10^{23}) of ions.

In all aqueous solutions, whether neutral, basic, or acidic, a chemical equilibrium or balance is established between these two concentrations, so that:

$$K_{eq}=\left[H^+\right]\left[OH^-\right]=10^{-14} \tag{9.2}$$

where K_{eq} = equilibrium constant. The numerical value for K_{eq} given above holds for room temperature and only when the concentrations are expressed in g-ion/L. In acid solutions, $[H^+]$ is > $[OH^-]$ in basic solutions, $[OH^-]$ predominates.

The pH is a direct measure of the hydrogen ion concentration and is given by:

$$\text{pH}=-\log\left[H^+\right] \tag{9.3}$$

Thus, an acidic solution is characterized by a pH below 7 (the lower the pH, the higher the acidity), a basic solution by a pH above 7, and a neutral solution by a pH of 7.

Partial pressure. In engineering practice, mixtures of gases are more often encountered than single or pure gases. The ideal gas law is based on the number of molecules present in the gas volume; the nature of molecules is not of significance. This law applies equally well to mixtures and to pure gases. Since pressure is caused by gas molecules colliding with the walls of a container, it seems reasonable that the total pressure of a gas mixture is made up of pressure contributions from each of the component gases. These pressure contributions are called *partial pressures*. Dalton defined the partial pressure of a component as the pressure that would be exerted if the same mass of the component gas occupied the same total volume alone at the same temperature as the mixture. The sum of these partial pressures equals the total pressure, *P*:

$$P=P_1+P_2+P_3+\cdots+P_n=\sum_{i=1}^{n}P_i \tag{9.4}$$

For ideal gases, partial pressure is equal to:

$$P_i=y_iP \tag{9.5}$$

where y_i is the mole fraction of Component I (Theodore et al. 2009).

Humidity. Curves showing the relative humidity (ratio of the mass of the water vapor in the air to the maximum water vapor the air can hold at that temperature, i.e., if the air were saturated) of humid air appear on *psychrometric charts* (Green and Southard 2019). The curve for 100% relative humidity is also referred to as the saturation curve. The abscissa of the humidity chart is air temperature, also known as the dry-bulb temperature (TDB). The wet-bulb temperature (TWB) is another measure of humidity. It is the temperature at which a thermometer with a wet wick wrapped around the bulb stabilizes. As water evaporates from the wick to the ambient air the bulb is cooled; the rate of cooling depends on how humid the air is. No evaporation occurs if the air is saturated with water; hence, at that point, the TWB and TDB are the same (Theodore and Ricci 2010).

9.4 Properties of Hydrogen

9.4.1 Chemical and Molecular Properties of Hydrogen

The elements in the universe are predominantly hydrogen, which contains 75% of the mass of all elements in the galaxies. In the Earth's atmosphere, its lightweight makes it less abundant, with a 0.14% by mass in the Earth's crust. Naturally, hydrogen exists as a diatomic molecule, called dihydrogen, which consists of two atoms of hydrogen represented by the chemical formula H_2 with a mass of 2.016 g/mol. In this form, they are unreactive because of the strong H—H bond between the atoms.

Subatomic particles like electrons and protons can be indexed by their spin, which arises from their orientation along a specific direction or their intrinsic angular momentum. When hydrogen molecules are formed, the constituent atoms' nuclear spins interact by aligning the spins of the protons in the nuclei. The hydrogen molecule formed when the proton spins are aligned in the same direction is called ortho-hydrogen. In contrast, a hydrogen molecule with a proton or nuclear spin aligned in the opposite direction is called para-hydrogen. These two forms are distinct in their properties and usually cannot be transformed from one form to another except under certain conditions. Ortho-hydrogen has a total spin of $S = 1$ with spin components of $m_s = 1, 0, -1$, thereby exhibiting a triplet state. They are unstable because of their excited state nature, thus possessing higher energy than para-hydrogen. On the other hand, para-hydrogen molecules are stable molecules with a total spin of $S = 0$, resulting in a singlet state formation. At equilibrium, the hydrogen molecule can contain different percentages of each form, which strongly depends on temperature. Equilibrium can be achieved by applying a catalyst, an electric discharge, or heating the gas molecules to a very high temperature. At absolute zero temperature, the hydrogen molecule is primarily para-hydrogen, which exhibits strong stability. Increasing temperature results in the variation of the ratios of both forms. At 80 K, the ortho-to-para ratio is 1:1, and at room temperature (293 K), it rises to 3:1 (Subramani et al. 2015). Furthermore, their somewhat different physical properties result in both forms having different melting points. Practically, pure para-hydrogen is obtained by adsorbing ordinary hydrogen on charcoal at approximately −225°C.

The hydrogen atom has an atomic number of 1 and is smaller than the molecules of most gases; as a result, hydrogen diffusion through impermeable materials is possible. Since the elements in the periodic table are arranged in such a way that the group numbers provide information about the number of electrons in their outermost shell, hydrogen becomes the first element in the periodic table.

The weight of a single atom or molecule of any substance can be determined by dividing the corresponding conventional atomic or molecular weight by 6.023×10^{23}, or Avogadro's number. A molecule of hydrogen gas, H_2, therefore weighs 3.346×10^{-24} g, and the weight of a single hydrogen atom is 1.673×10^{-24} g.

The hydrogen atom belongs to the alkali metals in Group 1 because of the single electron in its valence shell. It also exhibits similarities with the halogens in Group 17, whose valence shells contain seven electrons. Like alkali metals, hydrogen loses its single valence electron and becomes a positively charged ion (H^+), or gains an additional electron like the halogens and becomes a uni-negative ion (H^-), or hydride ion, necessitating several debates as to what group the hydrogen atom should belong to.

In gaining an electron, the hydrogen atom conforms to the duet structure (i.e., containing two electrons in their valence shells) in contrast to the octet structure acquired by the halogens. Like the bonding between hydrogen atoms to form a hydrogen molecule, a hydrogen atom can also react with other atoms/molecules by sharing its valence electron. An example is the bond between a hydrogen atom and a chlorine atom to form hydrochloric acid (HCl). The hydrogen atom has a single electron in its outer shell, and the chlorine atom has seven valence electrons. To become stable (H atom–duet structure, Cl atom–octet structure), both atoms must gain additional electrons, which is made possible by sharing one electron each from the atoms. This kind of reaction is termed covalent bonding. Other covalent compounds involving hydrogen include water (H_2O), ammonia (NH_3), and hydrogen sulfide (H_2S). By having only a single proton and a single electron, the electronic structure of the hydrogen atom makes it the element with the simplest atomic structure in nature. The proton's mass in the hydrogen nucleus is 1,800 times the mass of the orbiting electron; therefore, the mass of the hydrogen atom is concentrated in the nucleus. The atomic hydrogen nucleus can also accommodate an additional particle called the neutron: a neutral particle that has stabilizing effect on the nucleus.

Hydrogen reacts with many nonmetals. The reaction of oxygen and hydrogen takes place at room temperature only in the presence of a catalyst such as finely divided platinum. When hydrogen is mixed with air or oxygen and ignited, the mixture explodes. Hydrogen also combines with some metals such as sodium and lithium to form hydrides. It acts as a reducing agent on metallic oxides, such as copper oxide, removing the oxygen and leaving the metal in a free state. Hydrogen reacts with unsaturated organic compounds to form corresponding saturated compounds. Besides being a constituent in water, hydrogen is present in carbohydrates, proteins, and animal and plant tissues. With much research effort geared toward producing clean energy, hydrogen has been recently considered a secondary source of abundant clean energy.

Hydrogen burns readily with oxygen, releasing considerable amounts of energy as heat, producing only water as an end product:

$$2H_2 + O_2 \rightarrow 2H_2O \tag{9.6}$$

When hydrogen burns in air some oxides of nitrogen can be formed. Hydrogen is highly flammable with a wide flammability range, being flammable in mixtures of approximately 4 –74 vol% in air (Table 9.1). It has a high energy content by weight – nearly three times that of gasoline. By contrast, it has a low-energy density (by volume) at a given temperature and pressure (Table 9.2).

To summarize, hydrogen has the highest HHV (61,031 Btu/lb) relative to methane at 23,890 Btu/lb (natural gas is approximately the same value), gasoline at 20,300 Btu/lb, and jet fuel at 18,700 Btu/lb. Due to the lightweight and highest energy-to-weight ratio of hydrogen, it has found wide applications in outer space programs, particularly in multistage rockets. Due to the low specific gravity of liquid hydrogen, this ratio is reversed when the volume of the fuel is considered. Its HHV in units of Btu/gal is 36,100, while for LNG is 84,600 Btu/gal, and that of gasoline is 121,800 Btu/gal (Theodore et al. 2009; Theodore 2014; Green and Southard 2019).

Table 9.1 Upper and lower flammability limits for common gases.

Gas	% by Volume in Air
Hydrogen	4.15–75
Methane	4.9–15.4
Ethane	2.5–15.0
Propane	2.2–7.3
Ethylene	3.2–34.0
Propylene	2.2–9.7
Acetylene	1.5–80.5

Table 9.2 Specific energy and energy density comparison of commonly used fuels.*

Fuel	Specific Energy (MJ/kg)	Energy Density (MJ/L)
Hydrogen	142.0	0.01 (1 atm); 7.1 (1,000 bar); 10.0 (liquid)
Methanol	20.0	15.9
Ammonia	22.5	15.6
Gasoline	47.1	35.0
Diesel	42.8	40.4
Heavy Fuel Oil	42.4	40.7
Biodiesel	42.2	33.0
Natural Gas	50.0	0.04
LNG	50.0	22.2

*1 kWh = 3.6 MJ.

An ignition source is required for the combustion of any fuel. Ignition sources that are commonly used are sparks (from electrical equipment and valves), catalysts, heating equipment, atmospheric discharge, and flames. These sources provide ignition energy that must be applied to start combustion. Hydrogen has a lower ignition energy than other fuels at only 0.02 mJ. As a result, hydrogen is easily ignited when mixed with air. In fact, electrostatic discharge from the human skin has been reported to possess enough energy to ignite hydrogen. The highly localized nature of hydrogen fire makes it safer than fires from gasoline. This localization is attributed to the high buoyancy and diffusivity properties of hydrogen. For this reason, properties around hydrogen fires do not suffer severe damage compared to the damage resulting from the significant dispersion of gasoline fires. A relatively high autoignition temperature (584.85°C) is another property of hydrogen. Autoignition temperature describes the required minimum temperature that can initiate combustion when fuel and oxidizer are mixed without an ignition source. The high autoignition temperature of hydrogen suggests that combustion between hydrogen and an oxidizer will be difficult without an ignition source. Comparable values of the autoignition temperature of notable fuels and hydrogen are listed in Table 9.3.

Table 9.3 Autoignition temperature of hydrogen and other common fuels.

Gas	Autoignition Temperature (°C)
Hydrogen	584.85
Methane	537
Propane	450
Gasoline	246–480
Diesel	210
Methanol	433

9.4.2 Physical Properties of Hydrogen

At ambient temperatures, hydrogen is a colorless, tasteless, odorless, and nontoxic gas. The non-toxicity of hydrogen does not preclude the adverse effect of a high concentration of hydrogen in an environment. High hydrogen concentration will displace environmental oxygen and lead to hypoxia and asphyxiation (ACGIH 1991). Notable side effects associated with mild hypoxia include headache, short-term memory loss, mental fatigue, impaired judgment, etc. Sweating, coma, cyanosis, tachypnea, and cardiac arrhythmia are characteristics of rapid asphyxiation (NRC 2007).

Hydrogen melts at −259.2°C (−434.56°F) and boils at −252.77°C (−422.99°F). These boiling and melting temperatures are extremely low, and as such, the valves and vents of the storage vessels are usually covered with accumulated ice resulting from air moisture. Because of the low melting point, liquid hydrogen in contact with the human skin can lead to severe freezing, and splashes to the eyes can result in frostbite and hypothermia. Vessels containing liquid nitrogen must be air-locked as air or gases leaking into the vessel can result in serious hazards.

Physically, hydrogen has distinguishable properties from other substances in nature. Like most substances, hydrogen can exist in liquid, solid, and gas forms. Hydrogen can coexist in all three forms (solid, liquid, and gas) at the triple point temperature and pressure (13.8 K and 7.2 kPa) (Verfondern 2007).

At standard atmospheric conditions, hydrogen exists in the gaseous form and has a large volume, significantly more than its volume in liquid form. The volume of liquid hydrogen to gaseous hydrogen is 1:848; due to this large volume, difficulty in storing and transporting gaseous hydrogen is usually encountered since a large storage space is required for gaseous fuels. This ratio can be increased to 1:240 by storing it as a high-pressure gas at 3,600 psig (250 bar) and atmospheric temperature (Andre et al. 2001). For this reason, hydrogen storage is conveniently carried out in cryoliquid or pressurized gaseous forms.

In the gaseous form, it is diatomic but dissociates into free atoms at high temperatures. Hydrogen has a lower boiling and melting point than any other substance except helium. Due to the low boiling point of hydrogen, evaporation quickly occurs when exposed to air at normal ambient temperatures. Gaseous hydrogen combusts and explodes at normal atmospheric conditions; it burns in air and oxygen to produce water. During combustion, the flame appears pale blue and is almost invisible. Although hydrogen is a nontoxic and noncorrosive substance, high concentration in the atmosphere can displace oxygen in the air, which can cause severe health issues like unconsciousness, dizziness, headaches, vomiting, etc. In the absence of adequate care and treatment, exposure to a high concentration of hydrogen may lead to death.

Detection of hydrogen during leakage is difficult. Hydrogen gas rises and rapidly diffuses in the air due to its low molecular weight and low relative density (0.005 lb/ft^3, specific gravity of 0.071), which is approximately 14 times less than that of air. During leakage, unwanted consequences like fire outbreaks that would have occurred from the mixture with air are significantly reduced due to the molecules' high buoyancy, which stems from their very low density; therefore, hydrogen leakage results in a localized flammable region, which rapidly disperses. At cryotemperatures, saturated gaseous hydrogen becomes heavier than air and loses the ability to disperse into the atmosphere (Molkov 2012). Hydrogen has a negative Joule-Thomson coefficient, which describes the variation in temperature with respect to pressure change: the expansion of hydrogen gas when released from a pressurized container leads to an increase in the temperature of hydrogen gas.

Reports on the diffusion coefficient of hydrogen have been in the range of 6.1 to 6.8×10^{-5} m^2/s (Baratov et al. 1990; Alcock et al. 2001). Diffusion through certain solids can also be possible with gaseous hydrogen molecules. An example is the diffusion of hydrogen through gypsum panels with a coefficient of 1.4×10^{-5} m^2/s at 25°C (Molkov 2012). The high rate of leaks through the vents of vessels containing hydrogen can also be attributed to the very low viscosity of hydrogen in addition to the small molecular sizes. Gaseous hydrogen has a viscosity of 8.9 μP-s at normal temperature and pressure, and at normal boiling point, it reduces to 1.1 μP-s.

At constant pressure, the specific heat capacity of hydrogen gas is 14.85, 14.3, and 12.15 kJ/kg-K at normal temperature and pressure, standard temperature and pressure, and normal boiling point, respectively (Jordan 2004). Hydrogen is also physically characterized by thermal conductivity, which is a measure of a substance's heat-conducting ability. The significantly higher thermal conductivity of hydrogen than other gases is not surprising as the molecules are smaller and, as such, will possess high molecular velocities. At stoichiometric conditions, hydrogen also has a higher burning speed than other fuels (3 m/s). The burning speed is defined as the speed of a flame front propagating in a combustible mixture of gas. Hydrogen can cause cracks in metallic substances. This phenomenon is called hydrogen embrittlement, and it occurs when metallic and non-metallic substances are exposed to hydrogen for a long time. Responsible factors are hydrogen concentration, temperature, the purity level of hydrogen, stress level, metal composition and tensile strength, grain size, etc.

In gaseous form, hydrogen possesses a thermal conductivity of 0.01694 W/m-K, while in liquid form, it has a value of 0.09892 W/m-K at a normal boiling point (Molkov 2012).

Liquid hydrogen, the lightest of all liquids, is colorless (but light blue in thick layers) and has a density of 4.43 lb/ft^3 (specific gravity of 0.07 or 7% of that of water). When allowed to evaporate rapidly under reduced pressure, it freezes into a colorless solid. Liquid hydrogen has a viscosity value of 13.2 μP-s (Jordan 2004).

Crystalline hydrogen, with a density of 0.088 g/cm^3, is also the lightest of all crystalline substances.

9.5 Hydrogen Isotopes

An isotope is any of two or more forms of an element having the same or very closely related chemical properties and the same atomic number but different atomic weights (or mass numbers). For example, the three isotopes of uranium, U, are: U_{235}, U_{238}, and U_{239}.

The presence and number of neutrons in the nucleus determine the form of the hydrogen atom. The number of neutrons in the hydrogen nucleus can vary from 1 to 6, but naturally, existing forms of hydrogen have 1–2 neutrons in the nucleus; as a result, hydrogen can exist in three isotopic

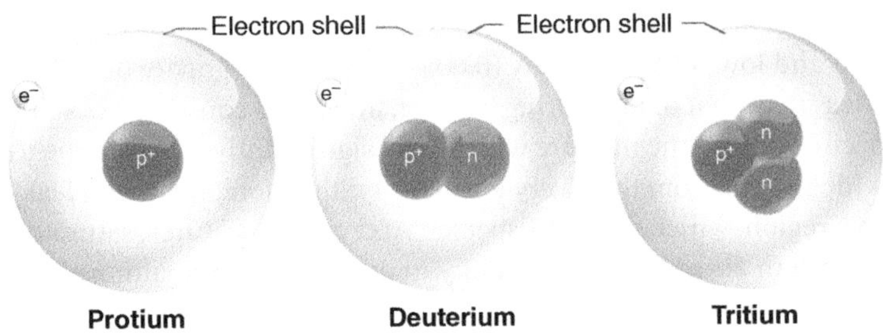

Figure 9.1 Isotopes of the hydrogen atom (Wikipedia contributors 2023).

forms. The isotopes are denoted as ^1_1H, ^2_1H, and ^3_1H. ^1_1H is called *protium*, or normal hydrogen, and contains only a proton in the nucleus. ^2_1H is *deuterium*, which contains a proton and a neutron and has an atomic mass of 2. Deuterium is the hydrogen isotope having an atomic weight of 2.014, a normal boiling point of −249.7°C, and is often referred to as *heavy* hydrogen. It reacts with oxygen to form deuterium oxide (D_2O). A deuteride is a compound analogous to a hydride in which ordinary hydrogen is replaced by deuterium. Water containing ^2_1H is called "heavy water," and it is used to cool nuclear reactors and as a neutron moderator. Deuterium can also be employed as fuel for commercial nuclear fusion reactions. Deuterium is present in ordinary hydrogen to the extent of 0.02%. The third isotope ^3_1H is called *tritium*, which contains two neutrons and one proton, is an unstable, radioactive isotope, and has an atomic mass of 3. The three isotopes are illustrated in Figure 9.1.

Tritium (^3_1H) is the only radioactive isotope, which decays into a non-radioactive helium isotope (helium-3) by emitting low-energy β^- particles with a half-life of 12.3 years (Molkov 2012). The radioactive process is described below.

$$^3_1H \xrightarrow{\beta^-} {}^3_2He + e^- \tag{9.7}$$

Besides the naturally occurring tritium atoms, bombarding lithium-6 with neutrons in a nuclear reactor is a common tritium production method. Heavy isotopes are synthesized from the laboratory. Bombarding tritium with fast-moving deuterium will result in the production of ^4_1H with an atomic mass of 4.026. The neutron from the fast-moving deuterium is captured by the tritium so that the number of neutrons increases to 3 in the ^4_1H nucleus. ^4_1H radioactively decays into ^3_1H by emitting the captured neutron. The half-life of this reaction is 1.39×10^{-22} s. ^5_1H is another radioactive isotope of hydrogen. This is synthesized by tritium bombardment. Here, tritium is bombarded by another tritium, and three neutrons are captured during the reaction resulting in four neutrons in the nucleus of ^5_1H. Radioactive decay occurs through the emission of two neutrons into ^3_1H with a half-life of 8.6×10^{-23}. Increasing the number of neutrons to five results in ^6_1H. During radioactive decay, three or four neutrons are emitted. Depending on the number of emitted neutrons, decay can lead to a tritium or deuterium formation. Bombarding hydrogen with helium-8 has been employed by scientists to produce ^7_1H. Here six neutrons are captured from helium-8 by hydrogen. The half-life of ^7_1H is 6.52×10^{-22} s, and it decays to ^3_1H by emitting four neutrons. The following reactions describe the radioactive decay of these heavy isotopes.

$$^4_1H \xrightarrow{1.39\times10^{-22}\ s} {}^3_1He + {}^1_0n \tag{9.8}$$

Table 9.4 Physical and atomic properties of hydrogen isotopes.

Properties	Protium (Normal Hydrogen)	Deuterium	Tritium
Density at STP (g/L)	0.089	0.18	0.27
Ionization energy (kJ/mol)	1,312	—	—
Relative abundance (%)	99.985	0.0145	10^{-15}
Half-life	Stable	Stable	12.32 yr
Relative atomic mass (g/mol)	1.008	2.014	3.016
Melting point (K)	13.99	18.73	20.62
Boiling point (K)	20.27	23.67	25.0
Magnetic ordering	Diamagnetic	—	—
Thermal conductivity (W/m-K)	0.1805	—	—
Enthalpy of vaporization (kJ/mol)	0.904	1.226	—
Enthalpy of fusion (kJ/mol)	0.117	0.197	—
Covalent radius (pm)	37	—	—
Ionic radius (H^-) (pm)	208	—	—
Internuclear distance (pm)	74.14	74.14	—
Van der Waals radius (pm)	120	—	—

$$^{5}_{1}H \xrightarrow{8.6\times10^{-23}\,s} {}^{3}_{1}He + 2\,{}^{1}_{0}n \tag{9.9}$$

$$^{6}_{1}H \xrightarrow{2.94\times10^{-22}\,s} {}^{3}_{1}He + 3\,{}^{1}_{0}n \tag{9.10}$$

$$^{7}_{1}H \xrightarrow{6.52\times10^{-22}\,s} {}^{3}_{1}He + 4\,{}^{1}_{0}n \tag{9.11}$$

Chemically, the hydrogen isotopes exhibit slightly different properties as reflected in their thermodynamics and also vary in physical properties owing to their different masses. Table 9.4 highlights the different properties of the three naturally occurring hydrogen isotopes.

9.6 The Hydrogen Bond

The unusual properties of water briefly mentioned above are due to the power of its molecules to attract one another especially strongly. This power is associated with a structural feature which is defined as the *hydrogen bond*. The concept of what was once called "coordinated hydrogen" or a "hydrogen bridge" is now referred to as the hydrogen bond. It is the presence of the hydrogen bond that accounts for the somewhat unusual physical properties of not only water but also liquid ammonia, alcohols, phenols, etc.

The H—H molecular bond is the strongest bond between two atoms of any element. In fact, due to the strength of the bond between the constituent atoms, the dissociation of hydrogen molecules into the individual atoms is less than 0.1% at 2,000 K. For this reason, hydrogen molecules are relatively inert at room temperature. As also discussed in the introductory part, hydrogen reaction with other compounds primarily depends on losing, gaining, or sharing the valence electron. In general, the reaction of hydrogen with other substances forms hydrides.

Any hydrogen atom can be associated with no more than two electrons, and hence in the formation of a hydrogen bond, it is not possible for both atoms that are linked in this manner to be attached to the hydrogen by an ordinary covalence. Because of the fact that the linked atoms must be strongly electronegative, the hydrogen bond is largely electrostatic or ionic in nature. The strength of the bond usually increases with an increase in the electronegativity of the atoms which are bridged by the hydrogen. These atoms may be the same or different, but they are limited almost exclusively to oxygen and nitrogen. In certain instances, hydrogen bonds are formed between carbon and oxygen, and between carbon and nitrogen. In these cases, the carbon atom is always attached to a strongly electronegative group.

In most cases, the hydrogen bond exists between two similar atoms derived from different molecules of the same substance. Other types of hydrogen bonding involving a single molecule only or molecules of two different substances are also known.

In summary, a hydrogen bond is formed when the hydrogen in a polar bond approaches the lone pair of electrons on an ion or atom. A polar bond to hydrogen occurs when the hydrogen is attached to an atom of high electronegativity, such as nitrogen, oxygen, or fluorine. Because hydrogen atoms have no inner core of electrons, the pull of electron density from them exposes a significant positive nuclear charge to interact electrostatically with nearby electron density. This is further strengthened by the very small size of the hydrogen atom. The electronegativity difference between carbon and hydrogen is small enough that a significant dipole is not produced, resulting in very weak hydrogen bonds involving C—H bonds. In any event, the strength of the hydrogen bond is determined by the polarity of the bond in which the hydrogen is covalently bound and the electronegativity of the atom to which the hydrogen is electrostatically attracted.

9.7 The Quintessential Energy Carrier

In spite of the many advantages of electricity at the stage of fuel utilization, its generation and transmission can give rise to many difficult problems. Electricity is not only difficult and costly to store for periods of peak demand, but it is also costly to transmit to the point of final use. Furthermore, the generation of electricity by the conversion of renewable energy such as solar, wind, geothermal, or thermal energy is a low-efficiency process. Keep in mind that electricity is a carrier of energy that has been converted from its original form in fossil fuels or renewable energy to electricity, which is transmitted over distances to cities or towns where it is used.

Other means for storing and transporting a host of energy types could be developed in the future. The qualities of a suitable energy carrier include one that is: efficiently obtained from the original energy source, is efficiently transported and distributed at low cost, is readily stored, readily used in industry and transportation, and is safe to handle and safe for the environment.

A large financial commitment to the generation, transmission, and distribution of electric power has already been made. Electric power is relatively safe to handle and is relatively straightforward to generate and transmit to customers. However, electricity storage remains expensive. Pumped water storage is a commonly used method of energy storage at the present time. Furthermore, while electric power has been used efficiently for electric streetcars and rapid transit systems, it cannot yet be readily used for car, truck, and bus transportation purposes without extensive efforts that are currently underway to develop a new generation of electric vehicles.

What is required for an ideal energy carrier is a liquid or gaseous carrier which can be obtained from abundant resources. This carrier should be capable of readily restoring the energy in the form of heat at the final consumption stage without introducing significant environmental pollution.

Table 9.5 Overall comparison of the characteristics of electricity and hydrogen as energy carriers.

Characteristic	Electricity	Hydrogen
Efficiency of conversion to carrier	35–40%	35–50%
Transportation and distribution	Expensive	Less Expensive
Storage	Expensive	Less Expensive
Methods of using the carrier	Good in industry	Good in Industry and Transportation
Safety	Safe	Minimum safety concerns

Hydrogen as an energy carrier can be favorably compared to electricity as indicated in Table 9.5 based on characteristics that are expected to be realized by the end of this decade. Hydrogen appears to be more efficiently converted to the carrier, transported readily and more inexpensively, stored less expensively, and readily used in industry and transportation compared to electricity. Hydrogen does have some safety concerns compared to electricity which are addressed in Chapter 17.

The economics of transmission of any energy carrier becomes a very important consideration as the new sources of energy become wind, solar, geothermal, and other renewable resources. Note that some of these energy resources are often not located near population centers; therefore, they may require transportation of the energy over long distances. However, hydrogen can be transmitted and distributed in liquid or gaseous form. In gaseous form, hydrogen can be transmitted in much the same ways that natural gas is today – primarily by means of pipeline.

References

ACGIH (1991). *Threshold Limit Values and Biological Exposure Indices for 1990–1991*, 6th Edition, Vol. II. Cincinnati, OH: American Conference of Governmental Industrial Hygienists.

Alcock, J.L., Shirvill, L.C., and Cracknell, R.F. (2001). Comparison of existing safety data on hydrogen and comparative fuels. Report of European FP5 project European Integrated Hydrogen Project, May. Brussels, Belgium: EIHP. http://www.eihp.org/public/documents/CompilationExistingSafetyData_on_H2_and_ComparativeFuels_S.pdf.

Andre, L., James, H., and Colin, M. (2001). *Hydrogen Fuel Cell Engines and Related Technologies*. Washington, DC: Federal Transit Administration, U.S. Department of Transportation, Report FTA-CA-26-7022-01.1. https://rosap.ntl.bts.gov/view/dot/14966.

Baratov, A.N., Korolchenko, A.Y., and Kravchuk, G.N. ed. (1990). *Fire and Explosion Hazards of Substances and Materials*. Moscow: Khimia. 496 p.

Green, D.W. and Southard, M. (2019). *Perry's Chemical Engineer's Handbook*, 9th Edition. New York: McGraw-Hill.

Jordan, T. (2004). Safety of hydrogen as an energy carrier. Final Activity Report, SES6-CT-20040502630. Karlsruhe, Germany: Forschungszentrum Karlsruhe GmbH. https://cordis.europa.eu/docs/results/502/502630/123655741-6_en.pdf.

Molkov, V. (2012). *Fundamentals of Hydrogen Safety Engineering, Part I*. London. www.bookboon.com. www.arma.org.au/wp-content/uploads/2017/03/fundamentals-of-hydrogen-safety-engineering-i.pdf.

National Research Council (2007). *Emergency and Continuous Exposure Guidance Levels for Selected Submarine Contaminants*, Vol. 1. Washington, DC: The National Academies Press.

Subramani, V., Basile, A., and Veziroglu, T.N. (2015). Compendium of hydrogen energy. In *Hydrogen Production and Purification*, Vol. 1. Swanston: Woodhead Publishing. https://www.perlego.com/book/1832731/compendium-of-hydrogen-energy-hydrogen-production-and-purification-pdf (accessed 28 October 2023).

Theodore, L. (2014). *Chemical Engineering: The Essential Reference*. New York: McGraw-Hill.

Theodore, L. and Dupont, R.R. (2012). *Environmental Health and Hazard Risk Assessment: Principles and Calculations*. Boca Raton, FL: CRC Press/Taylor Francis Group.

Theodore, L. and Reynolds, J. (1992). *Thermodynamics, a Theodore Tutorial*. East Williston, NY: Theodore Tutorials. Originally published Research Triangle Park, NC: USEPA/APTI.

Theodore, L. and Ricci, F. (2010). *Mass Transfer Operations for the Practicing Engineer*. Hoboken, NJ: John Wiley & Sons.

Theodore, L., Hyland, M., McGuinn, Y., Schoen, F., and Taylor, F. (1988). *Principles of Accident and Emergency Management*. Research Triangle Park, NC: USEPA/APTI.

Theodore, L., Ricci, F., and VanVliet, T. (2009). *Thermodynamics for the Practicing Engineer*. Hoboken, NJ: John Wiley & Sons.

Verfondern, K. Biennal (2007). Chapter I: Hydrogen fundamentals. In *Biennal Report on Hydrogen Safety*, Eggenstein-Leopoldshafen, Germany: International Association for Hydrogen Safety. http://www.hysafe.org/download/1196/BRHS_Chap1_V1p2.pdf.

Wikipedia Contributors (2023). Hydrogen. *Wikipedia*, The Free Encyclopedia. https://en.wikipedia.org/w/index.php?title=Hydrogen&oldid=1181926838 (accessed 28 October 2023).

10

Hydrogen-Bearing Compounds

10.1 Introduction

It should now be clear that the chemical hydrogen (H_2) must be derived from compounds containing hydrogen. One normally thinks of hydrocarbons when referring to these compounds, but there are others. This chapter provides information on some of the more important hydrogen-bearing compounds from which hydrogen can be produced. They include:

1) Water
2) Ammonia
3) Methane
4) Other hydrocarbons

It should be noted that the compound water (H_2O) will be revisited again later in this part of the book. A separate chapter is devoted to water as a by-product of the combustion reaction of hydrogen in air (or oxygen). As one might suppose, the bulk of the material in this chapter is also concerned with water.

A compound is composed of more than one atom, and the molecular weight of the compound is nothing more than the sum of the weights of the atoms of which it is composed. Thus H_2O consists of 2 hydrogen atoms and 1 oxygen atom, and the molecular weight of water is (2) (1.008)+ 16.000 = 18.02. These weights are all relative to the oxygen atom at 16.0000, and any mass unit can be attached to the molecular weight of water, such as 18.02 g/gmol, 18.02 lb/lbmol.

One can compute average molecular weights for mixtures of constant composition even though they are not chemically bonded if their compositions are known accurately. Of course, for a hydrocarbon such as fuel oil or coal whose composition may not be exactly known, one cannot determine an exact molecular weight, although an estimate of the approximate average molecular weight would be acceptable for a host of engineering calculations.

An individual atom can be represented by the symbol of the element. Most substances, however, are compounds, in that they are composed of combinations of atoms. The formula for water, H_2O, indicates that two atoms of hydrogen are present for every atom of oxygen. Similarly, the formula for carbon dioxide is CO_2, for gasoline is C_8H_{18}, for oxygen is O_2, and for candle wax is CH_2. The subscripts in each case (with a 1 understood if no subscript is given) show the relative number of each element in the substance. CO_2 has 1 C for every 2 Os, and CH_2 has 1 C for every 2Hs. A novice might ask why O_2 and C_8H_{18} rather than O and C_4H_9, which show the same atomic and

Hydrogen Energy: Principles and Applications, First Edition. Vincent J. DelGatto, Louis Theodore, R. Ryan Dupont, and Matthew C. Ogwu.

weight ratios. Experiments show that atmospheric oxygen consists not of single atoms (O) but of molecules made up of pairs of atoms (O_2); gasoline consists of carbon and hydrogen molecules of C_8 and H_{18} rather than any other combination of carbon atoms and hydrogen atoms. The formulas of atmospheric oxygen and gasoline are examples of molecular formulas. Water contains H_2O molecules, and carbon dioxide contains CO_2 molecules, so H_2O and CO_2 are also molecular formulas.

All formulas that are multiples of simpler ratios can be assumed to represent molecules. The formulas H_2, H_2O_2, and C_2H_6 represent hydrogen gas, hydrogen peroxide, and ethane, respectively. But formulas that show the simplest possible atomic ratios must be assumed to be empirical unless evidence exists to the contrary.

More than half a million pure chemical compounds have already been discovered and prepared. These compounds are composed of simpler substances, the elements. For compounds, the symbols are combined into formulas to denote the elementary atoms present in the compound. As noted, CH_4 is the formula for the gas methane. It shows that the gas is made by combining a carbon atom, C, with four hydrogen atoms, H. The law of *simple multiple proportions*, given by John Dalton in 1803, states that the atoms in a compound are in ratios of simple multiples of each other. Molecules are pure substances and, except for the fact that they can be decomposed into elements, show all the constancy of properties that elements show.

Section numbers and their accompanying titles for this chapter concerned with hydrogen-bearing compounds are listed below.

10.2 Water

Water is perhaps the key source of hydrogen. It deserves special consideration when discussing hydrogen characteristics. Water is a clear, transparent liquid, and colorless in thin layers. Thick layers of water have a bluish-green color. It is an especially suitable substance for hydrogen-bond formation because each molecule has two attached hydrogen atoms and two unshared electron pairs and hence can form four hydrogen bonds. The tetrahedral arrangement of the shared and unshared electron pairs causes these four bonds to extend in the four tetrahedral directions in space and leads to the crystal structure of ice. This structure, in which each molecule is surrounded by only four immediate neighbors, is a very open structure, and accordingly, ice is a substance with abnormally low density. When ice melts this tetrahedral structure is partially destroyed, and the water molecules are packed more closely together, causing liquid water to have greater density than ice. Many of the hydrogen bonds remain, however, and aggregates of molecules with the open tetrahedral structure persist in water at the freezing point. With an increase in temperature, some of these aggregates break up, causing a further increase in the density of the liquid; only at 4°C does the normal expansion due to the increase in molecular agitation overcome this effect and cause water to begin to show the usual decrease in density with increasing temperature.

Table 10.1 Key water properties.

Temperature (°C)	Density (kg/m^3)	Dynamic Viscosity (Ns/m^3 × 10^3)	Surface Tension (N/m)
0	999.8	1.781	0.0765
10	999.7	1.307	0.0742
20	998.2	1.002	0.0728
30	995.7	0.798	0.0712
40	992.2	0.653	0.0696
50	988.0	0.547	0.0679
60	983.2	0.466	0.0662
70	977.8	0.404	0.0644
80	971.8	0.354	0.0626
90	965.3	0.315	0.0608
100	958.4	0.282	0.0589

The abnormally large dielectric constant of water is due to its power to form hydrogen bonds (Pauling 1950). Some key water properties are provided in Table 10.1 (Vennard and Street 1975).

Two other characteristics of water also need to be noted:

1) Life forms on Earth consist primarily of water.
2) Water is most dense at 4°C so it expands below this temperature and becomes less dense. Thus water with a temperature of 4°C in a larger body of water is denser than the surrounding water and sinks; a characteristic that has prevented the Earth's lakes from freezing from the bottom up and providing liquid habitat for lake species during the coldest season of the year.

Finally, water contains molecules of several different kinds, built out of isotopic atoms in various ways. Since these molecules have almost identical properties except for mass, the density of a sample of water is proportional to the average molecular weight of the molecules in it. If the sample of water consisted of ordinary oxygen combined only with deuterium, its molecular weight would be 20 instead of 18, and its density would accordingly be over 10% greater than that of ordinary water. The term *heavy water* is used to refer to this form of water, which may also be called *deuterium oxide*.

Water vapor pressure data over a range of system temperatures are provided in Table 10.2. Specific volumes, enthalpy, and entropy data for saturated steam are provided in Table 10.3. Tables for superheated steam and steam-ice mixtures are available in the literature (Theodore and Dupont 2020).

10.3 Deuterium

Any discussion of the physical and chemical properties of hydrogen and hydrogen-containing compounds should also include information on deuterium, a hydrogen isotope. Because the atomic weights of the two isotopes of hydrogen are in the approximate ratio of 2 to 1, which is greatly in excess of that for any other element, the otherwise small differences in physical and chemical properties of isotopes are so greatly accentuated that it has been considered desirable to ascribe different names to them. The hydrogen isotope of mass number 2 is called deuterium (Greek: *the second*)

Table 10.2 Vapor pressure of water at 1 atm and various system temperatures

Temperature (°C)	Vapor Pressure (mm Hg)	Temperature (°C)	Vapor Pressure (mm Hg)
−15	1.436	35	42.17
−10	2.149	40	55.32
−5	3.163	45	71.88
0	4.579	50	92.51
1	4.926	55	118.0
2	5.294	60	149.4
3	5.685	65	187.5
4	6.102	70	233.7
5	6.543	75	289.1
10	9.209	80	355.1
15	12.79	85	433.6
20	17.54	90	525.7
25	23.76	95	633.9
30	31.82	100	760.0

and is denoted by the symbol D_2. The isotope of mass number unity is represented by H and is referred to as hydrogen. The nucleus of the deuterium atom is called a deuteron, by analogy with proton. The marked differences in some of the physical properties of the two isotopes and of their oxides, i.e., light and heavy water, are shown by the data in Tables 10.4 and 10.5 (Pauling 1950).

A "bimolecular" reaction of interest and of apparent order 1.5 is that between hydrogen and deuterium as shown in Equation 10.1:

$$H_2 + D_2 \rightarrow 2HD \tag{10.1}$$

For working with mixtures of the two gases at equal partial pressure, the time taken to complete half the reaction is inversely proportional to the square root of the initial pressure. The process *probably* takes place in the following stages:

$$H_2 \rightarrow 2H \tag{10.2}$$

$$D_2 \rightarrow 2D \tag{10.3}$$

which are rapid and then followed by the following reactions:

$$H + D_2 \rightarrow HD + D \tag{10.4}$$

$$H_2 + D \rightarrow HD + H \tag{10.5}$$

The hydrogen and deuterium atoms can either react with each other or combine in pairs. The overall rate of the reaction is evidently determined by the stages involving an atom and a molecule, for these would be kinetic of the order 1.5 (Theodore 2012); the other changes are rapid so that they may be assumed to have no effect on the kinetics of the overall reaction.

Table 10.3 Specific volume, enthalpy, and entropy data for saturated steam.

		Specific Volume (ft^3/lb)			Enthalpy (Btu/lb)			Entropy (Btu/lb)		
Temperature (°F)	Absolute Pressure (lb/in^2)	Saturated Liquid	Evaporation Difference	Saturated Vapor	Saturated Liquid	Evaporation Difference	Saturated Vapor	Saturated Liquid	Evaporation Difference	Saturated Vapor
32	0.08854	0.01602	3306	3306	0	1,075.8	1,075.8	0	2.1877	2.1877
35	0.09995	0.01602	2947	2947	3.02	1,074.1	1,077.1	0.0061	2.1709	2.177
40	0.1217	0.01602	2444	2444	8.05	1,071.3	1,079.3	0.0162	2.1435	2.1597
45	0.14752	0.01602	2036.4	2036.4	13.06	1,068.4	1,081.5	0.0262	2.1167	2.1429
50	0.17811	0.01603	1703.2	1703.2	18.07	1,065.6	1,083.7	0.0361	2.0903	2.1264
60	0.2563	0.01604	1206.6	1206.7	28.06	1,059.9	1,088	0.0555	2.0393	2.0948
70	0.3631	0.01606	867.8	867.9	38.04	1,054.3	1,092.3	0.0745	1.9902	2.0647
80	0.5069	0.01608	633.1	633.1	48.02	1,048.6	1,096.6	0.0932	1.9428	2.036
90	0.6982	0.01610	468	468	57.99	1,042.9	1,100.9	0.1115	1.8972	2.0087
100	0.949 2	0.01613	350.3	350.4	67.97	1,037.2	1,105.2	0.1295	1.8531	0.9826
110	0.2748	0.01617	265.3	265.4	77.94	1,031.6	1,109.5	0.1471	0.8106	0.9577
120	0.6924	0.01620	203.25	203.27	87.92	1,025.8	1,113.7	0.1645	0.7694	0.9339
130	0.2225	0.01625	157.32	157.34	97.9	1,020	1,117.9	0.1816	0.7296	0.9112
140	0.8886	0.01629	122.99	123.01	107.89	1,014.1	1,122.0	0.1984	0.691	0.8894
150	0.718	0.01634	97.06	97.07	117.89	1,008.2	1,126.1	0.2149	0.6537	0.8685
160	0.741	0.01639	77.27	77.29	127.89	1,002.3	1,130.2	0.2311	0.6174	0.8485
170	0.992	0.01645	62.04	62.06	137.9	996.3	1,134.2	0.2472	1.5822	0.8293
180	0.51	0.01651	50.21	50.23	147.92	990.2	1,138.1	0.263	1.548	1.8109
190	0.339	0.01657	40.94	40.96	157.95	984.1	1,142	0.2785	1.5147	1.7932
200	11.526	0.01663	33.62	33.64	167.99	977.9	1,145.9	0.2938	1.4824	1.7762

(Continued)

Table 10.3 (Continued)

Temperature (°F)	Absolute Pressure (lb/in^2)	Specific Volume (ft^3/lb)			Enthalpy (Btu/lb)			Entropy (Btu/lb)		
		Saturated Liquid	Evaporation Difference	Saturated Vapor	Saturated Liquid	Evaporation Difference	Saturated Vapor	Saturated Liquid	Evaporation Difference	Saturated Vapor
210	14.123	0.01670	27.8	27.82	178.05	971.6	1,149.7	0.309	1.4508	1.7598
212	14.696	0.01672	26.78	26.8	180.07	970.3	1,150.4	0.312	1.4446	1.7566
220	17.186	0.01677	23.13	23.15	188.13	965.2	1,153.4	0.3239	1.4201	1.744
230	20.78	0.01684	19.365	19.382.	198.23	958.8	1,157	0.3387	1.3901	1.7288
240	24.969	0.01692	16.306	16.323	208.34	952.2	1,160.5	0.3531	1.3609	1.714
250	29.825	0.01700	13.804	13.82	218.48	945.5	1,164	0.3675	1.3323	1.6998
260	35.429	0.01709	11.746	11.763	228.64	938.7	1,167.3	0.3817	J.3043	1.686
270	41.858	0.01717	10.044	10.061	238.84	931.8	1,170.6	0.3958	1.2769	1.6727
280	49.203	0.01726	8.628	8.645	249.06	924.7	1,173.8	0.4096	1.2501	1.6597
290	57.556	0.01735	7.444	7.461	259.31	917.5	1,176.8	0.4234	1.2238	1.6472
300	67.013	0.01745	6.449	6.466	269.59	910.1	1,179.7	0.4369	1.198	1.635
310	77.68	0.01755	5.609	5.626	279.92	902.6	1,182.5	0.4504	1.1727	1.6231
320	89.66	0.01765	4.896	4.914	290.28	894.9	1,185.2	0.4637	1.1478	1.6115
330	103.06	0.01776	4.289	4.307	300.68	887	1,187.7	0.4769	1.1233	1.6002
340	118.01	0.01787	3.77	3.788	311.13	879	1,190.1	0.49	1.0992	1.5891
350	134.63	0.01799	3.324	3.342	321.63	870.7	1,192.3	0.5029	1.0754	1.5783
360	153.04	0.01811	2.939	2.957	332.18	862.2	1,194.4	0.5158	1.0519	0.5677
370	173.37	0.01823	2.606	2.625	342.79	853.5	1,196.3	0.5286	1.0287	1.5573
380	195.77	0.01836	2.317	2.335	353.45	844.6	1,198.1	0.5413	1.0059	1.5471
390	220.37	0.0185	2.0651	2.0836	364.17	835.4	1,199.6	0.5539	0.9832	1.5371
400	247.31	0.01864	1.8447	1.8633	374.97	826	1,201	0.5664	0.9608	1.5272

410	276.75	0.01878	1.6512	1.67	385.83	816.3	1,202.1	0.5788	0.9386	1.5174
420	308.83	0.01894	1.4811	1.5	396.77	806.3	1,203.1	0.5912	0.9166	1.5078
430	343.72	0.0191	1.3308	1.3499	407.79	796	1,203.8	0.6035	0.8947	1.4982
440	381.59	0.01926	1.1979	1.2171	418.9	785.4	1,204.3	0.6158	0.873	1.4887
450	422.6	0.0194	1.0799	1.0993	430.1	774.5	1,204.6	0.628	0.8513	1.4793
460	466.9	0.0196	0.9748	0.9944	441.4	763.2	1,204.6	0.6402	0.8298	1.47
470	514.7	0.0198	0.8811	0.9009	452.8	751.5	1,204.3	0.6523	0.8083	1.4606
480	566.1	0.02	0.7972	0.8172	464.4	739.4	1,203.7	0.6645	0.7868	1.4513
490	621.4	0.0202	0.7221	0.7423	476	726.8	1,202.8	0.6766	0.7653	1.4419
520	812.4	0.0209	0.5385	0.5594	511.9	686.4	1,198.2	0.713	0.7006	1.4136
540	962.5	0.0215	0.4434	0.4649	536.6	656.6	1,193.2	0.7374	0.6568	1.3942
560	1133.1	0.0221	0.3647	0.3868	562.2	624.2	1,186.4	0.7621	0.6121	1.3742
580	1325.8	0.0228	0.2989	0.3217	588.9	588.4	1,177.3	0.7872	0.5659	1.3532
600	1542.9	0.0236	0.2432	0.2668	617	548.5	1,165.5	0.8131	0.5176	1.3307
620	1786.6	0.0247	0.1955	0.2201	646.7	503.6	1,150.3	0.8398	0.4664	1.3062
640	2059.7	0.026	0.1538	0.1798	678.6	452	1,130.5	0.8679	0.411	1.2789
660	2365.4	0.0278	0.1165	0.1442	714.2	390.2	1,104.4	0.8987	0.3485	1.2472
680	2708.1	0.0305	0.0810	0.1115	757.3	309.9	1,067.2	0.9351	0.2719	1.2071
700	3093.7	0.0369	0.0392	0.0761	823.3	172.1	995.4	0.9905	0.1484	1.1389

Table 10.4 Key properties of hydrogen isotopes.

Property	Hydrogen	Deuterium
Boiling point (K)	20.4	23.5
Triple point (K)	14	18
Heat of fusion (cal/mol)	28	47

Table 10.5 Key properties of isotopic forms of water.

Property	Hydrogen	Deuterium
Specific gravity at 25°C	1.0000	1.1079
Boiling point (°C)	100.00	101.42
Freezing point (°C)	0.00	3.82
Surface tension (dynes/cm)	72.75	67.8

10.4 Ammonia

Ammonia is a colorless, pungent gas, and highly soluble in water. A saturated aqueous (water) solution of ammonia contains 45% ammonia by weight at 0°C (32°F) and 30% at ordinary room, temperatures. In solution in water, ammonia becomes ammonium hydroxide, NH_4OH, which is a weak base and similar in chemical behavior to the hydroxides of the alkali metals. This ammonia in water solution is sometimes called *ammonia water* or *aqua ammonia* and contains the molecular species NH_3, NH_4OH, NH_4^+, and OH^-.

In the 19th century, the principal source of ammonia was the destructive distillation of coal; it was an important by-product of the manufacture of fuel gases. Today most ammonia is produced synthetically from hydrogen and nitrogen. For the equilibrium reaction:

$$\frac{3}{2}H_2(g)+\frac{1}{2}N_2(g)\rightarrow NH_3(g) \tag{10.6}$$

The maximum conversion of nitrogen and hydrogen into ammonia will be obtained when the reactants are present in their stoichiometric ratio, i.e., 1 mol of nitrogen to three of hydrogen (Theodore 2012). Any increase in pressure will move the equilibrium in the direction of a decrease in the number of molecules, i.e., it favors the formation of ammonia.

Ammonia melts at −77.7°C (−107.9°F), boils at −33.35°C (−28.03°F) and has a specific gravity of 0.68 at its boiling point and 1 atm (760 Torr) of pressure. The properties of liquid ammonia are compared with those of water in Table 10.6.

Ammonia resembles water in its general physical and chemical properties. Both ammonia and water are associated liquids. They form compounds with salts, are good solvents, and although virtual nonconductors themselves, affect the ionization of electrolytes. Table 10.7 illustrates the solubility of ammonia in water as a function of temperature at 1 atm pressure.

Ammonia is one of the fundamental raw materials employed in modern industry. The total inorganic chemical tonnage in the United States has exceeded 10 million tons per year for many

Table 10.6 Key properties of ammonia and water.

Property	Liquid Ammonia	Water
Melting point (°C)	−77.7	0
Boiling point (°C)	−33.35	100
Specific heat (cal/g at bp)	1.07	1.01
Specific gravity at 20°C	0.607	0.9982
Critical temperature (°C)	131	365
Critical pressure (atm)	112	195
Surface tension (dynes/cm)	22.0 at 20°C	72.8 at 20°C
Dielectric constant	22.7 at −50°C	81.1 at 18°C

Table 10.7 Solubility of ammonia and water at 1 atm pressure.

Temperature (°C)	0	20	40	60	80
g of NH_3/100 g of solution (%)	47.4	34.1	25.3	19.3	13.3

decades. Ammonia is an important refrigerant and is widely used in various chemical industries, especially in the manufacture of fertilizer, nitric acid, and explosives. The largest consumer of anhydrous ammonia is the fertilizer industry, which utilizes it in the production of calcium and sodium nitrate, ammonium sulfate, nitrate and phosphate, ammoniated superphosphates, urea, and aqueous ammonia.

10.5 Methane

Methane is perhaps the second major source of hydrogen. As such it deserves consideration in this chapter. Natural gas consists mostly of methane and other hydrocarbons including ethane, propane, butane, and pentane. It contains other gases including carbon dioxide, nitrogen, and helium, all of which reduce its net heating value. The density of natural gas at ambient conditions is approximately 0.60 lb/ft^3, compared to that of air at 0.75 lb/ft^3. Its heating value is approximately 1,000 Btu/scf. The high hydrogen content of natural gas, compared to oil or coal, results in more water vapor being produced in the products of combustion. This results in a correspondingly lower thermal efficiency. The important physical–chemical properties of CH_4 in engineering units are provided in Table 10.8. Properties of the constituents of natural gas can be found in Tables 10.9 and 10.10. The composition of typical natural gas and its gross heating value is provided in Table 10.10 (Theodore, Ricci, and VanVliet 2009; Theodore 2012, 2014).

As noted earlier, natural gas is a nonrenewable fossil fuel as natural gas (as with oil) was formed from the remains of tiny sea animals and plants that died 300–400 million years ago. When these sea animals and plants died, they sank to the bottom of the oceans where they were buried by layers of sediment that turned into rock. Over the millennia, these layers increased in height to several thousands of feet, subjecting the remains to high pressure, which, when combined with the high temperature of the Earth, converted this biological material into not only oil

Table 10.8 Key properties of methane.

Property	
Molecular weight (g/gmol)	16.04
Melting point at 1 atm (°F)	−296.5
Boiling point at 1 atm (°F)	−263.2
Heat capacity at 1 atm, 60 °F (Btu/lb-°F)	0.528
Density at 1 atm, 32 °F (lb/ft^3)	0.0448
Density at boiling point (lb/ft^3)	0.1124
Density of liquid at boiling point (lb/ft^3)	26.47
Critical temperature (°F)	−116.5
Critical pressure (atm)	45.8
Heat of vaporization at boiling point (Btu/lb)	248.4
Higher heating value (Btu/lb)	23,890
Higher heating value at 1 atm, 77 °F (Btu/ft^3)	978

Table 10.9 Key physical properties of natural gas components.

Component	Molecular Weight	Boiling Point at 1 atm (°F)	Melting Point at 1 atm (°F)
Methane, CH_4	16.04	−263.2	−296.5
Ethane, C_2H_6	30.07	−127.5	−297.9
Propane, C_3H_8	44.09	−43.7	−305.8
Ethylene, C_2H_4	28.05	−154.7	−272.5
Isobutane, C_4H_{10}	58.12	10.9	−255.3
n-Butane, C_4H_{10}	58.12	31.1	−217.0
Isopentane, C_5H_{12}	72.15	82.1	−255.8
n-Pentane, C_5H_{12}	72.15q	96.9	−201.5

Table 10.10 Typical natural gas composition.

Component	Mole Fraction	Gross Heating Value (Btu/scf)
N_2	0.0515	0
CH_4	0.8111	1,013
C_2H_6	0.0967	1,792
C_3H_8	0.0351	2,590
C_4H_{10}	0.0056	3,370
Σ	1.0000	1,105

but also natural gas. Over time, high concentrations of natural gas became trapped in the layers of rock, becoming the natural gas deposits being recovered around the world today. It built up in pockets when its upward progress toward the surface of this light gas was blocked by an upper layer of solid impervious rock. This process has been described as being analogous to trapping water with a sponge.

The gas accumulation can be reached by drilling wells in a manner similar to how oil is recovered from an oil field. It should be mentioned that natural gas is produced on all continents except Antarctica (Skipka and Theodore 2014).

10.6 Other Hydrocarbon Molecules

Many organic materials are hydrocarbons, i.e., they are entirely composed of hydrogen and carbon. They are the organic compounds of the simplest composition and may be considered the parent substances from which all other organic compounds are derived. The hydrocarbons are conveniently classified into two major groups, open chain and cyclic. In open-chain compounds containing more than one carbon atom, the carbon atoms are attached to each other to form an open chain; the chain may carry one or more side branches. In cyclic compounds, the carbon atoms form one or more closed rings.

The intramolecular bonds of these hydrocarbons are covalent. Each carbon atom has four electrons that may participate in covalent bonding, whereas every hydrogen atom has only one bonding electron. A single covalent bond exists when each of the two bonding atoms contributes one electron. Double and triple bonds between two carbon atoms involve the sharing of two and three pairs of electrons, respectively. For example, in ethylene, which has the chemical formula C_2H_4, the two carbon atoms are doubly bonded together, and each is also singly bonded to two hydrogen atoms.

10.6.1 Open-Chain Hydrocarbons

The saturated open-chain hydrocarbons form a homologous series called the paraffin (Latin, *parum affinis*, "little affinity") series or the alkane series. The composition of each of the members of the series corresponds to the formula C_nH_{2n+2} where n is the number of carbon atoms in the molecule. Among the members of the series are methane CH_4, ethane, C_2H_6, propane, C_3H_8, and butane, C_4H_{10}. All the members of the series are unreactive; that is, they do not react readily at ordinary temperatures with such reagents as acids, alkalies, or oxidizers. The first four members of the series are gases at ordinary temperature and pressure; intermediate members are liquids; and the heavier members are semisolids or solids. Petroleum contains a great variety of saturated hydrocarbons, and such petroleum products as gasoline, kerosene, heavy fuel oil, lubricating oils, petroleum jelly, and paraffin consist principally of mixtures of paraffin hydrocarbons, which range from the lighter liquid members to the solid members.

10.6.2 The Alkene Series

The unsaturated open-chain hydrocarbons include the alkene or olefin series, the diene series, and the alkyne series. The alkene series is made up of chain hydrocarbons in which a double bond exists between two carbon atoms. The general formula for the series is C_nH_{2n} where n is the number of carbon atoms.

As in the paraffin series, the lower molecular weight members are gases, intermediate molecular weight compounds are liquids, and the higher molecular weight members of the series are solids. The alkene series compounds are more active chemically than the saturated compounds. They easily react with substances such as halogens, adding atoms at the double bonds. They are not found to any extent in natural products but are produced in the destructive distillation of complex natural substances such as coal and are formed in large amounts in petroleum refining, particularly in the "cracking" process. The first member of the series is ethylene, C_2H_4. The dienes contain two double bonds between pairs of carbon atoms in the molecule. They are related to the complex hydrocarbons in natural rubber and are important in the manufacture of synthetic rubber and plastics; important members of this series are butadiene, C_4H_6, and isoprene, C_5H_8.

10.6.3 The Alkyne Series

The members of the alkyne series contain a triple bond between a pair of carbon atoms in the molecule. They are very active chemically and are not found free in nature. They form a series analogous to the alkene series. The first and most important member of the series is acetylene, C_2H_2.

10.6.4 Cyclic Hydrocarbons

The simplest of the saturated cyclic hydrocarbons, or cycloalkanes, is cyclopropane, C_3H_6, the molecules of which are made up of three carbon atoms each of which two hydrogen atoms are attached. Cyclopropane is somewhat more reactive than the corresponding open-chain alkane, propane, C_3H_8. Other cycloalkanes make up a part of ordinary gasoline.

Several unsaturated cyclic hydrocarbons, having the general formula $C_{10}H_{14}$ occur in certain fragrant natural oils that are distilled from plant materials. These hydrocarbons are called terpenes and include pinene (in turpentine) and limonene (in lemon and orange oils).

The most important group of unsaturated cyclic hydrocarbons is the aromatics, which occur in coal tar. Although aromatics sometimes exhibit unsaturation, that is, the addition of other substances, their principal reactions bring about the replacement of hydrogen atoms by other kinds of atoms or groups of atoms. The aromatic hydrocarbons include benzene, toluene, anthracene, and naphthalene.

10.6.5 Other Organic Compound Groups

There are numerous other organic groups, many of which are involved in polymer structures. Categories of the more common groups include:

1) Alcohols, e.g., methyl alcohol
2) Ethers, e.g., dimethyl ether
3) Acids, e.g., acetic acid
4) Aldehydes, e.g., formaldehyde
5) Ketones, e.g., acetone

10.7 The Alkane Series

The names of the higher numbers of this series consist of a numerical prefix and the termination "-ane." These are shown in Table 10.11.

Table 10.11 The higher alkane series.

Carbon No.	Name	Carbon No.	Name	Carbon No.	Name
11	Undecane	21	Heneicosane	31	Hentriacontane
12	Dodecane	22	Docosane	32	Dotriacontane
13	Tridecane	23	Tricosane	33	Tritriacontane
14	Tetradecane	24	Tetracosane	40	Tetracontane
15	Pentadecane	25	Pentacosane	50	Pentacontane
16	Hexadecane	26	Hexacosane	60	Hexacontane
17	Heptadecane	27	Heptacosane	70	Heptacontane
18	Octadecane	28	Octacosane	80	Octacontane
19	Nonadecane	29	Nonacosane	90	Nonacontane
20	Eicosane	30	Triacontane	100	Hectane

References

Merriam-Webster (1971). *Compound. Webster's New World Dictionary*, Second College Edition. Upper Saddle River, NJ: Prentice-Hall.

Pauling, L. (1950). *College Chemistry*. San Francisco, CA: W.H. Freeman and Company.

Skipka, K. and Theodore, L. (2014). *Energy Resources: Availability, Management, and Environmental Impacts*. Boca Raton, FL: CRC Press/Taylor & Francis Group.

Theodore, L. (2012). *Chemical Reactor Analysis and Applications for the Practicing Engineer*. Hoboken, NJ: John Wiley & Sons.

Theodore, L. (2014). *Chemical Engineering: The Essential Reference*. New York: McGraw-Hill.

Theodore, L. and Dupont, R.R. (2020). *Water Resources Management Issues: Basic Principles and Applications*. Boca Raton, FL: CRC Press/Taylor & Francis Group.

Theodore, L., Ricci, F., and VanVliet, T. (2009). *Thermodynamics for the Practicing Engineer*. Hoboken, NJ: John Wiley & Sons.

Vennard, J.K. and Street, R.L. (1975). *Elementary Fluid Mechanics*, 5th Edition. Hoboken, NJ: John Wiley & Sons.

11

Hydrogen Production Processes

11.1 Introduction

The word process conjures up two terms: production and manufacturing. Each of these terms is briefly defined below.

1) *Process*. From a technical perspective, it can be taken to mean a series of operations involving physical and/or chemical changes in some specified material. Merriam-Webster (2023a) offers: "a series of actions or operations conducing to an end; *especially*: a continuous operation or treatment especially in manufacture."
2) *Production*. Merriam-Webster (2023b) provides: "the act or process of producing: the creation of utility; *especially* the making of goods available for use; total output especially of a commodity or an industry."
3) *Manufacturing*. Production and processing of goods or merchandise including their design, treatment at various stages, and financial services contributed by bankers. Merriam-Webster (2023c) provides: "a productive industry using mechanical power and machinery."

Processes fall into two general categories: *flow processes* and *non-flow* or *static processes*. Flow processes imply a condition in which material, such as hydrogen, is actually flowing through a process similar to a fluid flowing in a pipe, or a continuous chemical reactor with constantly flowing inlet and outlet streams. Flow processes are of two types: *steady state* processes and *nonsteady state* processes. In a steady-state process, all the conditions of temperature, pressure, concentration, and flow rate are constant at any given point and do not change with respect to time. A nonsteady state process will involve changing conditions of temperature, pressure, concentration, and/or flow rate over time.

Despite hydrogen's abundance in the Sun, hydrogen is rarely found by itself on planet Earth, as it reacts readily with other elements. Most of the hydrogen on Earth is bound to other elements in various molecular compounds of various forms and is almost always extracted from another source. Hydrogen production typically comes from the extraction of hydrogen from water or hydrocarbons, most notably by using natural gas and coal. Hydrogen can be used for a wide array of energy and industrial applications and can be stored for long periods of time in various forms.

With respect to the energy required, it is easy to remove hydrogen from compounds that are at a higher energy state, such as fossil fuels. This process releases energy, reducing the amount of process energy required. It takes more energy to extract hydrogen from compounds that are at a lower

Hydrogen Energy: Principles and Applications, First Edition. Vincent J. DelGatto, Louis Theodore, R. Ryan Dupont, and Matthew C. Ogwu.

energy state, such as water, as energy has to be added to the process to bring the reactions to completion (Lanz, Heffel, and Messer 2001).

Several hydrogen manufacturing processes are available and currently in use. For example, any carbonaceous material can be used to make hydrogen by steam reforming, but many of these materials are likely to contain more contaminants than the preferable natural gas and would require cleanup before use. Natural gas is the main source of hydrogen production because it is a relatively clean source material, the supply of natural gas is abundant, and its price continues to remain relatively low. If prices or availability of natural gas becomes prohibitive, water is another abundant source of hydrogen. There are presently five major forms of energy that can assist in the making of hydrogen. These include:

1) *Thermochemical.* Hydrogen is produced mainly from natural gas by steam methane reforming (SMR). SMR is currently the most common and least expensive method of producing hydrogen.
2) *Electrochemical.* Water electrolysis passes a direct current between two electrodes in water. The water is made more conductive by adding an electrolyte such as potassium hydroxide. Hydrogen is produced at the negative electrode and oxygen is produced at the positive electrode. The gases are then collected separately.
3) *Thermal.* Thermal decomposition of water into hydrogen and oxygen occurs at temperatures around 2,500°C. In a plasma arc process, water is heated to 5,000°C by an electric field resulting in the cracking products H, H_2, O, O_2, OH, HO_2, and H_2O. The plasma gases are quenched with a cryogenic liquid to prevent the gases from recombining. This process consumes significant amounts of energy and is very expensive to operate.
4) *Photoelectrochemical.* Sunlight provides the source of energy for this process. Photons interact with dissolved chemicals to produce activated species, which are deactivated by reacting with water to release hydrogen. This is solar-powered electrolysis.
5) *Photobiological.* Sunlight provides the source of energy for this process. Living organisms, such as green algae, make enzymes; the pigments of algae absorb solar energy, and their enzymes act as catalysts to split water molecules into hydrogen and oxygen.

Finally, much has been written about the various colors of hydrogen: green, yellow, blue, black, etc. These terms are based on the processes employed and energy sources used to produce hydrogen. Figure 11.1 provides a summary of these hydrogen color definitions, their technology basis and electricity sources, and the overall greenhouse gas footprint of the corresponding hydrogen generation process.

Green hydrogen is produced when electricity from renewable energy sources such as wind, solar, or hydropower is employed to electrolyze water. Water is separated into hydrogen and oxygen during this process, and because no greenhouse gas is emitted as a by-product, this process is the most sustainable and climate-neutral process. Common electrolyzers employed are Proton Exchange Membrane (PEM) and Alkaline. Note that electrolyzers convert electrical energy into molecules with high potential energy densities and fuel cells are electrochemical cells that convert chemical energy into electricity. Polymer electrolyte membrane (PEM) electrolyzers and fuel cells are also called proton exchange membrane (PEM) electrolyzers and fuel cells. All PEM's use a proton-conducting polymer membrane as the electrolyte. Green hydrogen production is expensive because the cost of the electrolyzers falls within the range of \$4–\$6/kg, and, as such, green hydrogen constitutes a small proportion (<0.1%) of global hydrogen production. With the expanding areas of hydrogen application such as steel manufacturing, cement production, stabilization of the electricity grid,

	Terminology	Technology	Feedstock/ electricity source	GHG footprint
Production via electricity	Green hydrogen	Electrolysis	Wind, Solar, Hydro, Geothermal, Tidal	Minimal
	Purple/pink hydrogen		Nuclear	
	Yellow hydrogen		Mixed-origin grid energy	Medium
Production via fossil fuels	Blue hydrogen	Natural gas reforming + CCUS gasification + CCUS	Natural gas, coal	Low
	Turquoise hydrogen	Pyrolysis	Natural gas	Solid carbon (by-product)
	Grey hydrogen	Natural gas reforming		Medium
	Brown hydrogen	Gasification	Brown coal (lignite)	High
	Black hydrogen		Black coal	

Figure 11.1 The color spectrum of hydrogen (GEI 2021).

cleaning products, fertilizers from green ammonia, transportation, and heating systems in buildings, interest in green hydrogen production technologies has risen significantly. Further, with the advancement in technologies and the declining cost of renewable energy, it has been forecasted that production costs could fall from \$6/kg to \$2/kg by 2035.

Pink or *purple hydrogen* is climate-neutral, i.e., the by-products are not environmentally harmful. It also involves water electrolysis, but with a different source of electricity employed in the production process (nuclear energy). Producing hydrogen from nuclear energy has the advantages of low-cost, high-volume production, and easy access by industrial customers.

Yellow hydrogen might not be categorized into the climate-neutral class because although it also involves the electrolysis of water enabled by the electricity generated from solar energy, it emits fluctuating amounts of CO_2 because sometimes, the generated electricity comes from a mixture of alternative renewable energy, solar, and fossil fuel. Renewable energy as the feedstock reduces the hydrogen's carbon intensity (CI) score, and employing natural gas raises the score.

Blue hydrogen employs the SMR process, which breaks down natural gas into H_2 and CO_2 through the mixture of the natural gas with hot steam and catalysts. This process is believed to be environmentally friendly as the emitted CO_2 is captured by a carbon-capture and storage (CCS) technology to prevent entering the atmosphere. However, not all the emitted greenhouse gas is captured because roughly 10% of the produced CO_2 still leaks out to the surroundings. The 90% captured is transported by underground pipes and stored deep beneath the Earth's surface in empty oil and gas reservoirs.

Heating a natural gas to a very high temperature in the absence of water and air also produces hydrogen but with solid carbon as a by-product. This process is called methane pyrolysis and the hydrogen produced is called *turquoise hydrogen*. This is an inexpensive hydrogen production process as the energy employed to break down methane is much less than that required for water molecule splitting. Depending on the source of energy, turquoise hydrogen may be termed low-emission hydrogen if the heat energy is supplied from alternative renewables.

Gray hydrogen also employs methane as the feedstock and is also an endothermic process. The steam reforming process applies a steam temperature within the range of 700–1000°C at a pressure of 25 bar. The production is carried out in the presence of a catalyst.

The *brown* and *black hydrogens* are the opposite of the green hydrogen. The process involves high-emission coal gasification. Brown hydrogen is realized from lignite (brown coal), while black hydrogen is processed from bituminous (black coal). The environmental pollution resulting from the production of brown and black hydrogen is large as the carbon dioxide from the conversion of the coals into hydrogen is released into the atmosphere.

Lastly, a natural form of hydrogen exists called *white hydrogen*. This hydrogen is found deep in the Earth's crust and is released by geochemical processes in nature. Its nonpolluting form arises from the absence of greenhouse gas emissions during creation. A significant deposit of white hydrogen (up to 46 million T and the world's largest known supply to date) was recently discovered in Southeastern France (Pironon and de Donato 2023), opening up significant possibilities to exploit this carbon-free energy source in that region of the world.

Details of these processes are provided in the sections below. Section numbers and their accompanying titles for this chapter concerned with hydrogen processes are listed below.

11.2 Overview of Hydrogen Production Processes

Today's dependence on natural gas and coal as the source of hydrogen means that any hydrogen production process will generate significant amounts of CO_2 emissions, most of which are released directly into the atmosphere. It can be seen from the above that hydrogen can be produced by employing a variety of processes. The three major categories of hydrogen feedstocks are:

1) Fossil fuels
2) Water (by direct splitting)
3) Biomass

A successful hydrogen energy economy will ultimately be based on the efficient and clean production of hydrogen. As an energy carrier, hydrogen has to be manufactured from a primary energy source. There are many specific industrial methods currently available for the production of hydrogen, but all of them are relatively expensive compared to the cost of supplying the same amount of energy with fossil fuels. The distribution and storage systems for hydrogen to be supplied on a large scale are also much more expensive because of the low volumetric energy density of hydrogen. Bringing all of these costs down sufficiently to make hydrogen a viable competitor to existing fuels will require important technological advances to be realized. It should also be noted that almost all of the hydrogen presently produced involves the *steam reforming of fossil fuels* using a nickel catalyst. This is a proven commercial technology and is the most economical way of making hydrogen on a large scale. Natural gas is normally employed as the energy source for this process, and this SMR process carried out in large central plants fed by the existing natural gas infrastructure is how almost all of the hydrogen is produced in the United States. This is an important pathway for the near-term production of hydrogen that will need to be supplemented with other sustainable production pathways in the future.

Direct water splitting processes involve the use of electricity, high-temperature heat, or light energy photons to generate hydrogen and oxygen from water. Electrolysis can be driven by electricity from the traditional fossil-fuel-based grid or a variety of diverse domestic and sustainable resources including solar and wind. Thermally driven water splitting processes can utilize heat from concentrated solar power or waste heat from advanced nuclear reactors.

Biomass is an abundant sustainable resource that can be produced domestically and converted to hydrogen and other by-products through a number of methods. In addition to using biomass directly via biomass gasification or co-fed in coal gasification, bio-derived liquid reforming, or fermentation, there are also opportunities to produce hydrogen from waste streams (e.g., municipal solid waste or agricultural residue containing biomass), which offer the additional benefit of waste cleanup.

Since hydrogen is an energy carrier (it can store and deliver energy in a usable form) but is not an energy source, it must be produced from hydrogen-containing compounds. This can be accomplished at large central plants, at medium-scale semi-central plants, or in small, distributed units located at or very near the point of use, such as at refueling stations or stationary power sites.

11.3 Fossil Fuels

Fossil fuels can be reformed to release hydrogen from the constituent hydrocarbon molecules through three main reforming technologies: steam reforming, partial oxidation, and autothermal reforming.

1) Steam reforming is typically the preferred process for hydrogen production in industry. It is a proven commercial technology that is the most favorable way to produce hydrogen and will no doubt remain so for the foreseeable future. In the steam reforming processes, a hydrocarbon is reacted with steam at high temperature over a catalyst to produce syngas ($H_2 + CO$), which is then further reacted with steam in a Water-Gas-Shift (WGS) reactor to convert CO into additional hydrogen ($CO + H_2O \rightarrow CO_2 + H_2$). The overall reaction sequence can be described as:

$$CH_4 + H_2O \rightarrow CO + 3H_2 \tag{11.1}$$

$$CO + H_2O \rightarrow CO_2 + H_2 \tag{11.2}$$

for the overall reaction

$$CH_4 + 2H_2O \rightarrow CO_2 + 4H_2 \tag{11.3}$$

Following this, pure H_2 is separated. This is explained under hydrogen purification in Section 11.6. The separation usually applies Pressure Swing Adsorption (PSA) technology. When applied to solid fuels, the reforming process is defined as gasification. The resulting hydrogen-rich gas mixtures are defined as the reformate and the equipment as a reformer and/or first processor.

The reforming reactions for hydrocarbons (in general) and methanol are presented below:

$$C_mH_n + 2mH_2O \rightarrow mCO_2 + (2m + 0.5n)H_2 \tag{11.4}$$

$$CH_3OH + H_2O \rightarrow CO_2 + 3H_2 \tag{11.5}$$

2) Partial oxidation (POX) converts hydrocarbons (usually methane) to syngas by partially reacting (combusting) the hydrocarbon with oxygen in an exothermic reaction. The major advantage of POX is that it does not require a catalyst, and therefore the POX reactor is not susceptible to catalyst poisoning that is encountered in catalytic reformers (e.g., steam and autothermal reforming). The challenges for POX include high reaction temperatures, which require more expensive reactor construction materials, some soot formation, and a low H_2/CO ratio, meaning less hydrogen is produced per molecule of hydrocarbon fuel consumed compared to steam and autothermal reforming.
3) Autothermal reforming (ATR) combines steam reforming with partial oxidation. POX provides the heat for the endothermic steam reforming process, in theory, resulting in a thermally neutral process. Since the heat for the steam reforming is provided by the exothermic POX reaction, an external heat source for the reactor is not required. However, POX and ATR both require expensive and complex oxygen separation units. Since a large amount of CO is produced, a WGS reactor is required to maximize the hydrogen production. While ATR produces a higher amount of hydrogen per hydrocarbon molecule than POX, it is still less than what steam reforming would produce, i.e., its overall conversion efficiency is somewhat less than that for steam reforming.

A total of 95% of the hydrogen produced in the United States today is made by steam SMR of natural gas (NG) in large central plants. This is an important pathway for near-term hydrogen production. An abundant supply of inexpensive NG makes hydrogen produced in central facilities through existing SMR processes cost-competitive with gasoline on an energy basis. The main barrier is the high cost of hydrogen transportation from the central plant to a dispensing station owing to the lack of distribution infrastructure.

Small-scale modular distributed natural gas reforming (DNGR) units have been developed that can be configured to achieve the desired scale of distributed production of hydrogen from NG. DNGR offers a potentially economical and technically viable near-term approach to supply hydrogen to vehicles, offering early availability, which is key to promoting acceptance of fuel cell electric vehicles (FCEVs) and hydrogen fuels and creating the market demand for them. It is anticipated that the industry will continue to make incremental improvements to the standard SMR technologies as part of base business operations.

Large-scale SMR typically uses a low-activity nickel-based catalyst. Although inexpensive, the nickel-based catalysts are prone to coking, require complicated start-up and shut-down protocols, and owing to their low activity, the reactors must be relatively large (Holiday et al. 2008). Most DNGR applications tend to use a precious metal-based catalyst. Some precious metal catalysts have higher activity and are less prone to coking; however, their cost is higher. Lower-cost catalysts are needed that can provide high yields to help increase the energy efficiency of distributed production.

Gasification of coal (coal gasification) or other carbonaceous raw materials is a mature, well-established, and the oldest technology for production of syngas and hydrogen that has been practiced for nearly two centuries, particularly in regions where coal is abundant. It has also been used by the chemical and fertilizer industries for the production of ammonia. Gasification generally occurs in large scale, high-temperature/pressure vessels where oxygen (or air) and steam are directly contacted with the coal or other solid feed material, causing a series of chemical reactions to occur that convert the feed into syngas composed primarily of the colorless, odorless, highly flammable carbon monoxide (CO), hydrogen (H_2), and ash/slag mineral residues. To produce pure H_2, CO in the syngas can be further converted in a WGS reactor and then pure hydrogen gas separated, generally through a PSA process as described above. CO_2 can be efficiently captured from syngas, preventing emission to the atmosphere and enabling its utilization or underground storage. A typical reaction for this process can be described as:

$$C(s) + H_2O \rightarrow CO + H_2 \tag{11.6}$$

This coal gasification process suffers from high impurity levels, low hydrogen yields, and the usual solids handling problems.

SMR and coal gasification extract hydrogen from carbonaceous fuel; however, both release CO_2. Efficiency for both technologies has been improved over time to the point where CO_2 emissions are close to the theoretical minimum. Additional reduction of CO_2 can be achieved by integration of these processes with Carbon Capture, Utilization and Storage (CCUS) technologies.

One CCUS option is to remove CO_2 from the high-pressure syngas stream, which is present immediately after the carbonaceous fuel conversion in the first stage of the process before the separation of H_2. This is somewhat analogous to the pre-combustion CO_2 removal in a coal power plant with CCUS. The advantage of this strategy is that the syngas stream is at high pressure and CO_2 is present at high concentration, requiring lower-cost equipment but limiting CO_2 capture to only 50–60% of the total produced in the overall process (IEAGHG 2017). Another CCUS option is to remove CO_2 from combustion in the low concentration, low pressure post-combustion stream at the exhaust of the overall process (analogous to post-combustion capture in power plants). As much as 90% of the overall CO_2 can be removed in the post-combustion capture, but this requires higher capital cost equipment. Recent techno-economic analysis of six different cases of CCUS integration with a large-scale SMR plant suggests that the levelized cost of H_2 increases by 18–45% from the base case without CCUS depending on the carbon capture option, cost of fuel, etc. (U.S. DOE 2010). Also, the cost of CCUS installation with a coal gasification plant is about 40% higher than for an SMR plant (U.S. DOE 2017).

Most commercially distributed MeOH is produced from NG; however, it can also be produced from renewable sources such as biomass and even from renewable hydrogen and captured CO_2. As more renewable MeOH enters the global mix, the developing infrastructure may gradually transition from fossil sources to renewable sources for hydrogen production. Most barriers to the widespread application of distributed MeOH reforming technology are the same as for DNGR. Distributed MeOH reforming would need to compete with DNGR on economics and availability.

MeOH toxicity may create an additional barrier to widespread utilization. While MeOH is often cited as a hazardous material, it is less toxic than many common liquid fuels and household chemicals.

MeOH reforming technology is sufficiently advanced so that private industry should be able to tackle the remaining technical and cost barriers that may arise and complete the commercial development of this production technology in the near future.

11.4 Water Splitting Production Processes

Direct water splitting (U.S. DOE 2017) is used to decompose water into elemental components hydrogen and oxygen. It encompasses both near- and long-term pathways, including both low- and high-temperature water electrolysis as well as emerging solar to hydrogen pathways based on direct photoelectrochemical (PEC) and solar thermochemical (STCH) water splitting. Unfortunately, this direct water splitting currently requires large amounts of electricity, which is a major problem, particularly from an economic and energy perspective.

The near-term pathway for hydrogen production through water electrolysis calls for using the existing infrastructure for water and electricity. Emission impacts of this technology will be greatly reduced when the electricity to power electrolysis is supplied by near-zero emissions sources such as wind, solar, or nuclear energy. Water electrolysis is being pursued for distributed, semi-central, and central production.

A long-term goal for large-scale, centralized hydrogen production is solar water splitting via PEC and/or STCH production pathways. These pathways use solar energy directly, providing enhanced energy security independent of the grid while offering the potential for meeting economic targets.

11.4.1 Water Electrolysis Production Process

Water electrolysis is a promising option for hydrogen production that uses electricity to split water. The purity of hydrogen produced by this process is a significant benefit of this option. Electrolyzers (the unit in which the electrolysis reactions take place) can range in size from small, appliance-size equipment that is well-suited for small-scale distributed hydrogen production to large-scale, central production facilities that can be tied directly to the grid or renewable or other non-greenhouse-gas-emitting forms of electricity production. The electrolyzer functions similarly to a fuel cell – only in reverse. This process is described by Equation 11.7:

$$H_2O + \text{electricity} \rightarrow H_2 + {}^1\!/\!_2\,O_2 \tag{11.7}$$

There are several types of electrolyzers. In a polymer electrolyte membrane (PEM) electrolyzer water reacts at the anode to form oxygen and positively charged hydrogen ions (protons). The electrons then flow through an external circuit, and the hydrogen ions selectively move across the PEM to the cathode, where the hydrogen ions combine with electrons from the external circuit to form the desired hydrogen gas.

Alkaline electrolyzers operate via transport of hydroxide ions (OH^-) through the electrolyte from the cathode to the anode, with hydrogen being generated on the cathode side. Electrolyzers using a liquid alkaline solution of sodium or potassium hydroxide as the electrolyte have been commercially available for many years.

Solid-oxide electrolyzers are another option. They use a solid ceramic material as the electrolyte that selectively conducts negatively charged oxygen ions at elevated temperatures, but with lower electricity requirements (i.e., higher electrical efficiency). Water at the cathode combines with electrons from the external circuit to form hydrogen gas and negatively charged oxygen ions, which then pass through the solid ceramic membrane and react at the anode to form oxygen gas and generate electrons for the external circuit.

Distributed commercial hydrogen production via water electrolysis is considered a near- to mid-term technology and is being currently deployed on-site at distributed/forecourt hydrogen refueling stations, where it could stimulate market acceptance. In the longer term, centralized production has the potential to substantially expand the commercial supply of hydrogen by water electrolysis; larger semi-central and central production via wind and nuclear heat and power are being actively investigated by DOE.

O&M costs for distributed hydrogen production from water electrolysis remain high at this time. For distributed production, some of the O&M issues are related to durability and scheduled maintenance.

Central hydrogen production entails O&M costs, and even though the specific needs and constraints may vary from the distributed case, the issues are similar. All system components must be considered in O&M, including power conditioning/management, feed pre-conditioning (e.g., water purification), controls, utilities, QA/QC, compression, storage, dispensing, and safety. Commercial PEM electrolyzers reportedly can run for tens of thousands of hours; however, there are some potential durability concerns with more advanced cell components, which are still at the lab scale.

Control and safety barriers associated with water electrolysis include the efficiency of start-up and shut-down processes, turn-down capability, and the capability for rapid on-off cycling. Control and safety system costs remain high owing to complex system designs and necessary high-cost sensors. For distributed production, the permitting process relies on the proven reliability and safety of these units.

11.4.2 Photoelectrical Hydrogen Production Process

The Photoelectrical Hydrogen Production (PEC) water splitting process uses semiconductor materials to convert solar energy directly into chemical energy in the form of hydrogen. The semiconductor materials used in the PEC process are similar to those used in photovoltaic solar electricity generation, but for PEC applications the semiconductor is immersed in a water-based electrolyte, where sunlight provides the energy for the water splitting process.

PEC is one of the more attractive among solar to hydrogen (STH) conversion technologies because efficient STH conversion can potentially be achieved at low operating temperatures using cost-effective thin-film and/or particle-based materials. It is an early-stage development technology that may prove suitable for semi-central and central hydrogen production. PEC reactors can be constructed in panel form (similar to photovoltaic panels) as electrode systems, or as slurry-based particle systems, with each approach having its own advantages and challenges. To date, panel systems have been the most widely studied because of their similarities with established photovoltaic panel technologies.

11.4.3 Thermochemical Water Splitting Production Process

Thermochemical water splitting uses high temperatures – from concentrated solar power or from the waste heat of nuclear power reactions – and chemical reactions to produce hydrogen and oxygen from water. The chemicals used in the process are reused within each cycle, creating a closed loop that consumes only water and no chemicals, and produces hydrogen and oxygen. This is a long-term technology pathway and calls for producing hydrogen in semi-central and central facilities. Numerous solar thermochemical water splitting cycles have been investigated for hydrogen production, each with different sets of operating conditions, engineering challenges, and hydrogen production opportunities. It offers a potential technology for clean, sustainable, relatively safe, high-volume hydrogen production with current R&D priorities focused on material development for both the reactive material as well as the reactor material and design.

The high temperatures necessary to split water via the "direct" STCH processes require the use of concentrated solar power. One approach is the deployment of a central STCH reactor in a solar receiver tower surrounded by heliostat fields (fields of sun-tracking mirrors) of a suitable size. Another approach is the use of multiple smaller-scale STCH reactor modules, each coupled to a tracking dish concentrator. Each approach has unique advantages and disadvantages; however, both can incur significant capital expenses.

With solar radiation as the driving energy source, overcoming these barriers is made even more difficult, primarily because of the source's known transient nature and relatively low power density. The low power density characteristic of solar power requires large collector areas and efficient concentrators to drive energy-intensive processes such as water splitting. Additionally, reactors must endure daily cycling from low to high temperatures resulting from the Sun's day/night cycles, as well as weather effects. One option is decreasing reactor cycling by storing the thermal energy in salts or other materials to enable continuous production. A combination of thermal storage and/or backup power might be needed to ensure continuous operation.

11.5 Biomass Production Processes

Another major resource that the United States can tap for energy is biomass (U.S. DOE 2017). Biomass comprises plant and animal matter; it is organic in nature with the potential for over a billion dry tons of biomass available annually. In general, the two types of biomass feedstocks available for use in hydrogen production are:

1) Primary biomass such as energy crops like poplar, willow, and switchgrass
2) Biogas produced from anaerobic digestion of organic residues from sources such as animal waste and municipal solid waste

Primary biomass can be gasified producing a carbon-rich syngas using well-established technology similar to the coal gasification process previously described and via pyrolysis. It can also be processed into bio-derived liquids for subsequent reforming. Biogas, with additional cleanup requirements, can be reformed to produce hydrogen using the SMR process previously discussed. Waste streams can also be used to produce hydrogen through biological-based processes such as fermentation and microbial electrolysis.

Hydrogen can be produced by distributed or semi-central facilities that employ gas-phase or aqueous-phase reforming (APR) of bio-liquids such as sugars, cellulose slurries, ethanol, or bio-oils, which can be extracted from primary biomass feedstocks. Bio-liquid reforming is similar to NG reforming but is usually challenged by limited catalyst activity and durability. For APR, water-soluble organics are decomposed at high pressure (~2 MPa and relatively low temperatures, <300°C) to hydrogen. The high pressure keeps the components in the liquid phase and the relatively low temperatures favor CO_2 formation over CO, thus maximizing the hydrogen yield without the need for a WGS reactor. As APR is still in the developmental stages, researchers and developers are trying to address some of the issues related to catalysts and process development.

Feedstock costs, including transportation costs, are typically the single largest factor determining economic viability of producing hydrogen from bio-derived liquids (Holladay et al. 2008). Feedstock availability may be limited owing to competition for biomass or lack of crops in some areas, for example, and the lower-cost feedstock may change throughout the year. Without significant reductions in feedstock cost, hydrogen from bio-derived liquids would not be economically viable.

The reformer and WGS unit operations require large amounts of catalysts and considerable maintenance. Commercial catalysts are designed for continuous use, whereas in distributed production, the reactors are turned on and off almost daily, causing thermal cycling that almost certainly will degrade catalyst performance. Therefore, long-life, inexpensive, durable catalysts are required. Gas-phase reforming of liquids is very similar to NG reforming and has many of the same barriers. The main differences are that vaporization must occur, the catalysts may need to be adjusted, and the steam-to-carbon ratio typically needs to be higher to avoid catalyst deactivation caused by coking and charring. APR has been primarily proposed for reforming biomass or the aqueous phase of the pyrolysis oil. However, the aqueous phase of pyrolysis oil often contains significant amounts of carboxylic acids, such as acetic acid, which are very difficult to decompose to the desired products, making catalyst development a primary need for APR's future. For current systems, necessary repairs are frequent, and repair costs can be significant. Downtime may adversely affect co-located businesses (e.g., convenience store, car washes).

Biogas can be produced by anaerobic digestion of organic matter, such as municipal solid waste, discards from food processing, animal manure, sewage, stillage, and glycerin from biofuel production, as well as energy crops and agricultural residues. In the anaerobic digestion process, organic matter is broken down in an oxygen-free environment producing significant amounts of methane. The U.S. total methane potential in raw biogas is estimated at about 16 million tons per year, with the net availability calculated at about 6.2 million tons (U.S. DOE 2017). Landfills represent the largest potential source of methane and hydrogen from both a total perspective as well as current net availability. Wastewater treatment plants (WWTPs) can also be a significant source of methane for hydrogen production. The total potential from industrial, institutional, and commercial sources is slightly less than from the other biogas sources, yet it still represents a significant source of methane and hydrogen. Animal manure is a more dispersed resource found in rural areas. However, it can be a significant source of biogas if some of the economic issues of collection and aggregation can be overcome. It can provide a source of hydrogen for bridging transportation corridors between major metropolitan areas and bringing economic opportunities to rural areas.

The geographic distribution of the biogas sources suggests that the largest biogas resources are located in close proximity to population centers, as would be expected as the majority of it is the result of human activity. This is also where the demand for hydrogen in industrial and vehicle refueling applications is expected to be highest. Initial estimates suggest that biogas and hydrogen from biogas produced from local, sustainable resources can supply upwards of 5% of the current U.S. vehicle fleet if that portion were replaced by FCEVs and, therefore, can aid with early FCEV rollout. In addition, reducing waste through the use of anaerobic digestion and utilization of produced biogas can help decrease greenhouse gas emissions and other environmental pollution in the air and water.

As methane is the main component of biogas, it can be converted to hydrogen through the same processes as NG, as described in the fossil resources section. Though, unlike NG, which consists mainly of methane and has relatively few impurities, biogas contains methane at only 50–70% with 30–50% CO_2 and large amounts of nitrogen and other particulates and contaminants, which can complicate processing. Sulfur, siloxanes, and halogens are the most damaging impurities that need to be removed to sub-ppm levels, which adds to the cost of cleanup as well as fuel quality monitoring. Therefore, additional cleanup steps are required in biogas processing before it can be converted into hydrogen. Various chemical and biological purification processes are available on the biogas market.

Once the biogas has been upgraded to NG quality, it can be used as a substitute for NG in an SMR process to produce hydrogen. Most technical barriers to converting biogas to hydrogen are similar to those for the conversion of NG using SMR-based systems. In addition, the need to pre-condition and cleanup biogas before reforming constitutes an additional set of challenges to the widespread application of hydrogen production from biogas.

Pathways for fermentative hydrogen production include traditional dark fermentation and microbial-aided electrolysis. In dark fermentation, bacteria decompose biomass into hydrogen and by-products without the need for sunlight. Microbial-aided electrolysis combines the energy from the microbial decomposition of organic matter with an additional small electric current to produce hydrogen. Fermentation uses microorganisms that break down biomass to produce hydrogen as a metabolic product or through microbe-aided electrolysis. This is a mid- to long-term technology that will most likely be suitable for distributed, semi-central, and central hydrogen production facilities, depending on the feedstock used. Dark fermentation uses anaerobic bacteria on carbohydrate-rich substrates grown, as the name indicates, without the need for light. As the microbes break down the biomass substrate, a number of pathways can result in hydrogen production. Many different metabolic steps are needed to either reduce or oxidize these electron carriers, depending on the initial feedstock and other conditions. Some by-products of this metabolism, such as organic acids, may be broken down further by microbes for energy but would not produce the necessary reduced ferredoxin or NADH to produce additional hydrogen.

Microbial Electrolysis Cells (MECs) are a variation of microbial fuel cells. In microbial fuel cells, a microbial culture decomposes organic matter, excreting protons and transferring electrons to the anode; the electrons travel to the cathode and combine with oxygen and the protons to produce water, along with a low voltage. In an MEC, a small voltage is added, resulting in the recombination at the cathode to produce hydrogen gas instead of water. With the bacteria decomposing the organic matter into electrons and protons and generating power, less external electric power is required as compared to standard water electrolysis. MECs can use feedstocks that would not support fermentative hydrogen production. Although CO_2 is produced at the anode through the metabolic processing, reactor designs can be made to ensure that the gas collected at the cathode is nearly pure hydrogen.

Both fermentation and MEC systems have the potential, alone or in combination, to be integrated with waste treatment systems in which a bioavailable waste product (food waste, sewage, etc.) could act as the feedstock. Such a system could reduce the net costs of hydrogen production by replacing costly wastewater treatment systems with a system that could both treat waste and produce a usable by-product. In this case, production could occur on a distributed or semi-central scale with the reactor systems installed at the site of waste production. Different waste feedstocks would have different challenges with respect to feedstock utilization and may have lower stoichiometric hydrogen yields, but the economic and sustainability considerations may allow this production pathway to reach commercial viability earlier than other biological production pathways.

There are other biological-based options. Photobiological hydrogen production uses microorganisms to convert solar energy into hydrogen. Photolytic production uses microorganisms, such as green microalgae or cyanobacteria, and sunlight to split water through direct or indirect photolysis routes. In photofermentative hydrogen production, sunlight provides the energy for photosynthetic bacteria to break down organic compounds (either generated by the microbes themselves or supplied to them), releasing hydrogen. By using sunlight and water (added organic compounds are also an option), photobiological hydrogen production offers a potential for clean, sustainable hydrogen production. These biological hydrogen production methods are in the early stage of research and present many technical challenges, beginning with the bioengineering of microorganisms that can efficiently produce hydrogen at high rates. Some additional challenges are related to the need for increased light utilization efficiency, increased rate of hydrogen production, improved continuity of photoproduction, and increased hydrogen molar yield. Another biological pathway showing potential for increased hydrogen yield is in vitro synthetic enzymatic biosystems to convert sugar to hydrogen.

11.6 Hydrogen Purification

Merriam-Webster (2023d) defines purification as ". . . the act or an instance of purifying . . . to clear from material defilement . . . to free from undesirable elements." To engineers, purification involves processes that can be defined as *unit operations* (Theodore and Ricci 2010; Theodore, Dupont, and Ganesan 2017). The three major purification unit operations are:

1) Distillation
2) Absorption
3) Adsorption

Theodore (2014) has also listed and provided a brief review of others that he refers to as *novel* processes including:

1) Freeze crystallization
2) Ion exchange
3) Liquid ion exchange
4) Resin adsorption
5) Foam fractionation
6) Dissociation extraction
7) Electrophoresis

All of these separation processes may be classified into two categories: phase separation and component separation. In the latter, a particular species is separated from a single-phase, multi-component system. Most component separation processes remove particular ionic or molecular species, generally without the use of chemicals. Phase separation processes are employed to separate phases, reduce feed volume, and concentrate a feed into one phase before further treatment and material recovery are performed. Feed streams which include slurries, sludges, and emulsions that contain more than one phase, are the usual candidates for this category.

Many factors need to be considered when selecting a particular type of purification process. These include: the characteristics of the feed stream and the desired characteristics of the product stream, the technical feasibility of the different physical processes when applied to a particular case, and economic, environmental, and energy considerations.

There are many special techniques for the isolation and purification of substances such as hydrogen. Some substances occur very nearly pure in nature, such as diamonds, and some minerals. Most natural materials, however, are mixtures, which must be separated or purified if pure substances are desired, and most manufactured materials also require purification (as is the case with hydrogen).

The separation of two different *phases* is often rather easy. Particles of a solid phase mixed with a liquid phase may be separated from the liquid by filtration. Often the solid is present because it has been produced from the liquid by a chemical reaction or by change in conditions (such as cooling); the solid is then called the *precipitate*. The precipitate can be removed simply by pouring the mixture onto a folded filter paper in a funnel. The liquid (called the *filtrate*) passes through and the grains of precipitate (the *residue*) larger than the openings in the filter paper are retained.

A mixture of two or more solid phases can occasionally be profitably separated by picking out the crystals or particles by hand; this is the way in which diamonds are picked out of diamond-bearing rock. Such a mixture may also be separated by the use of some physical property of one of the phases. Particles of a strongly magnetic material can be separated from other particles by use of a magnet. A difference in density between the two phases may permit separation by stirring the mixture into a liquid of intermediate density. A difference in solubility in water or in some other solvent may also permit easy separation of the solid phases.

An impure liquid substance may also be purified by *fractional freezing*. The impure liquid substance is cooled until part of it has crystallized, and the remaining liquid, which usually contains most of the impurities, is then poured off, leaving the purified crystals. A closely related and highly important method of purification is *fractional crystallization* from a solution. The impure substance is dissolved in a suitable solvent, the solution is filtered, and the filtrate is then allowed to evaporate until some of the substance has crystallized. These crystals are usually purer than the original material.

An impure liquid can be similarly purified by *distillation* in a still. The liquid is boiled in a flask or other container, and the vapor is condensed in a condenser, forming a liquid distillate that is collected in a receiver. The first portions (fractions) of the distillate contain larger amounts of the more-volatile impurities and the residue in the flask tends to retain the less-volatile impurities. Stills of special design have been developed that are very effective in separating liquid mixtures into their constituent parts. Distillation, the workhorse of the chemical industry, is an important method of obtaining specialty materials, such as gasoline and lubricating oils, from natural sources such as petroleum.

The situation with hydrogen is unfortunately more complicated simply because the purification step is a strong function of the process employed to produce the hydrogen in the first place. Four processes employed in the past are briefly described below. Additional details are available in the literature (IEA 2019).

11.6.1 Carbon Dioxide and Hydrogen Sulfide Removal

Many processes for the removal of acid gases have been employed commercially (Riesenfeld and Kohl 1959). These include the use of monoethanolamine (MEA), hot potassium carbonate processes, and physical solvent processes.

MEA. A water solution of MEA is reacted with CO_2 in an absorber vessel under pressure at room temperature. A major problem associated with the use of MEA solutions is their corrosive effect on process equipment. Corrosion is most severe at elevated temperatures and where the acid gas concentration in solution is highest.

Hot potassium carbonate process. This process, developed originally by the Bureau of Mines, is particularly useful for removing large quantities of CO_2. Although it can remove CO_2 down to 0.1% by volume in the gas being purified, it is generally more economical at purity levels of 1% or greater.

Physical solvent processes. These are processes in which CO_2 is removed by physical solution in a solvent, which is often proprietary. Examples of this class of process are the Rectisol process using cold (approximately −60°C) methanol; the fluor solvent process using a nonaqueous organic solvent such as propylene carbonate; the Sulfinol process using an organic solvent, sulfolane tetrahydrothiophene dioxide); the Selexol process using the dimethyl ether of polyethylene glycol; and the Purisol process using *N*-methyl-2-pyrrolidone.

11.6.2 Adsorptive Purification

Fixed-bed adsorption can remove such impurities as CO_2, H_2O, CH_4, C_2H_6, CO, Ar, and N_2, among others. One type of process is the thermal-swing process in which the impurity is adsorbed at a low temperature and desorbed thermally by raising the temperature and passing a nonabsorbable purge gas through the bed to aid desorption and carry the desorbed gas from the bed. For continuous operation, two beds are necessary; while one bed is on-stream, the second bed is being regenerated.

11.6.3 Cryogenic Liquid Purification

An impure hydrogen stream can be partially purified by simply cooling to the appropriate cryogenic temperature (Lanz, Heffel, and Messer 2001), where the impurity condenses and can be separated as a liquid stream. This is often used for bulk removal of light hydrocarbons from hydrogen in a refinery off-gas. Product purity obtainable depends upon the vapor pressure of the impurity, and high purities are not readily attainable in most cases using this method.

11.6.4 Carbon Monoxide Removal

High concentrations of carbon monoxide are commonly reduced by conversion to hydrogen via the WGS reaction, while low concentrations can be converted via catalytic methanation. Removal by scrubbing in solutions of complex copper ammonium salts has also been used commercially (Riesenfeld and Kohl 1959).

11.7 Hydrogen Laboratory Processes

In the laboratory, hydrogen may be prepared via the reaction of an acid such as sulfuric acid, H_2SO_4, with a metal such as zinc. The equation for the reaction is:

$$H_2SO_4 + Zn \rightarrow ZnSO_4 + H_2 \uparrow \qquad (11.8)$$

The vertical arrow placed beside the formula of hydrogen in this equation indicates that hydrogen is a gas, which escapes from the region of reaction. Hydrogen can also be prepared by the reaction of a host of metals with water or steam.

Sodium and its congeners react very vigorously with water, so vigorously as to generate enough heat to ignite the liberated hydrogen. An alloy of lead and sodium, which reacts less vigorously, is sometimes used for the preparation of hydrogen. The equation for the reaction of sodium with water is as follows to produce sodium hydroxide and hydrogen gas:

$$2Na + 2H_2O \rightarrow 2NaOH + H_2 \uparrow \qquad (11.9)$$

Calcium also reacts with water, with less vigor. The reaction of calcium with cold water provides a simple and safe way of preparing hydrogen in the laboratory. The equation for this reaction is:

$$Ca + 2\,H_2O \rightarrow Ca(OH)_2 + H_2 \uparrow \qquad (11.10)$$

The calcium hydroxide (slaked lime) product is not very soluble in water, and in the course of the reaction it precipitates as a white solid.

It should be noted that both the sodium and calcium reactions liberate only half of the hydrogen contained in the water with which it reacts.

Hydrogen can also be made by the reaction of a *metallic hydride* (a compound of a metal and hydrogen) with water. Thus, calcium hydride, CaH_2, produces hydrogen according to the following reaction:

$$CaH_2 + 2H_2O \rightarrow Ca(OH)_2 + 2H_2 \uparrow \qquad (11.11)$$

The greatest yield (i.e., the greatest ratio of weight of hydrogen produced to weight of reacting material) is obtained by the use of the hydride of the lightest metal, lithium:

$$LiH + H_2O \rightarrow LiOH + H_2 \uparrow \qquad (11.12)$$

Much of the hydrogen that was used in industry, especially for the hydrogenation of vegetable oils to convert them into solid fats, was produced by the reaction of iron with steam. Steam from a boiler was passed over iron filings heated to a temperature of about 600°C. The reaction that occurred is:

$$3Fe + 4H_2O \rightarrow Fe_3O_4 + 4H_2 \uparrow \tag{11.13}$$

After a mass of iron had been used in this way, and was converted into iron oxide, Fe_3O_4, the iron could then be regenerated by passing carbon monoxide, CO, over the heated iron oxide to form carbon dioxide and elemental iron, allowing the iron to be used over and over again:

$$Fe_3O_4 + 4CO \rightarrow 3Fe + 4CO_2 \tag{11.14}$$

This method was used in the past at an industrial level for small- to moderate-scale operations. However, it has been supplanted by the modern hydrogen production methods described earlier in this chapter.

11.8 Emerging Hydrogen Technologies

Future development of other hydrogen-generating processes is under way. A currently promising method of hydrogen production is pyrolysis or co-pyrolysis. In this process, raw organic material is heated and gasified at pressures of 0.1–0.5 MPa, and temperatures in the 500–900°C range. The process takes place in the absence of oxygen, and therefore the formation of dioxins can be almost ruled out. Since no water or oxygen is present, no oxides of carbon, i.e., CO or CO_2, are formed, eliminating the need for purification steps. This process offers potentially significant emission reductions. However, if air or water are present in the feed material due to insufficient drying, significant CO_2 emissions will be produced. This reaction can be generally described by the following equation:

$$C_nH_m \rightarrow nC + 0.5mH_2 \tag{11.15}$$

The application of the co-pyrolysis of a mixture of coal with organic wastes has also recently received interest in industrially advanced countries, as it provides an option for diverting organic waste from landfills for beneficial use. Pyrolysis and co-pyrolysis are well-developed processes that could be used at a commercial scale.

Methane splitting offers another potential new way to produce hydrogen from NG. The main technology is based on alternating-current three-phase plasma and uses methane as a feedstock and electricity as an energy source. It produces hydrogen and solid carbon, but no CO_2 emissions. Methane splitting requires high-temperature plasma and significant thermal losses reduce its efficiency advantage, but it uses three to five times less electricity than electrolysis for the same amount of hydrogen produced. It has very low CO_2 formation and creates solid carbon in the form of carbon black. It requires more NG than electrolysis but could create additional revenue streams from the sale of carbon black for use in rubber, tires, printers, and plastics (IEA 2019). A thermodynamic analysis suggests that methane pyrolysis is an attractive production process for generating hydrogen (Theodore, Personal Notes, East Williston NY, 2024).

Alternative process designs for SMR are also being explored. While NG would still be required as feedstock, other energy sources could be used to produce the necessary steam, and this could facilitate the capture of the more concentrated "process" CO_2 stream. Electricity is a potential candidate to produce the necessary high-temperature steam, while concentrating solar heat could be used in areas with sufficient solar resources.

In summary, hydrogen is produced today from diverse domestic feedstocks using a variety of process technologies. Hydrogen-containing compounds such as fossil fuels, biomass, or even water can be a source of hydrogen. Thermochemical processes are used to produce hydrogen from fossil fuels such as NG and coal or from renewable sources such as biomass. Electricity generated from sunlight, wind, and nuclear sources can be used to electrolytically produce hydrogen from water. Sunlight by itself can also be used to directly produce hydrogen from water using advanced photo-electrochemical, thermochemical, and photobiological processes.

References

Global Energy Infrastructure (2021). Hydrogen – data telling a story. https://globalenergyinfrastructure.com/articles/2021/03-march/hydrogen-data-telling-a-story/ (accessed 4 November 2023).

Holladay, J.D., Hu, K., Kim, L., and Yang, Y. (2008). An overview of hydrogen production technologies. *Catalyst Today* 139: 244–260.

IEA (2019). *The Future of Hydrogen: Seizing today's opportunities*. Paris, France: OECD. https://doi.org/10.1787/1e0514c4-en.

IEA Greenhouse Gas R&D Programme (2017). Techno-Economic Evaluation of SMR Based Standalone (Merchant) Hydrogen Plant with CCS. Technical Report 2017-02. Cheltenham, UK: IEAGHG.

Lanz, A., Heffel, J., and Messer, C. (2001). *Hydrogen Fuel Cell Engines and Related Technologies*. Palm Desert, CA: College of the Desert. https://www.energy.gov/eere/fuelcells/articles/hydrogen-fuel-cell-engines-and-related-technologies-course-manual.

Merriam-Webster (2023a). Process. Merriam-Webster.com. https://www.merriam-webster.com/dictionary/process?src=search-dict-box (accessed 4 November 2023).

Merriam-Webster (2023b). Production. Merriam-Webster.com. https://www.merriam-webster.com/dictionary/production (accessed 4 November 2023).

Merriam-Webster (2023c). Manufacturing. Merriam-Webster.com. https://www.merriam-webster.com/dictionary/manufacturing (accessed 4 November 2023).

Merriam-Webster (2023d). Purification. Merriam-Webster.com. https://www.merriam-webster.com/dictionary/purification (accessed 13 November 2023).

Pironon, J. and de Donato, P. (2023). How we chanced upon what may be the world's largest white hydrogen deposit. The Conversation. https://theconversation.com/how-we-chanced-upon-what-may-be-the-worlds-largest-white-hydrogen-deposit-212499 (accessed 4 November 2023).

Riesenfeld, F.C. and Kohl, A.L. (1959). *Gas Purification*. Houston, TX: Gulf Publication Company.

Theodore, L. (2014). *Chemical Engineering: The Essential Reference*. New York: McGraw-Hill.

Theodore, L. and Ricci, F. (2010). *Mass Transfer Operations for the Practicing Engineer*. Hoboken, NJ: John Wiley & Sons.

Theodore, L., Dupont, R.R., and Ganesan K. (2017). *Unit Operations in Environmental Engineering*. Hoboken, NJ: John Wiley & Sons.

U.S. Department of Energy (2010). Assessment of hydrogen production with CO2 capture volume 1: baseline state-of-the-art plants. DOE/NETL-2010/1434.

U.S. Department of Energy (2017). *Hydrogen Production Tech Team Roadmap*. Washington, DC: U.S. Department of Energy, DRIVE Partnership. https://www.energy.gov/eere/vehicles/articles/us-drive-hydrogen-production-technical-team-roadmap.

12

Hydrogen Storage

12.1 Introduction

Most hydrogen is used locally where it is produced or generated, and it is, generally speaking, technically difficult and expensive to not only store but also transport hydrogen safely other than by pipeline.

Many have argued that hydrogen storage is a key technological barrier to the widespread use of fuel cell power technologies. The ability to store hydrogen economically and safely will play a major role in the future of hydrogen energy. At the present time, hydrogen storage, especially at large scale or using "bulk" storage approaches, needs further development and investment before they are ready for broad deployment. This necessary development is currently underway, and there are several pathways for the development of a hydrogen storage infrastructure in the coming years. Existing regulatory programs apply to traditional storage approaches, but additional regulatory activity in the future will help assure that this industry understands and complies with any new legal and regulatory requirements.

As indicated earlier, hydrogen has the lowest gas density and the second lowest boiling point of all known substances, creating a challenge to store as either a gas or a liquid. As a gas, it requires very large storage volumes and pressures. Storage of hydrogen as a gas typically requires high-pressures, 350–700 bar (5,000–10,000 psi) tank pressure (U.S. DOE 2020). As a liquid, it requires a cryogenic storage system as the boiling point of hydrogen at 1 atm pressure is −252.8°C.

Hydrogen's low density, both as a gas and a liquid, also results in very low energy density. This increases the relative storage tank size, as more hydrogen is required to meet a particular application. Despite its low volumetric energy density, hydrogen has the highest energy-to-mass ratio of any fuel. Unfortunately, this advantage is usually overshadowed by the high weight of the hydrogen storage tanks and other associated storage equipment. Thus, most hydrogen storage systems are considerably bulkier and/or heavier than those used for other fuels, e.g., gasoline or diesel fuels.

For all practical purposes, hydrogen can be stored as either a high-pressure gas, a liquid in cryogenic containers, or a gas chemically bound to certain metals, e.g., hydrides. However, it appears that the optimal way to store hydrogen is in the form of hydrocarbon fuels, although it requires additional processes to extract hydrogen from these hydrocarbons.

In the final analysis, storing hydrogen therefore requires the input of energy to compress it as a gas. For electric power generation purposes, storage can potentially be accomplished in a large pressure vessel or a cavern. "Very large amounts of hydrogen can be stored in constructed underground salt caverns of up to 500,000 m^3 at 2,990 psi, which would mean about 100 GWh of stored

Hydrogen Energy: Principles and Applications, First Edition. Vincent J. DelGatto, Louis Theodore, R. Ryan Dupont, and Matthew C. Ogwu.

electricity" (ESA 2020). Thus, it is fair to say that developing an acceptable system to store hydrogen could entail substantial costs, whether for a completely new system or a gradually upgraded and modified system.

Section numbers and their accompanying titles for this chapter concerned with hydrogen storage are listed below.

12.2 Chemical Industry Storage Options

Storage of materials, particularly gases and liquids, requires the use of tanks and reaction vessels for a wide variation of temperature and pressure conditions. A few principles employed in industrial settings, particularly in chemical process industries, are presented in this section.

12.2.1 Gas Storage

The distinction between pressure vessels, bottles, and pipes is arbitrary. They can all be used for storing gases under pressure. A storage pressure vessel is usually a permanent installation. Storing a gas under pressure not only reduces its volume but also, in many cases, liquefies it at ambient temperature. Some gases in this category are carbon dioxide, several petroleum gases, chlorine, ammonia, sulfur dioxide, and some types of Freon. Pressure tanks are frequently installed underground for safety reasons.

Atmospheric pressure gas can be stored in vertical cylindrical tanks commonly known as wet- or dry-seal gas holders. The wet gas holder maintains a liquid seal of water or oil between the top movable inside tank and the stationary outside vertical tank. A dry-seal holder maintains a seal between the inner and outer tanks by means of a flexible rubber or plastic sheet; it is specified where dry gas storage is a necessity. Earlier developments in bulk natural gas or gas product storage showed that pumping the gas into porous underground strata is the cheapest method available. High-pressure gas must be stored in a bank of spherical or horizontal pressure vessels.

12.2.2 Liquid Storage

Bulk storage of liquids is generally handled by closed tanks to prevent the escape of volatiles and contamination. In some instances, such as water storage, where contamination and dilution are not a factor, large open reservoirs can be employed. Natural terrain, concrete-walled excavations, or concrete tanks without tops are typical construction. Reinforced wall design is required, and the concrete must be waterproofed with a suitable paint to prevent any possibility of leakage. Storage of liquid materials in a typical process industry is carried out in tanks classified as spherical or vertical and horizontal cylindrical. Since safety is an important consideration in storage tank design, the National Fire Protection Association (NFPA 2021) and the American Petroleum Institute (API 2020) publish rules for safe design and operation. *Vertical tanks* are most commonly used for outdoor storage of such materials as petroleum products. Water towers are typical of elevated vertical tank outdoor construction. They are used for maintaining a uniform head of water and to store water for temporary emergencies and fire protection. *Horizontal tanks* are most frequently found inside buildings where floor loading and allowable headroom are primary considerations. A series of horizontal pressure vessels of 2–6 ft in diameter are used for bulk storage of liquids under pressures greater than 250 psig. Typical piping includes individual pressure relief valves and a common header. *Spherical tanks* using the thin-steel construction of vertical storage tanks are economically designed for pressures between 2 and 250 psig.

Venting of storage tanks is an important design consideration. In the low vapor-pressure, non-flammable category, an inverted U or downcomer cap of the open type should be specified. The venting capacity in terms of scf air should be twice the pumping rate. Storage of flammable liquids requires flame arrestors (Theodore and Dupont 2012) on all vapor openings. Flame arrestors consist of a number of thin slotted plates in a small chamber of 40- to 60-mesh wire screen of such a design that rapid heat dissipation prevents sustained combustion. Tanks for conserving valuable fluids of significant vapor pressure are vented via pressure-vacuum-controlled openings. These are commercially available and work on an open-close pressure range tolerance. This avoids loss by normal tank "breathing" caused by ambient temperature cycling and excursions.

12.2.3 Tank Details

Storage is commonly accomplished in industrial plants, as noted above by the use of cylindrical, spherical, or rectangular tanks. These tanks were constructed in earlier days of wood and are now constructed of concrete or metal, with metal being the most common modern material of construction. The design of storage vessels involves consideration of details such as size and number of openings, shape of heads, necessary temperature and pressure controls, and corrosive action of the contents. One of the major factors in the design is the wall thickness. The wall must be sufficiently thick to permit safe usage under all operating conditions. The necessary wall thickness for metal vessels is a function of the ultimate tensile strength or the yield point of the metal at the operating temperatures, the operating pressure, the diameter of the tank, and the joint or welding efficiencies. Various codes are available, which specify the conditions that must be met for different vessels. A summary of design equations and other information is available in the ASME Code for Unfired Pressure Vessels (2023).

12.2.4 Storage Batteries

Storage batteries may reach economical use for the storage of large amounts of energy before the middle of this century. Batteries have the advantage of minimal siting problems and can be placed near any anticipated load, thus reducing transmission costs. Efforts are needed to produce a battery with higher specific energy. In addition, a longer lifetime is desired. Furthermore, the efficiency of storage and discharge should be relatively high – normally more than 70% in total. Unfortunately, no presently available storage batteries meet all of the above criteria.

12.3 Hydrogen Storage Overview

Once hydrogen is produced via a particular process, it can be stored in gaseous, liquid, solid, or bound (e.g., physical and/or chemical) states. Methods for storage currently available or under study are:

1) High-pressure gas tanks
2) Cryogenic liquid tanks
3) Underground storage salt caverns
4) Adsorption in metal hydrides
5) Bonding with liquid organic hydrogen carries (LOHCs)

Each form of storage has advantages and disadvantages that are presented below.

12.3.1 Compressed Gas

High-pressure gas containers minimize the volume of storage. These typical approaches are well established, as they differ very little from the storage of any other industrial gas. Recent developments have also reduced the cost of high-pressure storage for stationary applications. Design and material research is ongoing to make high-pressure gas containment more affordable. This has the added benefit of increasing the energy density of the storage system. However, gas compression can be energy intensive, so storage at lower pressures should be considered if the system volume is not a major concern.

12.3.2 Liquid Storage

Hydrogen can also be stored as a liquid for many applications. Cryogenic liquid hydrogen is stored at 20K (−423°F; −253°C). Liquefying hydrogen further increases the volumetric energy density of hydrogen, but it requires a considerable amount of energy to cool hydrogen to such low temperatures, which reduces the roundtrip storage efficiency. Furthermore, over long periods of storage time, the system would be particularly susceptible to boil-off losses due to unavoidable temperature excursions and ambient heating of the liquid. Recent cryogenic design advancements have reduced the rate of heat transfer and resulting hydrogen loss, but it is still not ideal for storing hydrogen for long periods of time. This typically makes liquid hydrogen an undesirable storage method for some applications, but it is relevant for the transportation of hydrogen from production facilities to load centers. Cryocompression is a newer approach to mitigate boil-off losses in hydrogen systems. In cryocompressed storage, a high-pressure tank is modified to operate at liquid nitrogen temperatures (about 77K).

12.3.3 Underground Storage

In addition to manufactured containers, gas pipelines and geologic storage are attractive methods for storing large volumes of hydrogen. Hundreds of miles of pipelines carrying pure hydrogen are near refineries in South Texas and Southern California. Unused or abandoned pipelines could make excellent storage vessels, depending on the material with which they are made. A number of studies have recommended new pipelines as economically feasible for both long and relatively short distances.

Geologic storage refers to the use of underground caverns to store large volumes of gas. Although rock caverns, aquifers, and abandoned mines have been considered for this purpose, the most viable caverns for hydrogen are created by solution mining in salt deposits. This results in large, impermeable salt caverns deep underground. (The petroleum and chemical industries use aquifers and underground salt caverns to store hydrogen, natural gas, and other fuels near refining operations, although the problems associated with the permeability of rock are significantly more serious for hydrogen storage.) Similar to the proposed use case for grid-scale hydrogen storage, these caverns are often used to help rectify the variable consumption of natural gas throughout the year with the constant production rate from refineries.

Hydrogen storage in geologic formations requires careful siting to limit stability concerns, and the requisite salt deposits are not available everywhere. Although geologic hydrogen storage may have limited siting locations, this mode of storage is incredibly inexpensive for a very large storage capacity, making it an attractive storage option.

12.3.4 Metal Hydrides

Hydrogen can also be stored in the chemical bonds of various compounds. Solid materials such as metal hydrides are sometimes used. In these systems, a metal (M) absorbs hydrogen (H) to form a metal hydride (MH). Hydrogen can then be released by heating the MH as needed. These materials increase the volumetric energy density of the storage system even beyond that of liquid hydrogen. These systems also offer increased safety because they operate at lower pressures, and the release of hydrogen is slowed by desorption processes in the event of a leak.

12.3.5 Liquid Organic Hydrogen Carriers

Liquid Organic Hydrogen Carriers (LOHCs) such as methylcyclohexane provide another way to store hydrogen. These energy-rich liquids can be easily stored and transported without the need for cryogenic temperatures necessary to produce liquefied pure hydrogen. Hydrogen can then be released from the LOHC for use through dehydrogenation processes.

12.4 Gaseous Hydrogen Storage

The most common method to store hydrogen in gaseous form is in steel tanks, although lightweight composite tanks designed to endure higher pressures are also becoming more and more common. Another novel method to store hydrogen gas at high-pressure is to use glass microspheres. Both these methods are reviewed below.

12.4.1 Composite Tanks

There are advantages to composite tanks. Their low weight meets key targets, and the tanks are already commercially available, well-engineered, and safety-tested. A composite tank requires no internal heat exchange and may be usable for cryogas. Their main disadvantages are:

1) The large physical volume required
2) The fact that the ideal cylindrical shape makes it difficult to conform storage to available space
3) Their high cost
4) The energy penalties associated with compressing the gas to very high pressures

There are also some safety issues that still have not been resolved, such as the problem of rapid loss of hydrogen in the event of an accident. The long-term effect of hydrogen on the materials under temperature excursion conditions is also not fully understood.

12.4.2 Glass Microspheres

The basic concept for how glass microspheres can be used to store hydrogen gas on board a vehicle can be described by three steps: charging, filling, and discharging. First, hollow glass spheres are filled with H_2 at high pressure (350–700 bar) and high temperature (300°C) by permeation in a high-pressure vessel. Next, the microspheres are cooled down to room temperature and transferred to the low-pressure vehicle tank. Finally, the microspheres are heated to 200–300°C for controlled release of H_2.

The main problem with glass microspheres is the inherently low volumetric density that can be achieved, and the high pressure required for filling. The glass microspheres slowly leak hydrogen at ambient temperatures. Another practical challenge is that there is potentially too much breakage during temperature or pressure excursions. The main operational challenge is the need to supply heat at temperatures higher than those available by the vehicle.

Glass microspheres have the potential to be inherently safe as they store H_2 at a relatively low pressure onboard and are suitable for conformable tanks. This allows for low container costs. The significant technical advantage is the demonstrated storage density. Additional studies are needed to reduce the hydrogen liberation temperatures to less than 100°C for the microspheres. Improvements are also needed to develop stronger glass and coating techniques for the optimization of H_2 permeability at lower temperatures.

12.5 Liquid Hydrogen Storage

Liquid hydrogen storage systems overcome some of the weight and size problems associated with high-pressure gas storage systems. Perhaps the most common way to store hydrogen in liquid form is to cool it down to cryogenic temperatures, while other options include storing hydrogen:

1) As a constituent in other liquids
2) As a rechargeable organic liquid
3) As anhydrous ammonia (NH_3)

12.5.1 Cryogenic Liquid Hydrogen

Cryogenic hydrogen, usually simply referred to as liquid hydrogen (LH_2), has a density of 70.8 kg/m^3 at its normal boiling point (−253°C). Liquid hydrogen has a much better energy density than pressurized gas solutions. However, it is important to note that approximately one-third of the energy of hydrogen is lost when LH_2 is produced. The main disadvantage of LH_2 is the boil-off loss during dormancy plus the fact that super-insulated cryogenic containers are needed. In addition, the general public's perception of LH_2 is that it is unsafe. The main advantage of LH_2 is the high energy storage density at even low pressures. Liquid hydrogen has already been demonstrated in commercial vehicles; it can also be co-utilized as aircraft fuel since it provides the best mass advantage of any H_2 storage option. Improvements are expected in the future, including lower cost improved insulated containers, and systems to capture and re-liquefy the boil-off.

12.5.2 Storage as a Constituent in Other Liquids

Borohydride ($NaBH_4$) solutions can be used as a liquid storage medium for hydrogen. The catalytic hydrolysis reaction for hydrogen release is:

$$NaBH_4(l) + 2H_2O(l) \rightarrow 4H_2\uparrow + NaBO_2(s) \tag{12.1}$$

The main advantage of using $NaBH_4$ solutions is that they allow for safe and controllable onboard generation of H_2. The main disadvantage is that the reaction product $NaBO_2$ must be regenerated back to $NaBH_4$. Although use of $NaBH_4$ solutions in vehicles may be prohibitively expensive due to the cost of $NaBO_2$ regeneration, there are a few commercial companies that promote the technology. The required cost reduction is unlikely because of the unfavorable thermodynamics. However, $NaBH_4$ solutions may be usable in high-value portable and stationary applications.

12.5.3 Rechargeable Organic Liquids

Organic liquids can also be used to indirectly store hydrogen in liquid form. The following three steps summarize the basic concept.

1) An organic liquid is catalytically dehydrogenated to produce H_2 gas.
2) The dehydrogenated product is removed from the tank while simultaneously refilling the tank with fresh H_2-rich liquid.
3) The dehydrogenated project is transported to a central processing plant where it is re-hydrogenated back to the starting compound for reuse.

One example of a rechargeable organic liquid process is the dehydrogenation and re-hydrogenation of methylcyclohexane (C_7H_{14}) and toluene (C_7H_8):

$$C_7H_{14}(l) \rightarrow 3H_2 \uparrow + C_7H_8(l) \tag{12.2}$$

$$C_7H_8(l) + 3H_2 \rightarrow C_7H_{14}(l) \tag{12.3}$$

It should also be noted that any of these organic liquids must be handled with great care because of fire and explosion hazards (Theodore and Dupont 2012). It is therefore necessary to perform detailed safety and toxicity studies when considering the use of these materials. Future improvements need to include developing systems that can be dehydrogenated at low temperatures to produce H_2 at useful pressures, developing optimal metal dehydrogenation catalysts and onboard systems, and improving the re-hydrogenation process. This process compares favorably with LH_2 and borohydride solutions.

12.6 Solid Hydrogen Storage

Storage of hydrogen in the solid state has the *potential* to become a safe and efficient way to store energy, both for stationary and mobile applications. There are four main categories of suitable materials:

1) Carbon and other high surface area materials
2) Rechargeable metal hydrides
3) Water-reactive chemical hydrides
4) Thermal chemical hydrides

Details are provided below.

12.6.1 Carbon and Other High Surface Area Materials

12.6.1.1 Carbon-Based Materials

Carbon-based materials, such as nanotubes and graphite nanofibers, have received significant attention in the research community and the public press over the last decade. It is believed that high H_2 storage capacities (50 wt%) are impossible. Pure H_2 molecular physisorption has been clearly demonstrated, but is useful only at cryogenic temperatures, and extremely high surface area carbons are required. Pure atomic H-chemisorption has been demonstrated, but the covalent-bound H is liberated only at impractically high temperatures (above approximately 400°C). Room temperature adsorption up to a few wt% H_2 is occasionally reported but has not been reproducible. This requires a new bonding mechanism with energies between physisorption and strong covalent chemisorption. The surface and bulk properties needed to achieve practical room temperature storage are not clearly understood, and it is far from certain that useful carbon can be economically and consistently synthesized. Hence the potential for H_2 storage in carbon-based materials is questionable at this time.

12.6.1.2 Other High Surface Area Materials

The most predominant examples of other high surface area materials are zeolites, metal oxide frameworks (MOFs), and clathrate hydrates. These definitions and main features are listed below.

1) *Zeolites.* Complex aluminosilicates with engineered pore sizes and high surface areas—Well-known as "molecular sieves." The science of capturing non-H_2 gases is well-known.
2) *Metal oxide frameworks (MOFs).* Typically ZnO structures are bridged with benzene rings. These materials have an extremely high surface area, are highly versatile, and allow for many structural modifications.
3) *Clathrate hydrates.* H_2O (ice) cage structures often contain "guest" molecules such as CH_4 and CO_2. The cage size and structure can often be controlled by organic molecules.

These materials are all characterized by extremely high surface areas that can physisorb molecular H_2. They have been shown to store a few wt% H_2 at cryogenic temperatures. The main concern with these materials remains whether they can be engineered to reversibly store high levels of H_2 near room temperature. These materials, particularly MOFs, and clathrate hydrates, represent new storage ideas and should be studied to determine the potential for their development and use in the near future.

12.6.2 Rechargeable Metal Hydrides

Metal hydride storage systems are based on the principle that some metals readily adsorb gaseous hydrogen at high-pressure and moderate temperatures to form metal hydrides. These metal hydrides release the hydrogen gas when heated at low pressure and relatively high temperature. In essence, the metals soak up and release hydrogen like a sponge. The bottleneck is that heat energy is released as the hydrogen atoms fill the interstitial sites, and to remove the stored hydrogen atoms, heat must be supplied, which might lead to the inefficient removal of the hydrogen atoms. Also, compared to other fuels, the storage capacity of these hydrides is low (Sakintuna, Lamari-Darkrim, and Hirscher 2007; Lim et al. 2010). Complex hydrides still provide hope for the future, particularly the nontransition metal types such as alanates and borohydrides.

12.6.2.1 Alanates

The key properties of the most well-known alanate, $NaAlH_4$, are being intensely studied on many fronts. The low-temperature kinetics and reversibility of these alanates are improved by adding a catalyst (e.g., Ti). The research on catalyzed $Mg(AlH_4)$ shows that this type of alanate is not nearly as reversible as $NaAlH_4$. Hence, the near-term applicability of $Mg(AlH_4)$ is unlikely. The main interest today is to extend the catalyst concept to other alanates beyond $NaAlH_4$.

12.6.2.2 Borohydrides

The key properties of the most well-known borohydrides show that the borohydrides have much higher potential capacities for H_2 adsorption than the alanates. However, they are also much less studied than alanates. In general, borohydrides are too stable and are not as reversible, but are at the same time considered safer than the alanates. There has been some recent preliminary progress on $LiBH_4$, particularly on reversibility and destabilization.

12.6.3 Water-Reactive Chemical Hydrides

Chemical hydrides can be handled in a semi-liquid form, such as mineral oil slurry. In this form, hydrides can be pumped and safely handled. Controlled injection of H_2O during vehicle operation, for example, can be used to generate H_2 via hydrolysis reactions. The liberation of H_2 is exothermic and does not require waste heat from the vehicle power source. One approach is to encapsulate small spheres of NaH in polymeric shells. In general, MgH_2 probably offers the best combination of H_2 yield and affordability. The major interest today is to lower the cost of processing the spent hydroxide back into the starting hydride.

12.6.4 Thermal Chemical Hydrides

Ammonia borane is another group of chemical hydrides that potentially could be used to store hydrogen in a solid state. Preliminary results show that NH_4BH_4 can be thermally decomposed with very high H_2 yields. However, the reactions are not reversible, and regeneration is required. The possibility of gaseous boranes in the evolved H_2 must be carefully studied since they are toxic.

12.7 The Moon Project

An extensive literature search was conducted prior to writing this chapter, and several relevant DOE sites were found as reported above, but much of what has been reported is quite redundant. However, there was one truly exceptional and informative work, Dr. Donald Rapp's classic 2005 report to JPL *Transporting Hydrogen to the Moon or Mars* (Rapp 2005). This is a report that belongs in the hydrogen energy literature and is a must-read for those involved with this emerging field and emerging technologies. The following are a number of excerpts from this report.

> It would seem to be elementary that the first thing to do in studying hydrogen storage is to establish a baseline of characteristics of storage as a highly compressed gas at high pressure and room temperature. However. I have searched the Internet for available information on the masses of such tanks, and although there are many relevant sites, few deal in any detail

with tank masses. In fact, it is truly amazing how many websites discuss hydrogen storage at length but rarely say anything useful or interesting, and almost never mention the word "mass."

1) One problem (of many) with storing hydrogen as a cryogenic liquid for terrestrial use is that in order to cool and condense hydrogen starting with room temperature gas, it takes roughly 30% of the heating value of the hydrogen.
2) For most space applications, storage as cryogenic liquid hydrogen is clearly superior to any alternatives.
3) For longer-term missions, two approaches can be taken. One approach is to simply oversize the hydrogen tank covered with thick multilayer insulation (MLI) blankets and allow some of the hydrogen to boil off during the mission.
4) The other approach is to use active refrigeration to pump out heat that leaks into the hydrogen tank, thus achieving zero boil-off (ZBO).
5) Because of the longer trip time to get to Mars and the fact that the hydrogen tank would be embedded within an entry aeroshell, boil-off in transit to Mars is a much more serious problem than for the Moon.
6) At a pressure of 1 bar and a temperature of 20.4 K, the density of liquid hydrogen is 70.97 kg/m^3. If hydrogen liquid is stored at other pressures, the temperature and density will be as shown in Table 12.1. There may be good reasons to store liquid hydrogen at slightly elevated pressures if the system that utilizes the hydrogen also operates at an elevated pressure.
7) An alternate approach to using a larger tank and allowing boil-off is to use a tank that contains the ultimate amount of hydrogen needed from the start and provides an active cooling system to remove an equivalent amount of heat that continually leaks in during the storage period, thus avoiding boil-off. This is called the "zero boil-off" (ZBO) approach.
8) Solid hydrogen storage presents several complications: (1) difficulties in supplying hydrogen to users at a high enough pressure and flow rate, (2) mass penalties associated with a full-volume internal metal foam, and (3) additional complexity of the ground loading system.
9) The use of slush hydrogen (SLH) at the triple-point pressure (7.042 kPa or 1.02 psi) provides several advantages and several disadvantages compared to liquid hydrogen (Hardy and Whalen 1992). On the positive side, a 50% solid mass fraction in slush has 15% greater density and 18% greater heat capacity than normal boiling point liquid hydrogen. However, it is not clear how to perform vapor extraction with vapor at 1 psi.

Table 12.1 Pressure–temperature–density data for saturated liquid storage of hydrogen.

Temperature (K)	Pressure (psi)	Density (g/mL)
16	3.17	0.075
18	7.07	0.073
20	13.71	0.071
25	48.24	0.064
30	120.52	0.053

Source: Rapp (2005)/Springer Nature.

10) The production of densified hydrogen does not appear to be a difficult task. Several production methods for SLH have been demonstrated at the laboratory level, and densifying hydrogen to a triple-point liquid is easily achieved by pumping to reduce the fluid to triple-point pressure and temperature. The SLH production methods include evaporative cooling processes such as "freeze–thaw" . . . and refrigeration processes . . . The major issues related to the production of SLH include large-scale production, methods of production, safety concerns, energy efficiency, and costs . . . (Hardy and Whalen 1992).
11) Triple-point hydrogen (TPH) – liquid hydrogen at 1.02 psi and 13.8 K could offer increases in density (8%) and heat capacity (12%) in comparison to liquid H_2 at its normal boiling point. Although these benefits are not as large as those for slush, TPH does not have the added complication of solid particles.
12) Gas/sorbent pairs are characterized by whether sorption takes place by physical adherence due to polarization forces ("physisorption") or whether the gas molecules chemically combine with the sorbent ("chemisorption"). The great advantage of chemisorption is that stronger gas-surface forces are involved whereas physisorption involves weaker van der Walls forces. With these stronger forces, greater storage densities of gases can be achieved. However, it is difficult to find a suitable chemisorption sorbent material for a gas like hydrogen. Dissociative hydride alloys have been used, but they tend to be too massive and expensive for use in commercial storage systems. The advantage of physisorption sorbent materials is that they can be used for many gases, particularly hydrogen. All these choices are challenged by the demands of high gravimetric and volumetric density for any conceivable application.
13) Finely powdered "activated" carbon is a widely used physisorption material used for absorbing and desorbing gases. It derives its effectiveness from the very high surface area that can be produced if processed effectively.
14) Since their discovery, carbon nanotubes have become a focal point of the nanomaterials field, and many hopes have been pinned on them. Scientists have envisaged these molecular-scale graphitic tubes as the key to a variety of potentially revolutionary technologies ranging from super-strong composites to nano-electronics. It is widely conjectured that carbon nanotubes might provide the best medium for storing hydrogen.
15) Metal hydrides are metallic alloys that absorb hydrogen. These alloys can be used as a storage mechanism because of their ability to not only absorb hydrogen but also release it. The release of hydrogen is directly related to the temperature of the hydride. Typically metal hydrides can hold hydrogen equal to approximately 1–2% of their weight. The key tradeoff to utilizing a metal hydride storage system is whether there is sufficient heat available to extract the hydrogen from the hydride. The excess heat produced by the fuel cell (or other hydrogen-consuming component) must be greater than that required by the metal hydride to liberate the hydrogen at the desired flow rate and pressure.
16) According to Colozza (2002), glass microspheres can be used to store hydrogen in tiny hollow spheres of glass. If heated, the spheres' permeability to hydrogen will increase. This provides the ability to fill the spheres by placing the warmed spheres in a high-pressure hydrogen environment. The hoop stresses achievable for glass microspheres can range from 345 MPa (50,000 psi) to 1.034 MPa (150,000 psi) (Mueller and Durrant 1999). Once cooled, the spheres lock the hydrogen inside. The hydrogen is released by subsequently increasing the temperature of the spheres. This method of storage provides a safe, contamination-resistant method for storing hydrogen.
17) The use of hydrogen as a major feedstock for ISRU or a fuel for ascent does not appear to be practical on Mars.

Obviously, calculational details of the report have been omitted here, but it should be clear that this report is a "must-read" for those in the hydrogen energy field, particularly those working in the storage and transportation/transmission fields.

12.8 Summary of Hydrogen Storage Strategies

As noted above, hydrogen can be compressed or liquefied, and, in a compressed or liquid state, the same number of hydrogen molecules take up less volume, decreasing the footprint necessary for hydrogen storage facilities. Compressed hydrogen is generally defined as hydrogen at pressures in the 7,500 psi range. Liquefied hydrogen is generally defined as cryogenically cooled hydrogen at −253°C (or −423°F). Compressed hydrogen contained in high-pressure tanks is the technology of choice for a number of applications. Liquefied hydrogen storage is often used for bulk stationary hydrogen storage in above-ground tanks and truck transportation of liquefied hydrogen. Hydrogen can also be deposited on the surfaces of or within solid materials by adsorption. Storing hydrogen molecules in other compounds, e.g., ammonia, during transportation and storage, allows these compounds to be employed later by releasing the hydrogen closer to the point of use.

Hydrogen can be stored in bulk in a variety of ways. For end-use locations at small and medium scales, high-pressure cylinder tanks are employed since these tanks can be transported relatively easily and sized for specific applications. For larger-scale storage on-site, large cryogenic tanks store liquefied hydrogen since they have a higher volumetric density than pressurized gas storage systems. Most industrial, chemical, or commercial applications do not require large volumes of hydrogen and instead employ large cylindrical cryogenic tanks to store liquefied hydrogen.

To date, there are two salt cavern storage facilities for hydrogen in Texas and opportunities for further development of underground storage across the United States and around the world (Bellini 2020). But since hydrogen is the smallest of molecules compared to conventional hydrocarbons, it can leak into the walls of depleted wells, which would not only decrease the inventory of hydrogen and potentially introduce impurities to the gas. However, salt caverns or other geologic formations may not be located near every end-use market, so transportation problems no doubt will arise.

Although there are some regulatory programs addressing hydrogen in the United States, these programs likely will need to expand to address the potentially growing hydrogen market. Providing appropriate regulatory guidance in the future will aid in the development of a hydrogen energy economy. Technically sound codes and standards will provide an essential basis for the safe and consistent deployment and commercialization of hydrogen and related technologies. Along with widely shared safety information and best practices, they will also improve confidence in the commercial viability of the technologies among all stakeholders, which can further accelerate the adoption of the technology.

The most suitable type of storage will depend on a number of factors such as the end-use application, the amount of hydrogen needed, geographical and geological constraints, safety concerns, and the required flow rates. Most hydrogen today is used where it is produced.

To summarize the state of the practice of the three most common hydrogen storage options:

1) Gas storage is commercially available but generally not considered economically viable. Safety and the reduction of volume requirements remain major concerns for this hydrogen storage option.
2) Liquid storage is also commercially available but economic, boil-off, and safety concerns remain.

3) Solid storage is still in early development with many R&D questions remaining. Present concerns with this hydrogen storage technology include desorption temperatures, desorption kinetics, recharge time and pressure, energy management, economics, cycle life, container compatibility, and process optimization.

An analysis of the three basic storage options indicates that the advantages of solid H_2 storage compared to gaseous and liquid hydrogen storage are:

1) Smaller required volumes for equivalent energy storage
2) Lower pressure (greater energy efficiency)
3) Higher purity H_2 output

Compressed gas and liquid storage are the most commercially viable options today, but completely cost-effective storage systems are still in development. The safety aspects of all storage options, particularly the novel hydride storage options, remain a major concern. Developing a system to store hydrogen could entail substantial costs, whether for a completely new system or a gradually upgraded and modified one.

References

American Petroleum Institute (2020). *API Standard 650, Welded Tanks for Oil Storage*, 13th Edition. Washington, DC: American Petroleum Institute. https://www.api.org/products-and-services/standards/important-standards-announcements/standard650.

American Society of Mechanical Engineering (2023). *BPVC Section VIII-Rules for Construction of Pressure Vessels Division 1*. New York: ASME. https://www.asme.org/codes-standards/find-codes-standards/bpvc-viii-1-bpvc-section-viii-rules-construction-pressure-vessels-division-1/2023/print-book.

Bellini, E. (2020). Hydrogen storage in Salt Caverns. *PV Magazine*, June 16. https://www.pv-magazine.com/20206/16/hydrogen-storage-in-salt-caverns/.

Energy Storage Association (2020). Hydrogen energy storage. http://energystorageassociationarchive.org/why-energy-storage/technologies/hydrogen-energy-storage/ (accessed 18 November 2023).

Hardy, T.L. and Whalen, M.V. (1992). Technology issues associated with using densified hydrogen for space vehicles. *NASA TM 105642. AIAA 92-3079, ALAA/SAB/ASME/ASEE 28th Joint Propulsion Conference*, Nashville, TN.

Lim, K.L., Kazemian, H., Yaakob, Z., and Daud, W. R. W. (2010). Solid-state materials and methods for hydrogen storage: a critical review. *Chemical Engineering and Technology* 33:213–226.

Mueller, P. and Durrant, T.C. (1999). Cryogenic propellant production, liquefaction, and storage for a precursor to a human Mars mission. *Cryogenics* 39: 1021–1028.

National Fire Protection Association (2021). *Flammable and Combustible Liquids Code*. NFPA 30. Quincy, MA: NFPA. https://www.nfpa.org/product/nfpa-30-code/p0030code?Edition=2021&Language=English&Format=Softbound&type=digital.

Rapp, D. (2005). Transporting hydrogen to the Moon or Mars and storing it there. Report to Nasa's Johnson Propulsion Laboratory, Pasadena, CA. Appendix F In Rapp, D. (2013). *Use of Extraterrestrial Resources for Human Space Missions to Moon or Mars*. Berlin, Germany: Springer Praxis Books. https://doi.org/10.1007/978-3-642-32762-9. https://link.springer.com/content/pdf/bbm:978-3-642-32762-9/1?pdf=chapter%20toc.

Sakintuna, B., Lamari-Darkrim, F., and Hirscher, M. (2007). Metal hydride materials for solid hydrogen storage: a review. *International Journal of Hydrogen Energy* 32: 1121–1140.

Theodore, L. and Dupont, R.R. (2012). *Environmental Health and Hazard Risk Assessment: Principles and Calculations*. Boca Raton, FL: CRC Press/Taylor Francis Group.

U.S. Department of Energy (2020). *Hydrogen Storage*. Washington, DC: Office of Energy Efficiency and Renewable Energy, Hydrogen and Fuel Cell Technologies Office. https://www.energy.gov/eere/fuelcells/hydrogen-storage.

13

Hydrogen Transportation and Transmission

13.1 Introduction

This chapter discusses four primary modes of U.S. hydrogen transportation and transmission: motor carrier, rail, pipeline, and vessel. As the U.S. hydrogen economy continues to develop and grow, additional modes of hydrogen transportation/transmission may also develop. Merriam-Webster's definition of transportation (Merriam-Webster 2023a) is ". . . means of conveyance or travel from one place to another; public conveyance of passengers or goods especially as a commercial enterprise." In more general terms, transportation concerns the conveyance of people or property from one place to another. Modern commercial transportation serves the public interest and includes all the instrumentation and facilities used in the movement of people or property, and all services involved in the receipt, delivery, and handling of such property.

Merriam-Webster's definition of transmission Merriam-Webster (2023b) is ". . . an act, process, or instance of transmitting; . . . to send or convey from one person or place to another." In more general terms, transmission refers to electrical, hydraulic, or mechanical systems employed in most powered vehicles to transmit power from a prime mover such as an internal combustion engine, to the drive shaft. Transmission systems are also common in airplanes, automobiles, boats, ships, and locomotives, and the power is usually delivered through a gearing device.

According to DOE, the existing hydrogen gas pipeline network provides a low-cost option for delivering large volumes of hydrogen. Approximately 1,600 miles of hydrogen pipelines are currently operating in the United States. Owned by merchant hydrogen producers, these pipelines are located where large hydrogen users, such as petroleum refineries and chemical plants, are concentrated such as the Gulf Coast region (DOE 2020a). Beyond the initial high capital costs associated with pipeline construction, expanding the network of pipelines for hydrogen transportation presents unique technical barriers. DOE states that ongoing research is focusing on overcoming issues such as:

1) The potential for hydrogen to embrittle the steel and welds used to fabricate the pipelines
2) The need to control the potential for hydrogen permeation and leaks
3) The need for lower-cost, more reliable, and more durable hydrogen compression technology

DOE is investigating potential methods to address these issues, including the use of fiber-reinforced polymer (FRP) pipelines for hydrogen distribution. DOE estimates that the installation

Hydrogen Energy: Principles and Applications, First Edition. Vincent J. DelGatto, Louis Theodore, R. Ryan Dupont, and Matthew C. Ogwu.

costs for FRP pipelines are about 20% less than that of steel pipelines because the FRP can be obtained in sections that are much longer than steel, minimizing welding requirements (DOE 2020a).

Adapting the natural gas delivery infrastructure to accommodate hydrogen may be another option, with natural gas pipelines prospectively carrying a blend of natural gas and hydrogen (e.g., up to about 15% hydrogen content). DOE expects that converting existing natural gas pipelines to deliver 100% hydrogen would likely require "more substantial modifications" (DOE 2020a).

At this point, the movement of a gas by pipeline is one of the cheapest and most attractive ways of moving energy. Hydrogen can also be transported in liquid form, at a state which is formed at approximately −420°F. This "system" would require continual cooling of the hydrogen. Liquid hydrogen is currently used for fuel in rockets for spaceships since it has a relatively high energy per unit mass and could be a useful fuel for aircraft.

As the production of electricity from hydrogen increases, so will the need for energy storage and transportation. Hydrogen may be the perfect carrier for this energy. It can store the energy and distribute it to wherever it is needed. Transmitting electricity long distances has been estimated to be up to four times more expensive than shipping hydrogen by pipeline; in effect, transporting energy over long distances is easier and less expensive when the energy is a chemical fuel rather than in the form of electricity.

Section numbers and their accompanying titles for this chapter concerned with hydrogen transportation/transmission are listed below.

13.2 Hydrogen Transportation/Transmission Options

The low energy density of hydrogen means that it can be very expensive to transport over long distances in some forms. Nonetheless, a number of possible options are available to overcome this hurdle, including compression, liquefaction, or incorporation of the hydrogen into larger molecules that can be more readily transported as liquids.

Pipelines are the safest and cheapest way to move large quantities of either crude oil or refined petroleum across land. About 100,000 miles of small gathering lines and large trunk lines move crude oil from wells to refineries. Other transport methods include motor carriers, ships (and barges), and trains. Details on these transportation/transmission methods are provided by Skipka and Theodore (2014).

13.2.1 Motor Carriers

Hydrogen (and natural gas) can be transported in the United States by trucks in a compressed gaseous state using tube trailers, and in a liquid state employing cryogenic tank trucks. More common options for hydrogen transport involving pipelines, ships, and trains are provided below.

13.2.2 Pipelines

Oil (and natural gas) can be transported in their natural states through pipelines buried underground or even on the seabed. Hundreds of miles of underground pipelines have been laid to transport oil (and gas) ashore from important offshore producing areas such as the North Sea and the Gulf of Mexico. There are also land pipelines carrying both oil (and gas) products over thousands of miles between producing areas and centers of population in North America, as well as in the Middle East.

Some of the earliest pipelines used to move fuel were made from wood. In the United States during the 19th century, holes were drilled through the center of tree trunks to provide tubes for gas distribution with inside diameters of up to 3 in. Main distribution lines for crude oil (and gas) are now almost always made of steel. The largest one, crossing Alaska from Prudhoe Bay to Valdez, has a diameter of 48 in and would be able to carry 100 million tons of oil a year if used to its full capacity.

Most land pipelines are buried at least 3 ft underground, often under fields where crops are growing undisturbed by the fuel passing beneath. Underwater pipelines may also be buried in a trench on the seabed. All oil and gas pipelines are coated with a layer of bitumen or fiberglass to prevent corrosion while underwater pipelines have an additional coating of concrete for extra protection against the effects of seawater. It should be noted that pipelines have low operating costs and lifetimes in excess of 50 years. Their two main drawbacks are their high capital costs and potential difficulties in acquiring the rights of way for their installation.

Long pipelines usually need pumping stations every 50–150 miles to give the oil or gas an additional pressure boost along the way, and in remote areas these stations sometimes get their energy supply by using a minute quantity of the fuel being carried in the line. The flow velocity is approximately 5 mi/h. Although pipelines are a very efficient way of transporting oil and gas, they need to be cleaned regularly to remove deposits that the fuels leave on the inside walls, particularly where there are bends in the line. Cleaning is carried out by using the oil or gas flow, to push through a device known as a *pig*. This odd name comes from the initials of pipeline inspection gadget, which describes another of its uses. The pig has a diameter that exactly fits the inside dimensions of the pipe and will scrape away blockages with its outer edge as it moves along the pipe section being cleaned.

13.2.3 Ships

Much of the world's oil comes from areas such as the Middle East that are too distant from main markets to make transport by pipeline either economical or practical. Oil from these regions is shipped to North America, Europe, and Japan in specially built tankers. The first oil tanker, the Glückauf, was launched in 1866 and could carry just 300 tons of oil. Modern vessels can carry up to 500,000 tons. These supertankers are more than 1,300 ft long and hide their bulk beneath the surface like icebergs. While large-scale oil transport via ships is common, there are currently no known ships that can transport pure hydrogen. As one might suppose, for ship transport, hydrogen must be liquefied or converted to some other form, e.g., methylcyclohexane, prior to its transportation.

Liquid natural gas (LNG) carriers exist to transport large quantities of natural gas but are not easily modified for transporting hydrogen. While cryogenic tanks used to store LNG might endure low temperatures required for storage of liquid hydrogen, hydrogen's low density and low boiling temperature demand an extremely low heat flux through the tank walls. Due to its characteristics liquid hydrogen storage tanks would require much thicker cargo tank insulation than is on LNG carriers.

Due to its extremely low storage temperature of −252°C, it would liquefy nitrogen and oxygen in tank insulation membranes to dangerously low temperatures that could embrittle the ships' structure to dangerous levels and run the risk of hydrogen leaks and significant safety issues.

13.2.4 Trains

Oil is transported by train in small quantities, usually across short distances. This mode is employed when the receiver is not near pipelines or major terminals. Jobbers handle the wholesale distribution of most petroleum products. The retailer, i.e., a gasoline station or a home heating oil company, then receives the product. The last stage of fuel distribution is when a car receives gasoline and/or the home receives fuel oil for heating purposes.

13.3 Traditional Transportation Options

The traditional transportation system in the United States is an extensive, interrelated public/private network of roads, airports, railroads, transit routes, waterways, terminals, ports, and pipelines. Millions of people and businesses rely on this expanding system to get to work, conduct business, ship goods, and travel on vacation. The transportation system links regions, joins urban and rural areas, and connects the United States with the rest of the world. Tables 13.1 and 13.2 display the immensity of the system. The number of airplanes and road-based vehicles continues to increase. Rail transportation has declined as a result of the loss of market share to freight trucks. This is observed in Table 13.2.

The traditional transportation industry discussed here includes four different modes: air, rail, water, and road. The material on each mode provides only a brief overview. The different modes often work together to transport a single good or person. Only "domestic" transportation for these four modes is discussed here, although the transportation industry is global, and the impacts are not confined to national borders.

13.3.1 Air Transportation

There are five types of aircraft that compose the aviation industry: commercial, air taxi operations, commuter, general, and military. The airline industry provides transportation of passengers, cargo, mail, and perishable goods. The system of airports in the United States is the largest and most complex in the world. As of 2022, there were 5,069 public-use landing facilities and 14,784 private-use facilities in the United States (FAA 2022).

In 2022, 7.6 million domestic passenger flights were flown in the United States by U.S. air carriers (BTS 2022). Additionally, 14.7 million tons of domestic cargo were flown in 2021 (BTS 2022). The 2021 Federal Aviation Administration (FAA) forecast predicts that between 2020 and 2041 the number of jets in the U.S. mainline carrier fleet is forecast to grow approximately 1.1%/year, while the regional carrier fleet is expected to expand by 0.2%/year, while the cargo carrier fleet is expected to more than double during this forecast period (FAA 2021). Possible option is to transport hydrogen in large balloons.

Table 13.1 The U.S. transportation network (U.S. DOT 2023a).

Mode	Components (2020)
Air	Total airports: 19,853 (2022) Public-use: 5,069 Private-use: 14,784
Highway	Public roads: 4,172,562 miles Public road lanes: 8,790,746 miles
Pipeline	Gas distribution: 1,328,885 miles Gas transmission and gathering: 319,244 miles
Rail	Class I freight railroad: 91,774 miles Amtrak commuter rail: 20,787 miles Amtrak stations: 526
Transit	Commuter rail: 7,930 miles Heavy rail: 1,663 miles Light rail: 2,096 miles Commuter rail stations: 1,311 Heavy rail stations: 1,057 Light rail stations: 1,287
Water	Navigable waterways: 25,000 miles Ports: 192 Cargo handling docks: 8,334 Lock chambers: 237

Table 13.2 Number of aircraft, vehicles, railcars, and vessels (U.S. DOT 2023a).

Mode	Types (2020)
Air	Air carrier aircraft: 5,882 General aviation aircraft: 204,980
Highway	Light-duty vehicles: 253,121,228 Trucks: 13,479,382
Pipeline	LNG facilities: 165
Rail	Class I freight locomotives: 23,544 Class I freight cars: 252,400 Class I total rail cars: 1.66 million Amtrak locomotives: 384 Amtrak cars: 1,313
Transit	Commuter rail cars: 7,524 Heavy rail cars: 11,064 Light rail cars: 2,799
Water	Nonself-propelled vessel (barges): 34,168 Self-propelled vessel: 10,333 Oceangoing vessel: 180 Recreational boat: 11,838,188

13.3.2 Rail Transportation

In 2019, there were seven Class I railroads and 626 additional local and regional railroads operating 136,729 route miles (BTS 2022). The total Class I railroad employment was 122,792 in August 2023 (AAR 2023), with Class I railroads transporting more than $100 billion of freight revenue in 2022 (Statistica.com 2023).

In 2022 (U.S. DOT 2023b), the top five rail-carried commodities (in terms of ton-miles) included coal (31.9%), cereal grains (11.8%), natural gas and other fossil products (6.3%), basic chemicals (4.4%), and other ag products (4.4%). The top five rail-carried commodities in terms of value included motorized vehicles ($126 billion), natural gas and other fossil products ($45 billion), plastics/rubber ($41 billion), basic chemicals ($30 billion), and cereal grains ($26.6 billion).

Over the past 50 years, Class I railroads and connecting facilities have developed increasingly efficient ways to carry and transfer cargo (e.g., larger cars, double-stack container railcars, and on-dock rail), allowing more cargo to be carried with fewer railcars. The railroads, which are private companies, invested $10.8 billion in 2020 to improve their facilities (BTS 2022). This is down 17% from that in 2019, in line with traffic and revenue reductions, but is higher than the investment as recently as 2010. Rail freight ton-miles are projected to increase by 30% in 2050 compared to 2020 levels, while the value of these goods is expected to increase by more than 90% to more than $1 trillion by 2050 (U.S. DOT 2023a).

13.3.3 Water Transportation

The water transportation industry includes establishments engaged in freight and passenger transportation on the open seas or inland waters and establishments furnishing such incidental services as lighterage, towing, and canal operations. This category also includes excursion and sight-seeing boats, water taxis, and cargo handling operations. U.S. shipyards play an important role in supporting the nation's maritime presence by building and repairing the domestic maritime fleet.

Water transportation vessels consist of barges and self-propelled vessels. Self-propelled vessels have onboard propulsion that is either steam- or diesel-powered. Barges rely on external means for movement (i.e., tugboats and push boats). Vessels are classified as follows:

1) General cargo vessels, which carry nonuniform items that must be lifted off the vessel using cranes or rope slings.
2) Bulk carriers, which carry homogenous, unpacked cargo as either dry bulk carriers or tankers for liquid commodities.
3) Intermodal vessels, which include container vessels and roll-on/roll-off vessels. The preloaded containers are loaded from trucks or railroads efficiently.

As of 2020, there were 44,501 U.S. flag vessels registered in the United States, 77% barge and 23% self-propelled vessels (U.S. DOT 2023a). Water is the leading transportation mode for U.S. International trade, both in terms of weight and value. Ships accounted for more than 41% of the trade value in 2020, amounting to more than $1.25 trillion. Imports accounted for 43% of the total weight of shipments and 68.7% of the total freight value shipped by vessels in 2020 (BTS 2023).

The domestic trade includes cargo moved on the oceans (including trade between the 48 contiguous states and Alaska, Hawaii, Puerto Rico, and Guam), along the coasts, on the inland waterways, and the Great Lakes. Water transportation occurs in, and is defined by, three basic geographic areas (U.S. EPA 1997).

1) Coastal, from one coast port to another, either domestic or foreign
2) Great Lakes, either domestic or foreign
3) Inland, riverways, and lakes of the United States

13.3.4 Highway Transportation

More than 2.9 trillion vehicle miles were traveled in the United States in 2020, 87.6% of person miles were made using personal vehicles (U.S. DOT 2023a). In 2017, 87.7% of person trips were in private vehicles, 2.9% walked, and 6.9% used public transit modes. The average person took 2.7 trips daily per driver, an average distance of 25.8 miles (U.S. DOT 2023a).

In general, trucking establishments fall into two broad categories: private and for-hire. Private carriers move their own goods. For-hire carriers are compensated for providing transportation freight belonging to another entity. In terms of the value of shipments, trucking accounts for the vast majority of total U.S. freight services, approximately 76% in 2020 based on shipment value and 65% based on tons shipped (U.S. DOT 2023a). This reflects trucking's higher revenues per-ton and per-ton mile (1 ton of weight over a 1-mile distance), compared to the rail and barge sectors, which generally carry lower-valued, bulk commodities. Thus, the trucking industry's overall share of ton-miles (45%) is much lower than its share of revenues, and this is especially true for short-haul (<500 miles) distances (U.S. DOT 2023a).

13.4 Chemical Industry Transportation Options

Most chemicals in the United States are transported as liquids or gases. Each of these is briefly discussed below.

13.4.1 Transportation of Liquids

Tank cars, consisting of horizontal cylindrical tanks made of riveted or welded-steel plate construction, permanently mounted on a railway car chassis, are used for bulk shipment of liquids and liquefied gases. Construction must conform to specifications of the Interstate Commerce Commission (ICC). Tank-car capacities for general-service conditions range from 4,000 to 12,000 gal. Materials of construction of the tanks must conform to corrosion specifications. Liquids that attack carbon steel are handled in lined tanks. Alloy cladding, rubber or plastic coatings, and glass or lead linings are prevalent. Materials that are normally solid at room temperature but melt at elevated temperatures are handled by use of internal heating coils. During loading and unloading steam, Dowtherm, or hot oil is circulated to liquefy the tank contents, which can then be removed by pumping, siphoning, or blowing.

Tank cars can be either rented from railroads or purchased outright from fabricators. Rental rates can vary significantly for simple steel tanks to more expensive stainless-steel tank cars with special fittings. The purchase price of tank cars is approximately 50% higher than the tank itself.

Tank trucks moving on highways furnish a second method of transportation, generally for short-haul distances. Design considerations are similar to those for tank cars. The maximum tonnage is limited by state and ICC regulation with a maximum loading of 15 tons in most states.

In recent years, pipeline pumping has been used for moving large-volume raw materials, such as petroleum, for thousands of miles across the country at a cost lower than for tankers and competitive with water transportation options.

13.4.2 Transportation of Gases

Insulated tank cars are available for shipping liquefied oxygen and nitrogen. Where high-pressure gas is transported locally, trucks carrying racks of horizontal cylindrical pressure vessels are available. Smaller volume deliveries are made in vertical gas cylinders. All designs must conform to ICC regulations.

Transportation of gases, particularly from petroleum operations, via pipelines over a range of several thousand miles is a common practice in the gas industry.

13.5 Hydrogen Transportation: Pipelines

In circumstances where hydrogen cannot be produced on-site at the point of consumption, or in a scenario where hydrogen production develops in a centralized rather than decentralized manner, safe, quick, and cost-effective transport will be of critical importance to end-users and project developers alike. Although pipelines for water have been used since ancient times, the present pipeline industry did not come into existence until the middle of the 19th century. Soon after, pipelines were a major element in the oil business, providing specialized transportation for liquefied products, including natural gas and pulverized coal.

Hydrogen transportation issues are directly related to hydrogen storage issues. In general, compact forms of hydrogen storage are more economical to transport, and diffuse forms are more costly. Note that pipelines can carry hydrogen as gas or as a liquid.

Gaseous hydrogen can be transported by pipeline in a similar fashion as natural gas. There are 305,000 miles of natural gas transmission pipelines throughout the continental United States with a daily movement of over 50 billion scf and a capacity of nearly 100 billion scf/day.

Hydrogen, being less dense than natural gas, results in less mass transport for a given pipeline size and operating pressure. In addition, the energy density of hydrogen is only one-third that of natural gas on a volumetric basis; hence, three times the amount of hydrogen gas must be pumped through a pipeline to transmit an equivalent amount of energy. To compensate for both of these properties, hydrogen pipelines need to be designed to operate at higher pressure in order to be practical. All pumps and other equipment must be hydrogen-compatible. Furthermore, hydrogen pipelines must be resistant to hydrogen embrittlement in order to prevent cracking.

Existing hydrogen gas pipelines operate in some parts of the world. In the United States, there are approximately 1,600 miles (2,575 km) of pipelines, 90% of which are located along the Gulf Coast in Texas, Louisiana, and Alabama, primarily serving refineries and ammonia plants in the region (Congressional Research Service 2021). Several hydrogen pipelines exist in Canada. Europe currently has approximately 1,600 km of hydrogen pipelines in operation within Germany and between Belgium and France, among several others. Compared to pipelines of other gases, these lengths are very short, however, they indicate that the high cost of transporting hydrogen by gas pipeline is already worth it in some areas and expanding (DOE 2020a).

It should also be noted that hydrogen pipelines are often used in regions where there is significant demand, and that demand is expected to remain stable into the middle of this century.

Hydrogen pipelines are capital intensive, but when the quantity of hydrogen demand is high, they have a lower levelized cost over time. While pipelines are the most energy-efficient approach to transporting hydrogen, their deployment is challenged by their high capital costs (DOE 2020b).

Hydrogen pipelines exist, but they are used to transport a chemical commodity from one production site to another. The energy required to move the gas has little relevance because energy consumption is part of the production costs. This is not so for hydrogen energy transport through pipelines. Normally, pumps are installed at regular intervals to keep the gas moving. These pumps are energized by energy derived from the delivery stream. It has been estimated that approximately 0.3% of the natural gas is used every 150 km to energize a compressor to move that gas.

As noted, hydrogen can be transported in a pipeline as a liquid, but liquid hydrogen pipelines have been accomplished over very short distances only. These pipelines require thorough insulation in order to maintain cryogenic temperatures and prevent the formation of a two-phase (liquid/gas) flow within them. If exposed to air, the low pipeline temperature would cause air to liquefy and oxygen to concentrate on the surface of the pipe, increasing the fire hazard of this form of hydrogen transport (DOE 2020a).

Existing domestic natural gas pipeline infrastructure has the potential to expand the transportation of hydrogen. Blending hydrogen into natural gas pipeline networks may be an option for delivering pure hydrogen to markets, using separation and purification technologies downstream to extract hydrogen from the natural gas blend near the point of end-use. It is necessary to assess multiple factors to safely integrate hydrogen blending into the existing natural gas pipeline systems (e.g., gaseous hydrogen embrittlement). Nationally, pipelines can handle from 15% to 30% hydrogen blends without modifications or significant detrimental effects. While hydrogen compression can be utilized for transport and storage, this compression comes with energy penalties (U.S. DOE 2020c).

One of the major considerations for the robust development of a hydrogen economy in the United States is the potential use of existing natural gas pipelines for the transportation of hydrogen. Since there are hundreds of thousands of miles of natural gas transportation pipelines and over a million miles of natural gas distribution pipelines across the United States, this presents a lower-cost opportunity for the hydrogen industry to deliver commercial-scale volumes of hydrogen. This possibility of using existing gas pipelines will require the hydrogen industry to partner with the natural gas industry to try and develop some successful path forward to avoid potential and protracted problems and delays.

As the hydrogen economy in the United States continues to develop, the need for new dedicated hydrogen pipelines is certain to emerge. These pipelines can offer an attractive economic advantage to mobile transport options described below.

13.6 Hydrogen Transportation: Mobile

Mobile transport includes transport by truck, rail, or barge. The hydrogen being transported is stored in tanks appropriate to hydrogen storage method and transport medium. Hydrogen as a high-pressure gas can be transported in cylinders at pressures ranging from 2,200 to 5,800 psig. For trucks, specially designed tube trailers carry a number of large, high-strength steel tubes linked together through a common manifold. This design works well in providing small quantities of hydrogen but is very inefficient in terms of transport energy. The weight of cylinders required is such that the gas is only 2–4% of the cargo weight (U.S. DOE 2014).

Since hydrogen is not dense enough for long-distance transport to be economically viable, generators need to consider liquefaction. It is usually more economical to transport hydrogen in a liquid state since a liquid hydrogen carrier can hold substantially more hydrogen than pressurized gas-state hydrogen. Hydrogen can be transported by truck, rail, or barge as a cryogenic liquid in double-walled, super-insulated vacuum-lined tanks. Transporting liquid hydrogen is far more efficient than a high-pressure gas, particularly where larger quantities are needed. On the downside, maintenance costs are much higher for liquid transportation (U.S. DOE 2014).

As a point of reference, 150 1b (70 kg) of liquid hydrogen occupies a volume of 35.3 ft^3 (1,000 L) and requires a transport container weighing between 440 and 660 lb (200–300 kg). The same weight of hydrogen, transported as a gas or hydride would require between 2 and 5 tons of containers. The liquid option is obviously much more efficient. Liquid hydrogen transport trailers include standard 40 and 80-foot truck trailers.

These high-pressure (≈2,600 psi) tube trailers are used when the quantity of hydrogen delivery is small (<1 ton/day) and the delivery distance is within 200 miles of the source when the hydrogen is transported as a compressed gas. Tube trailers are likely to remain an essential technology to supply growing markets that use relatively small amounts of hydrogen or do not yet have predictable demand. Today, tube trailers are usually filled using liquid hydrogen for transport distances up to 600 miles – the liquid hydrogen is pressurized and then vaporized into gaseous form. As markets of small-scale hydrogen consumers expand, this approach will be less efficient than filling tube trailers directly at a gaseous hydrogen production facility (U.S. DOE 2020b).

Compared to trucks, it could be more economical to transport hydrogen by railcars, barges, or ships for larger quantities of hydrogen. By utilizing the same style of tube or liquid tanks used by trucks, these modes of transport are not restricted by weight requirements encountered on roadways. The first liquid hydrogen transport ship was launched as a pilot in Japan in 2019, with an anticipated storage capacity of 1,250 m^3 (less than 1% of typical LNG carriers). Future hydrogen transport ships are expected to have much greater capacity (U.S. DOE 2020c). Even where shipping is utilized, rail and road transport is almost always necessary for the final delivery of hydrogen to the point of use, especially to serve remote locations or to provide backup fuel to filling stations in times of peak demand.

From a safety perspective, about one in 100 trucks is a gasoline or diesel tanker. For surface transportation of hydrogen, one may see 115 trucks on the road, 15 of them transporting hydrogen. One out of seven accidents involving trucks would therefore involve a hydrogen truck. Every seventh truck–truck collision would occur between two hydrogen carriers. This scenario is certainly unacceptable from a risk perspective (Theodore and Dupont 2012, 2023).

13.7 On-Site Hydrogen Production

One option for providing clean hydrogen at filling stations and dispersed depots is the on-site generation of the gas by electrolysis. On-site manufacture of hydrogen sidesteps the issue of transporting it. The energy for the manufacture of hydrogen is moved to the site, rather than the hydrogen itself. The energy may be in the form of electrical power to an electrolysis system or as a fossil fuel piped or trucked to a reformer. By manufacturing hydrogen at or near the location where it is required, the high cost and energy inefficiency of transport is avoided, and advantage is taken of the high transport efficiency of other forms of energy.

The energy needed to generate and compress hydrogen by this scheme must be compared to the HHV energy content of the hydrogen delivered to local customers, however. When calculations are done for the total energy needed to generate hydrogen by electrolysis (70% efficiency) and compress hydrogen (to 200 bar) at filling stations to service 100–2,000 vehicles/d, between 3 and 57 MW of power would be required. This total energy needed exceeds the HHV energy of the delivered hydrogen by 50–70%. The energy output of a 1 GW nuclear power plant would be required to serve 20–30 hydrogen filling stations on heavy-trafficked highways.

13.8 Transportation via Chemical Hydrogen Carriers

Another emerging method to transport large amounts of hydrogen is the use of chemical hydrogen carriers, which offer the potential to carry more hydrogen than tube trailers, and at a lower-cost. Chemical hydrogen carriers are liquid- or solid-phase materials that can chemically bond with hydrogen to "carry" it at low-pressure and near-ambient temperatures but can then release the hydrogen on demand. They may be ideally suited for applications where hydrogen demand is substantial, but not stable enough to warrant pipeline construction. They also offer the potential for significantly higher energy density compared with gaseous or even liquid hydrogen transport, thereby reducing hydrogen delivery costs. Chemical hydrogen carriers can be broadly classified as one-way or two-way carriers. One-way carriers are materials that do not release a by-product for reuse or disposal after the hydrogen is released (such as ammonia). Two-way carriers are those whose by-products are typically returned for processing for reuse or disposal after the hydrogen is released (such as methylcyclohexane/toluene). The use of chemical hydrogen carriers is in the early stages of commercialization and R&D efforts are needed to increase the hydrogen-carrying capacity of these materials and improve the charge-and-discharge rates, reversibility, and overall round-trip efficiency. Some carriers, like ammonia, can also be used for direct power generation without intermediate hydrogen release such as in turbines, internal combustion engines, and direct fuel cells (U.S. DOE 2020b).

13.9 International/Global Hydrogen Transportation

Since the costs of hydrogen production differ significantly between regions, long-distance transmission, and international trade in hydrogen can be attractive. This is particularly true for countries that are expected to have large hydrogen demand but lack suitable locations to deploy power generation from wind and solar, such as Japan and South Korea (Hydrogen Council 2020).

A global supply chain will likely consist of long-distance pipelines as well as shipping routes. Even existing natural gas pipelines can transport hydrogen, often with only modest upgrades. This could be an option, for example, to transport low-carbon hydrogen derived from natural gas from Russia or Norway into Central Europe, or renewable hydrogen from electrolysis from Northern Africa into Southern Europe.

For hydrogen shipping to become economically feasible, the industry needs to scale up its infrastructure to reach similar levels that LNG has today in the mid to long term. Since ship-based imports will always directly compete with domestic production, domestic hydrogen costs and their potential to decline will be decisive. In the case of Japan, the analysis shows that ship-based imports are economically competitive versus locally produced renewable hydrogen from electrolysis as long as local renewables levelized cost of energy (LCOE) are above $60/MWh.

There are several technology options for shipping hydrogen globally although there are no ships currently available to transport hydrogen. As indicated above, the three major options are liquid hydrogen, ammonia, and liquid organic hydrogen carriers. Liquid hydrogen shipping delivers hydrogen in pure form at the location of import. Today liquid hydrogen shipping costs are high (e.g., for the route from Saudi Arabia to Japan about $15/kg in 2020), but with enough scale, they could fall to $1.7/kg by 2030 (Hydrogen Council 2020). Indeed, the technology is similar to that used for LNG, which supports quick scale-up given similar conditions as for LNG – namely large enough demand in locations that might warrant investment into hydrogen production and transmission. Imported hydrogen offers options for countries to diversify their energy imports, and one result of this is significant interest in using ships to transport hydrogen. However, shipping supply chains require the necessary infrastructure, including storage tanks, liquefaction and regasification plants, and conversion and reconversion plants, to be built at the loading and receiving terminals as appropriate.

Using ammonia as the hydrogen carrier is an option that has the benefit of leveraging existing infrastructure for global distribution. Additionally, the conversion from hydrogen to ammonia is a well-established technology. In cases where the end use is ammonia, shipping ammonia is the preferred option (but requires careful handling by certified operators due to its toxicity). However, if the end use requires pure hydrogen, an additional reconversion step is required, which is currently at an early stage of development. In addition, reconversion would require access to low-cost clean energy at the arrival port if the desire is to have a low-carbon or renewable hydrogen product, the absence of which is exactly why hydrogen was being shipped into the demand center in the first place. Depending on the technology evolution and local conditions, this reconversion step could add another $1–$2/kg on top of conversion and shipping costs (Hydrogen Council 2020).

Liquid organic hydrogen carrier shipping based on a range of different chemical compounds, such as toluene or methyl-cyclohexane, has the benefit of leveraging existing shipping infrastructure and allows hydrogen to be transported and stored as a liquid. Like ammonia, a challenge lies in the dehydrogenation step, which requires scale-level development as well as significant energy input at the import destination. In contrast to ammonia, it also requires the return of the carrier to the port of origin. Further, there are several different technologies, hindering economies of scale unless global standardization is achieved. Given the low maturity of the technology, cost estimates for liquid organic hydrogen carriers, as well as cycling rates, are still highly uncertain and require additional research and development.

Which shipping technology will become the least cost option will be site-specific. If hydrogen is the end use, liquid hydrogen seems to be the closest to maturity across the value chain and the lowest-cost alternative in the immediate future, although significant scale-up will be a critical prerequisite for cost reductions.

13.10 Regulation Issues

In France, Great Britain, Italy, and many other countries of Western Europe, railroad, steamship, and air transportation facilities are government-owned. Motor carriers are nationalized only when operated in conjunction with a railroad or water carrier. Public carriers, privately owned, are regulated in all nations, with the minister of transport performing this function in most European countries. In the United States, specialized agencies are created by Congress for this purpose.

Key among these agencies is the Department of Transportation, an executive department of the U.S. federal government, created in October 1966 by the combination of transportation agencies or functions formerly dispersed throughout the government. The purpose of the department is to promote and develop safe, rapid, efficient, and convenient transportation in the United States. It is administered by a Secretary appointed by the President and approved by the Senate.

The major divisions of the department are the U.S. Coast Guard; the Federal Aviation, the Federal Highway, and Federal Railroad Administrations; the Saint Lawrence Seaway Development Corporation; the National Highway Traffic Safety Administration; and the Urban Mass Transportation Administration. The FAA regulates air traffic and commerce in the United States; the Federal Highway Administration is responsible for highway construction, administration, and safety; the Federal Railroad Administration consolidates government support of rail transportation activities, advises the Secretary of Transportation on matters relating to rail transportation, and administers the federally owned Alaska Railroad. The St. Lawrence Seaway Development Corporation oversees activities relating to the St. Lawrence Seaway. The National Highway Traffic Safety Administration determines the causes of motor vehicle accidents and makes recommendations to promote highway safety. The Urban Mass Transportation Administration assists in the planning and development of improved mass transportation facilities and equipment.

Here are some details of three other transportation-related agencies.

1) *The Interstate Commerce Commission* is an independent agency, that was established by Congress in 1887 and implemented subsequent legal powers. The function of the Commission is the regulation of domestic common carriers engaged in interstate commerce comprising railroads, inland, coastwise, and intercoastal water carriers, motor carriers, pipelines, and domestic freight forwarders. It is the largest, oldest, and broadest in scope of the regulatory agencies, and its decisions have greatly affected the U.S. economy and economic development.
2) *The Federal Maritime Commission* was established by reorganization as an independent agency, effective August 12, 1961, for the economic regulation of common carriers by water and other persons engaged in the foreign commerce of the United States, such as foreign freight forwarders and terminal operators.
3) *The Civil Aeronautics Board* is vested with economic regulatory powers over civil aviation within the United States and foreign countries. The board exercises its powers independently and its decisions are not subject to review by any department or agency, except for the approval of the President in granting or affecting certificates in overseas and foreign air transportation. It is composed of five members appointed by the President for 16-year terms.

There are also regulatory issues that may arise in the future. The development of new hydrogen gas pipelines could be subject to federal permitting regimes, which could include permits under the Clean Water Act, the Rivers and Harbors Act, the Endangered Species Act, as well as environmental review under the National Environmental Policy Act, if federal funding or permits are required. A state water quality certification may also be required if construction of a pipeline requires a federal permit and has the potential to discharge into state waters. Other state and local permits may also be required for construction. Ultimately, the regulation of existing, and future, hydrogen pipelines will depend, in part, on whether the pipelines are intrastate or interstate (K&L Gates 2020).

Federal regulation of hazardous material transportation is applicable to all intrastate, interstate, and foreign commerce. State regulation may also apply, however, pursuant to the Hazardous Materials Transportation Act of 1975, which provides that state directives that are inconsistent with federal law are preempted unless they afford equal or greater protection to the public. Advancements in tube trailers and cryogenic tanker technologies may push reconsideration or revision of these regulatory requirements (K&L Gates 2020).

The Pipeline and Hazardous Materials Safety Administration (PHMSA) is responsible for regulatory oversight of the carriage of hazardous materials by rail, aircraft, vessel, or public highway under its Hazardous Materials Regulations (49 C.F.R. Parts 171-180, Subtitle B, Chapter I, Subchapter A). More specifically, PHMSA provides material hazard class guidance for Class 2.1 flammable gases and Class 3 flammable and combustible liquids. It also outlines general requirements for loading, inspection, and lading pressure associated with hazardous materials in cargo tank motor vehicles (K&L Gates 2020).

In the absence of a consistent national regulatory framework for hydrogen transport, a patchwork of state regulatory regimes and standards for the construction and operation of export facilities will emerge and will no doubt mandate consideration of a wide range of safety issues (Stander and Theodore 2008).

13.11 New Hydrogen Transmission Options

A number of new options are available to transport hydrogen. Like natural gas, pure hydrogen can be liquefied before it is transported to increase its density. However, liquefaction requires hydrogen to be cooled to −253°C. If the hydrogen itself were used to provide this energy it would consume approximately one-third of the initial quantity of hydrogen. This is considerably more energy than is required to liquefy natural gas, which consumes approximately 10% of the initial quantity of natural gas.

An alternative possibility is to incorporate the hydrogen into larger molecules that can be more readily transported as liquids. Options include ammonia and liquid organic hydrogen-containing compounds which are both much easier to transport than hydrogen. The downside of using these alternative materials to transport hydrogen is that they often cannot be used as final products, and a further step is needed to liberate the hydrogen from them before final consumption. This entails extra energy and cost, which must be balanced against the lower transport costs for these materials.

As noted earlier, the transport of hydrogen as a gas by pipeline is generally the cheapest option if the hydrogen needs to be transported for distances of less than about 1,500 km. For longer distances, transmission as ammonia or liquid organic hydrogen-containing compounds could be a more cost-effective option, especially if the hydrogen needs to be moved overseas. For local distribution, pipelines are cost-effective for distributing high volumes of hydrogen over longer distances; in other cases trucks are likely to be the cheaper option.

These two options are briefly discussed below (IEA 2019).

1) Converting hydrogen to ammonia requires energy equivalent to between 7% and 28% of the energy contained in the hydrogen, depending on the size and location of the system. A similar level of energy is lost if the ammonia needs to be reconverted back to high-purity hydrogen at its destination. Nevertheless, ammonia liquefies at −33°C, a much higher temperature than is the case for hydrogen, and contains 17 times more hydrogen per cubic meter than liquefied hydrogen, making it much cheaper to transport than hydrogen. While ammonia already has a well-established international transmission and distribution network it is a toxic chemical, and this may limit its use in some end-use sectors. There is also a risk that some un-combusted ammonia could escape, which can lead to the formation of particulate matter (an air pollutant) and acidification.
2) Making a liquid organic hydrogen carrier involves "loading" a "carrier" molecule with hydrogen, transporting it, and then extracting pure hydrogen again at its destination. These compounds

have similar properties to crude oil and oil products, and their key advantage is that they can be transported as liquids without the need for cooling. However, as with ammonia, there are costs associated with the conversion and reconversion processes involved. These processes would require energy equivalent to approximately one-third of the hydrogen itself. In addition, the carrier molecules are often expensive and are not used up when hydrogen is created at the end of the process, requiring that they be shipped back to their place of origin. Several different liquid organic hydrogen carrier molecules are under consideration today including methylcyclohexane.

References

American Association of Railroads (2023). Freight rail facts and figures. https://www.aar.org/facts-figures#3-workforce (accessed 23 November 2023).

Bureau of Transportation Statistics (2022). *Transportation Statistics Annual Report 2022*. Washington, DC: U.S. Department of Transportation, Bureau of Transportation Statistics. https://rosap.ntl.bts.gov/view/dot/65841/dot_65841_DS1.pdf.

Bureau of Transportation Statistics (2023). *National Transportation Statistics Section 1.D Travel and Goods Movement. Table 1-51 – Top U.S. Foreign Trade Freight Gateways by Value of Shipment; Table 1-56 – U.S. Waterborne Freight*. Washington, DC: U.S. Department of Transportation, Bureau of Transportation Statistics. https://www.bts.gov/topics/national-transportation-statistics.

Congressional Research Service (2021). *Pipeline Transportation of Hydrogen: Regulation, Research, and Policy*. R46700. Version 1. Washington, DC: Congressional Research Service. https://www.everycrsreport.com/files/2021-03-02_R46700_294547743ff4516b1d562f7c4dae166186f1833e.pdf.

Federal Aviation Administration (2021). *FAA Aerospace Forecast Fiscal Years 2021-2041*. Washington, DC: FAA. https://rosap.ntl.bts.gov/view/dot/59861.

Federal Aviation Administration (2022). *National Plan of Integrated Airport Systems (NPIAS) 2023–2027*. Washington, DC: FAA. https://www.faa.gov/sites/faa.gov/files/npias-2023-2027-narrative.pdf.

Hydrogen Council (2020). *Path to hydrogen competitiveness. A cost perspective*. Brussels, Belgium: The Hydrogen Council. https://hydrogencouncil.com/wp-content/uploads/2020/01/Path-to-Hydrogen-Competitiveness_Full-Study-1.pdf.

IEA (2019). *The Future of Hydrogen: Seizing today's opportunities*. Paris, France: OECD. https://doi.org/10.1787/1e0514c4-en.

K&L Gates (2020). *The H2 Handbook. Legal, Regulatory, Policy, and Commercial Issues Impacting the Future of Hydrogen*. Washington, DC: K&L Gates LLC. https://marketingstorageragrs.blob.core.windows.net/webfiles/Hydrogen-Handbook-UNITEDSTATES.pdf.

Merriam-Webster (2023a). Transportation. Merriam-Webster.com. https://www.merriam-webster.com/dictionary/transportation (accessed 23 November 2023).

Merriam-Webster (2023b). Transmission. Merriam-Webster.com. https://www.merriam-webster.com/dictionary/transmission (accessed 23 November 2023).

Skipka, K. and Theodore, L. (2014). *Energy Resources: Availability, Management, and Environmental Impacts*. Boca Raton, FL: CRC Press/Taylor & Francis Group.

Stander, L. and Theodore, L. (2008). *Environmental Regulatory Calculations Handbook*. Hoboken, NJ: John Wiley & Sons.

Statistica.com (2023). Leading North American railroad companies in 2022, based on operating revenue. https://www.statista.com/statistics/271613/leading-north-american-railroad-companies-based-on-revenue/ (accessed 23 November 2023).

Theodore, L. and Dupont, R.R. (2012). *Environmental Health and Hazard Risk Assessment: Principles and Calculations*. Boca Raton, FL: CRC Press/Taylor Francis Group.

Theodore, L. and Dupont, R.R. (2023). *Chemical Process Industries: Environmental and Health Risk Calculations*. Boca Raton, FL: CRC Press/Taylor & Francis Group.

U.S. Department of Energy (2014). *Hydrogen Fuel Cell Engines and Related Technologies Course Manual. Module 2: Hydrogen Use*. Washington, DC: Office of Energy Efficiency and Renewable Energy. Hydrogen and Fuel Cell Technologies Office. https://www.energy.gov/sites/default/files/2014/03/f10/fcm02r0.pdf.

U.S. Department of Energy (2020a). *Hydrogen Pipelines*. Washington, DC: Office of Energy Efficiency and Renewable Energy, Hydrogen and Fuel Cell Technologies Office. https://www.energy.gov/eere/fuelcells/hydrogen-pipelines.

U.S. Department of Energy (2020b). *Hydrogen Program Plan*. DOE/EE-2128. Washington, DC: U.S. Department of Energy. https://www.hydrogen.energy.gov/docs/hydrogenprogramlibraries/pdfs/hydrogen-program-plan-2020.pdf?Status=Master.

U.S. Department of Energy (2020c). *Hydrogen Strategy. Enabling a Low-Carbon Economy*. Washington, DC: Office of Fossil Energy. https://www.energy.gov/sites/prod/files/2020/07/f76/USDOE_FE_Hydrogen_Strategy_July2020.pdf.

U.S. Department of Transportation (2023a). *Pocket Guide to Transportation 2023*. Washington, DC: Bureau of Transportation Statistics, U.S. Department of Transportation. https://doi.org/10.21949/1527467.

U.S. Department of Transportation (2023b). *Freight Analysis Framework Version 5 (FAF5)*. Washington, DC: Bureau of Transportation Statistics, U.S. Department of Transportation. https://faf.ornl.gov/faf5/dtt_total.aspx (accessed 23 November 2023).

U.S. Environmental Protection Agency (1997). *Profile of the Water Transportation Industry*. EPA 310-R-97-003. Washington, DC: Office of Enforcement and Compliance Assurance, U.S. Environmental Protection Agency. https://nepis.epa.gov/Exe/ZyPDF.cgi?Dockey=50000GOR.PDF.

14

Hydrogen Conversion

14.1 Introduction

As noted in earlier chapters, hydrogen can be viewed as a secondary source of energy as is the most common source of secondary energy, electricity. Secondary sources of energy are also known as energy carriers. Since it is not available in the "free" state, hydrogen must be produced using a hydrogen-rich primary source. Replacing fossil fuels with hydrogen in providing energy services could bring about major environmental benefits if the carbon dioxide, and noxious gases emitted in the hydrogen-production process are suitably controlled or reduced. Although hydrogen can be burned in conventional devices such as boilers, turbines, and internal combustion engines, fuel cells appear to be a promising technology to exploit hydrogen because of their high efficiency.

Following the production of hydrogen as detailed in Chapter 11 – Hydrogen Processes, is the all-important step of hydrogen *conversion* to energy in the form of heat and/or electricity.

This Chapter 14 – Hydrogen Conversion is concerned with the hydrogen conversion steps. Following in Chapter 15 – Hydrogen Uses are the details of the applications of heat or electricity generation and hydrogen as an alternative carbon-free feedstock to produce other end products.

Merriam-Webster's definition of conversion is ". . . the process of being converted . . . to alter the physical or chemical nature or properties of especially in manufacturing . . . to alter for more effective utilization . . . to exchange for an equivalent."

To generate electricity from hydrogen, it needs to be reacted with oxygen and the resulting energy either captured for heating purposes and/or used by an electric generator. The simplest form of this reaction is

$$2H_2 + O_2 \rightarrow 2H_2O + \text{energy} \tag{14.1}$$

where oxygen and hydrogen combine to produce water and release energy in the process. As a result, the only direct emission from a hydrogen-based power unit is clean water.

Two main classes of technologies are available for electric power production from hydrogen:

1) Hydrogen "engines" that combust the fuel and produce electricity with a conventional generator.
2) Fuel cells that directly use the reaction energy liberated in an electric circuit.

Hydrogen Energy: Principles and Applications, First Edition. Vincent J. DelGatto, Louis Theodore, R. Ryan Dupont, and Matthew C. Ogwu.

Chapter 14 is concerned with the conversion of hydrogen as secondary carbon-free energy source. Section numbers and their accompanying titles for this chapter concerned with hydrogen conversion are listed below.

14.2 Energy Conversion Technical Details

The hydrogen energy conversion dilemma involves the concept of *entropy* and energy. Its application allows calculations of energy relationships associated with all kinds of processes and conversions. Applications involve calculations for maximum power outputs from power plants.

Regardless of the types of energy involved in processes – thermal, mechanical, electrical, elastic, magnetic, etc. – the change in the energy of a system is equal to the difference between energy input and energy output. The first law of thermodynamics allows free convertibility from one form of energy to another, as long as the overall quantity is conserved. Thus, this law places no restriction on the conversion of work into heat, or on its counterpart – the conversion of heat into work (Theodore, Ricci, and VanVleet 2009; Theodore 2014).

The unrestricted conversion of work into heat is well-known to most technical individuals. Frictional effects are frequently associated with mechanical forms of work and the transformation of heat into work is of greater concern. In nations with a partially developed or developing technological society, the ability to produce energy in the form of work takes on prime importance. Work transformations are necessary to transport people and goods, drive machinery, pump liquids, compress gases, and provide energy input to so many other processes that are taken for granted in highly developed societies. Much of the work input in such societies is available in the form of electrical energy, which is then converted to rotational mechanical work. Some of this electrical energy (work) is produced by the combustion of fossil fuels or nuclear fission. Hence the study of the conversion of heat to work is extremely important – especially in the light of developing environmental problems, particularly with global warming (Theodore, Ricci, and VanVleet 2009; Theodore 2014).

The second law of thermodynamics implies that thermal energy loses some of its "quality" or is degraded when it is transferred by some process or conversion. Other important forms of energy degradation include energy transformations due to frictional effects and electric resistance. Such effects are highly undesirable if the use of energy for practical purposes is to be maximized. The second law provides some means of measuring this energy degradation (Theodore, Ricci, and VanVleet 2009) through a term referred to as *entropy* and it is the second law that serves to define this important thermodynamic property that can provide quantitative information on the "quality" of energy and energy degradation.

An example of the limitation of heat to perform work is that of the steam in a steam engine. The hot vapor in a steam engine's cylinder contains a significant amount of thermal energy. The random thermal motion of the water molecules contributes to the motion of the rigid metallic piston, thereby transforming some of the steam's internal energy into useful pressure–volume work. However, a process may never be created such that all of the steam's energy content can be transformed *completely* into useful work.

Returning to the chapter's main theme of hydrogen conversion processes, there are two primary paths to produce hydrogen that can be converted into energy in the form of heat or electricity.

1) Combustion of methane. This process involves the reaction of methane with oxygen to produce hydrogen as shown in Equation 14.2.

$$CH_4 + O_2 \rightarrow CO_2 + 2H_2 \tag{14.2}$$

with $\Delta H° = -212{,}800$ cal/gmol (liberated).

The hydrogen energy conversion process involves the combustion of hydrogen according to Equation 14.1.

$$2H_2 + O_2 \rightarrow 2H_2O + \text{energy} \tag{14.1}$$

with $\Delta H° = -2(68{,}300) = -136{,}600$ cal/gmol (liberated).

In effect, the 212,800 cal/gmol initially available for energy purposes has been reduced to 136,600 cal/gmol – providing an energy conversion efficiency of less than 65%.

The above analysis should be viewed as follows. The generation of one source of energy to another – in this case hydrogen from methane – has resulted in a loss of available and/or useful energy. And this does not consider the energy "losses" associated with friction, irreversibility, etc.

2) Combustion of hydrogen obtained from water. This occurs via the reaction shown in Equation 14.3, followed by the hydrogen energy conversion shown again in Equation 14.1.

$$2H_2O \rightarrow 2H_2 + O_2 \tag{14.3}$$

with $\Delta H° = 2(68{,}300) = +136{,}600$ cal/gmol (required).

$$2H_2 + O_2 \rightarrow 2H_2O + \text{energy} \tag{14.1}$$

with $\Delta H° = -2(68{,}300) = -136{,}600$ cal/gmol (liberated).

In effect, the hydrogen energy converted in Equation 14.1 was arrived at by employing the *same* amount of energy required to produce the hydrogen. The frictional effects, irreversibilities, etc., must once again be considered in this or any energy analysis.

The above analysis should provide a wake-up call to practicing engineers and applied scientists who are committed to a hydrogen energy economy for the future.

14.3 Electric Power Systems

Electric power systems involve the transformation of other types of energy into electrical energy and the transmission of this electrical energy to the point of consumption. The economic production and transmission of energy in the form of electricity have important implications. Such systems permit power generation by equipment that is comparatively cheap to build and operate in terms of cost per unit of power delivered.

Alternating current (AC) is generally used in modern power systems, because it has the important advantage of being easily converted to higher or lower voltages by means of transformers.

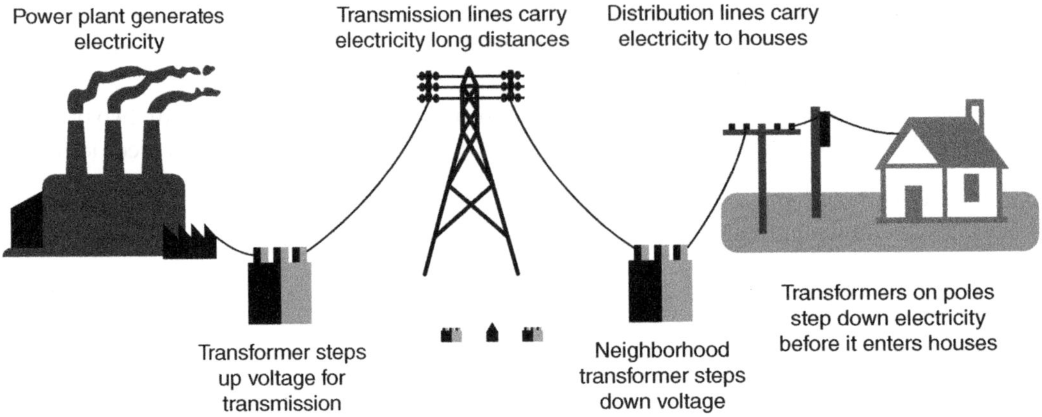

Figure 14.1 Electricity generation, transmission, and distribution. *Source:* EIA/ Public Domain (https://www.eia.gov/energyexplained/electricity/delivery-to-consumers.php.

Thus, each stage of the system can be operated at an appropriate voltage. An electric power system consists of six main elements (Figure 14.1):

1) Electric generation stations.
2) Transformer Substations to raise the generated power to the high voltages used on the primary transmission lines.
3) Transmission lines.
4) Terminal Substations at which the power is stepped down to the voltage of the secondary transmission lines or distribution lines.
5) Distribution lines.
6) Transformers that lower the secondary voltage to levels used by the consumer.

In a typical system, the electric generators at the central power stations produce electricity with a voltage of around 13,000 V. This voltage is stepped up by means of transformers to values ranging from 138,000 to 500,000 V for primary transmission lines. At the terminal substations, the voltage is stepped down again by transformers to a distribution level in the order of 13,000 V. The voltage is stepped down once or twice again by transformers near the point of use. Meanwhile, the current changes inversely proportional to voltage. At the lower voltages, the higher currents require more costly, heavier conductors that lessen the hazards of fire and electric shock. High-Voltage Direct Current (HVDC) transmission is now being built in place of high-voltage AC for very long-distance transmission and high-power system interconnections.

The central power-generating stations of an electric power system consist of a prime mover, e.g., a turbine, which operates an electric generator. An electric generator is a device that converts a form of energy into electricity. There are many different types of electricity generators. Most of world's electricity generation is from generators that are based on scientist Michael Faraday's discovery in 1831 that moving a magnet inside a coil of wire makes (induces) an electric current flow in the wire. He made the first electricity generator called a Faraday disk, which operates on this relationship between magnetism and electricity, and led to the design of the electromagnetic generators that are in use today.

Electromagnetic generators use an electromagnet – a magnet produced by electricity – not a traditional magnet. A basic electromagnetic generator has a series of insulated coils of wire that form a stationary cylinder – called a stator – surrounding an electromagnetic shaft – called a rotor.

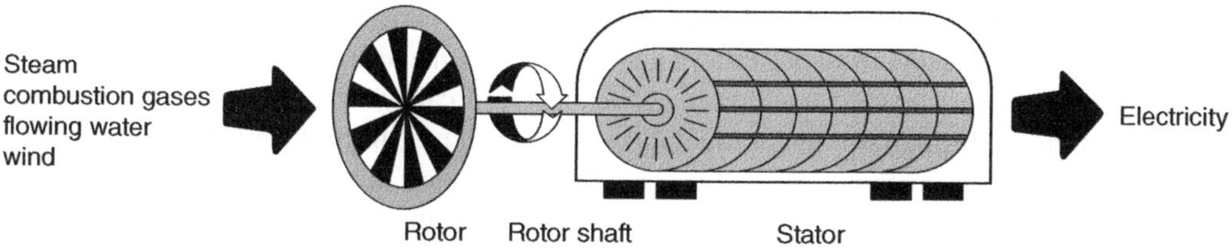

Figure 14.2 Electricity generation from an electric turbine. *Source:* EIA/ Public Domain (https://www.eia.gov/energyexplained/electricity/how-electricity-is-generated.php).

Turning the rotor makes an electric current flow in each section of the wire coil, which becomes a separate electric conductor. The currents in the individual sections combine to form one large current. This current is the electricity that moves from generators through power lines to consumers. Electromagnetic generators driven by kinetic (mechanical) prime movers account for nearly all of U.S. electricity generation.

Most electricity generation is from electric power plants that use a turbine to drive electricity generators. In a turbine generator, a moving fluid – water, steam, combustion gases, or air – pushes a series of blades mounted on a rotor shaft. The force of the fluid on the blades spins/rotates the rotor shaft of a generator. The generator, in turn, converts the mechanical (kinetic) energy of the rotor to electrical energy (Figure 14.2). Different types of turbines include steam turbines, combustion (gas) turbines, hydroelectric turbines, and wind turbines.

Steam turbines are used to generate much of the world's electricity. Most steam turbines have a boiler where fuel is burned to produce hot water and steam in a heat exchanger, and the steam powers a turbine that drives a generator. Nuclear power reactors use heat from nuclear fuel rods to produce steam. Solar thermal power plants and most geothermal power plants also use steam turbines for power production.

There are many different types of electricity generators that do not use steam turbines to generate electricity such as hydroelectric turbine generators, wind turbines, fuel cells, solar photovoltaic (PV) systems, and internal combustion engines (ICE). Solar photovoltaic cells convert sunlight directly into electricity. The panels can be connected together in arrays to power individual homes or into thousands of arrays to form large power plants. Internal combustion engines, such as diesel engines, are used all around the world for electricity generation including in many remote villages around the world. They are also widely used for mobile power supply at construction sites and for emergency or backup power supply for buildings and power plants. Internal combustion engine generators can use a variety of fuels including petroleum, diesel, biomass-based liquid fuels, biogas, natural gas, and propane. Small internal combustion engine generators fueled with gasoline, natural gas, or propane are commonly used by construction crews and tradespeople and for emergency power supply for homes.

Any electric power system involves a large amount of supplementary equipment for the protection of generators, transformers, and the transmission lines themselves. In addition, the electric power system often includes devices that are designed to regulate the voltage delivered to consumers and to correct the power factor of the system. To protect all elements of a power system from short circuits and overloads, and for normal switching operations, *circuit breakers* are employed. These are large switches that are actuated automatically in the event of a short circuit or other condition that produces a sudden rise of current. Because an arc is formed across the terminals of the circuit breaker at the moment when the current is interrupted, some large breakers (such as those used to protect a generator or a section of primary transmission line) are immersed in a liquid such as oil to quench the arc. In large air-type circuit breakers, as well as in oil breakers, magnetic fields are used to break up the arc. Small air-circuit breakers are used for protection in shops, factories, and modern home installations.

Long transmission lines have considerable inductance and capacitance as well as resistance. When a current flows through the line, inductance and capacitance have the effect of varying the voltage on the line as the current varies. Thus, the supply voltage can vary with the load. Several kinds of devices are used to overcome this undesirable variation in an operation called regulation of the voltage. They include induction regulators and three-phase synchronous motors called synchronous condensers), both of which adjust the effective amount of inductance and capacitance in the transmission circuit. Inductance and capacitance react with a tendency to nullify one another. When a load circuit has more inductive than capacitive reactance, as almost invariably occurs in large power systems, the amount of power delivered for a given voltage and current is less than when the two are equal. The ratio of these two amounts of power is called the power factor. Because transmission-line losses are proportional to current, capacitance is added to the circuit when possible, thus bringing the power factor as nearly as possible to 1. For this reason, large capacitors are frequently inserted as part of power transmission systems.

14.4 The Grid System

World electric power production and consumption continue to rise almost exponentially. In most parts of the world, local or national electric utilities have joined in *grid* systems. The linking grids allow electricity generated in one area to be shared with others. Each pooling company gains through sharing an increased reserve capacity, use of larger, more efficient generators, and compensation for local power failures.

Electricity is generated at power plants and moves through this grid of electricity substations, transformers, and power lines that connect electricity producers and consumers. Most local grids are interconnected for reliability and commercial purposes, forming larger, more dependable networks that enhance the coordination and planning of electricity supply. In the United States, the entire electricity grid consists of hundreds of thousands of miles of high-voltage power lines and millions of miles of low-voltage power lines with distribution transformers that connect thousands of power plants to hundreds of millions of electricity customers all across the country.

At the beginning of the 20th century, more than 4,000 individual electric utilities operated in isolation from each other. As demand for electricity grew, especially after World War II, utilities began to connect their transmission systems. These connections allowed utilities to share the economic benefits of building large and often jointly owned electric generating units to serve their combined electricity demand at the lowest possible cost. Interconnection also reduced the amount of extra generating capacity that each utility had to hold to ensure reliable service during times of peak demand. Over time, three large, interconnected systems evolved in the United States. These essentially operate largely independently from each other with limited transfers of electricity between them. These three systems are:

1) The *Eastern Interconnection* which encompasses the area east of the Rocky Mountains and a portion of the Texas panhandle.
2) The *Western Interconnection* which encompasses the area from the Rockies to the west.
3) The *Electric Reliability Council of Texas* (ERCOT) which covers most of Texas.

The Eastern and Western Interconnections in the United States are also linked with the Canadian power grid. The network structure of the interconnections helps maintain the reliability of the grid by providing multiple routes for power to flow and allowing generators to supply electricity to many load centers. This redundancy helps prevent transmission line or power plant failures from causing interruptions in service to retail customers.

The electric power system is the infrastructure that converts fuel and energy resources into electric power (thus generating electricity) and carries and manages that electric power from where it is generated to where it is used. It is a system of systems that comprise physical networks that include fuel and resources; power plants of many different varieties; electric transmission and distribution line networks and measurement; information and control systems; and virtual networks of money, business relationships, and regulation. Achieving balance among all of these elements is a fundamental challenge for the planning, engineering, and operation of the overall system because of the variability and uncertainty of load and unexpected equipment failures that affect the generation and delivery of electricity. This system of systems is loosely referred to here, once again, as the grid.

The major physical elements of the grid are generation, transmission, distribution, and load. Generation is the collection of power plants electrically connected to the grid and ranging in size from very small, distributed units to central stations rated at over 1,000 MW. The sources of U.S. electric generation in 2022 were 60% fossil fuel, 19% nuclear, 10% wind, 6% hydro, 4% solar, and 1% biomass and others. Many small, distributed generators or small power units are installed at hospitals, fire stations, and other critical facilities to provide power in case of emergencies or failure of the grid resulting in an interruption of the flow of electricity. These standby or emergency backup systems are not normally electrically connected to the grid. With the addition of proper control and switching systems, these units could be connected, however.

Transmission is the collection of networked high-voltage lines (above 100 kV) that tie generation to load centers. High-voltage lines also connect utilities to one another, reduce costs through sharing of resources, and provide enhanced reliability in case of events such as the loss of a large generator. The high-voltage transmission system also enables the wholesale marketplace for electricity. In general, the bulk power system refers to the network of interconnected generation and transmission lines, while the distribution system refers to the lower-voltage generally radial lines that deliver electricity to the final customer. The load created by the electrical equipment on the customer's side of the meter is electrically part of the overall power system and affects its operation; the load completes the system. The largest industrial and commercial customers may be served by transmission directly; the rest are served by the lower-voltage distribution system. The balance of generation to load must be maintained constantly every second.

Many entities – balancing authorities, regional entities, utilities, power pools, independent system operators (ISOs), regional transmission organizations (RTOs), and other transmission organizations – are involved in running the grid today. At the federal level, the Federal Energy Regulatory Commission (FERC) has regulatory authority over the interstate sale of electricity and the operation of regional markets. The North American Electric Reliability Corporation (NERC) has the responsibility, under FERC authority, for power system reliability, operating, and planning standards in the United States, and coordinates with Canada. Every utility in the United States and Canada participate in the NERC reliability assessments to ensure that the transmission system meets standards and will perform reliably. Most criteria for planning of transmission are based on the NERC standards.

In addition to spinning generator reserve power, short-term storage of electric energy can help counter the destabilizing effects on the grid from intermittent wind and solar renewable generation. Electric storage can consist of pumped hydro, batteries, and hydrogen. Variable hydrogen production from renewables including nuclear is being considered. Large-scale steam reforming and electrochemical production of hydrogen could be the controllable variable load to balance intermittent renewables. Figure 14.3 shows how the Hydrogen Economy and the Electric Grid could form a very positive and substantial synergistic relationship.

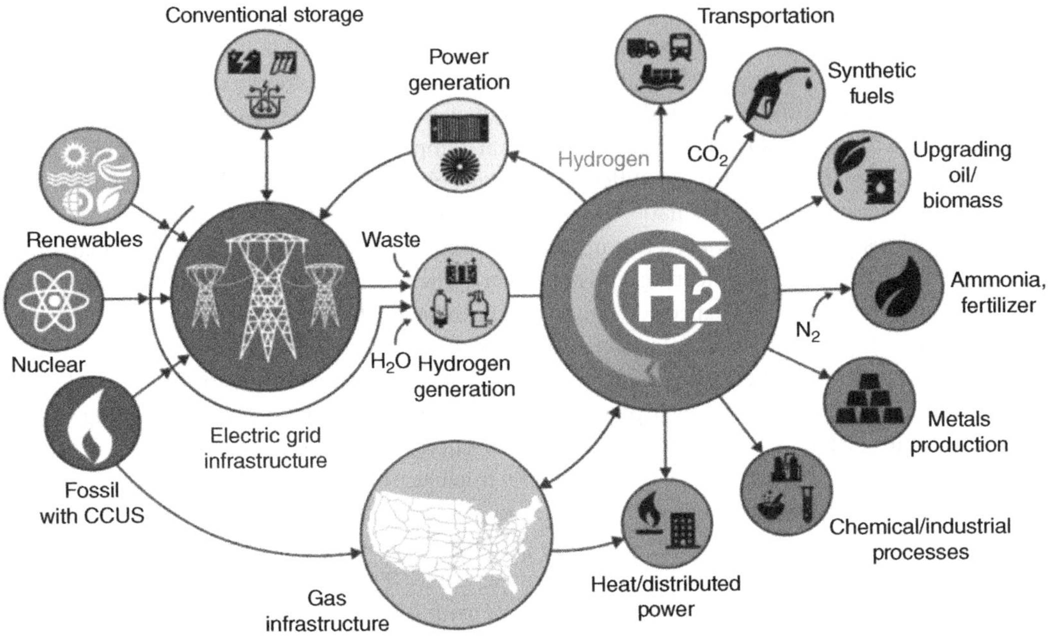

Figure 14.3 Synergy of Electric Grid and Hydrogen Economy. *Source:* Courtesy of University of Michigan.

A possible synergy of the Electric Grid and H2 is boosting the capacity factor (CF) of the intermittent wind and solar generation by storing surplus generation of renewables. The production of hydrogen for storage could also provide nuclear generation with a constant load (World Nuclear Association 2021). However, these capacity factor increases will greatly influence costs.

The traditional cost metric for generation has been the Levelized Cost of Energy (LCOE). However, LCOE does not account for the storage requirements of intermittent renewables. A newer cost metric called Levelized Full System Cost of Electricity (LFSCOE) includes storage to reach 95% capacity factor.

Two reports on the high cost of boosting capacity factor of solar with battery or hydrogen storage are here presented in full. Refer to Section 14.4.1 for the January 13, 2023 report by V.J. DelGatto titled "Storage Costs Multiply to Achieve 90% Capacity Factor for Large Solar PV". Refer to Section 14.4.2 for the September 1, 2023 report by V.J. DelGatto titled "Cost of Vogtle 3 and 4 Nuclear is Less than PV with Storage for 90% Capacity Factor." These costs were based on pricing for utility scale PV generation, battery storage and hydrogen storage from the NREL report: "Solar Photovoltaic System and Energy Storage Cost Benchmarks, With Minimum Sustainable Price Analysis: Q1 2022."

14.4.1 Storage Costs Multiply to Achieve 90% Capacity Factor for Large Solar PV

The report below (DelGatto 1-13-2023) proposes that there are costly implications to adding energy storage requirements to solar photovoltaic (PV) power plants to boost CF from 30% to 90%.

> "The Bhadla Solar Park in India is one of the largest PV solar farms in the world. The rated output is 2.25 GW and comprises 14,000 acres of photovoltaic solar panels. The Bhadla Solar Park will be assumed to have 30% CF based on 8 hours of full sunlight every day. A typical nuclear power plant has nearly 2.25 GW output at 90% CF. The goal here is to quantify

the amount of battery or hydrogen storage and the additional solar PV generation required to reach 90% CF. The intent is to increase the CF of a PV system with storage to match CF of a nuclear plant. In summary, the simple calculations show that solar PV generation for 2.25 GW at 90% CF could require either a battery storage solar PV plant with a total of 7.25 GW PV generation and 36 giga-watt hours (GWh) battery storage or a hydrogen storage solar PV plant with 13.5 GW PV generation and a 72GWh hydrogen storage system for an ideal 24-hour storage cycle.

The definition of Levelized Cost of Energy (LCOE) is lifetime costs of generation plant divided by energy production. Therefore, the added costs of generation and/or storage energy to backup 30% CF of solar or 50% CF of wind energy are not reflected in LCOE. The LCOE of a PV generating plant with storage to achieve a CF of 90% would be a better cost comparison to a nuclear or sequestered fossil fuel generating station. [The Levelized Full System Cost of Electricity (LFSCOE) metric is similar in that it uses 95% capacity factor and only one renewable generation type.]

The storage scenario considered was based on the Bhadla solar PV power station operating at full rated 2.25 GW power for 8 hours producing 18 GW hours of electric energy per day (2.25 GW × 8 hours = 18 GWh). Additional electric power output of 36 GW hours from battery storage or hydrogen storage would provide the electrical energy for the remainder of day (2.25 GW × 16 hours = 36 GWh). Therefore, Bhadla PV with storage for 90% CF on a 24 hour cycle would approximate the daily energy output of a typical 2.25 GW nuclear power plant. Other energy storage scenarios could scale down storage to 2.25 GWh for intermittent minutes of backup or scale storage up to 378 GWh for full one-week backup.

A 2.25 GW nuclear power plant could produce 54 GWh of energy in a 24 hour period (2.25 GW × 24 hours). The 2.25 GW Bhadla solar PV plant could produce 18 GWh of energy in a 24 hour period (2.25 GW × 8 hours = 18 GWh). The amount of the additional solar PV generating plant plus battery or hydrogen storage to provide the additional 36 GWh each day (54–18 GWh) are simple approximate calculations. The round trip efficiency of storage mediums is one major factor in sizing as shown below. The other major factor is the daily window of sunshine, which is the only time for all PV generation. Therefore, solar generation has to be increased multiple times to increase CF three times from 30% to 90%.

Battery storage scenario:

7.25 GW solar PV + 36 GWh battery storage = 2.25 GW nuclear. Battery storage round trip efficiency from charging to discharging onto grid is 90% efficient. 36 GWh battery output requires 40 GWh electric energy input. Daily PV generation of 40GWh over 8 available hours requires 5 GW additional generation for storage. Therefore, a total of 7.25 GW (2.25 GW + 5 GW) solar PV power plant with 36 GWh battery storage would be equivalent to 2.25 GW nuclear power plant.

Hydrogen storage scenario:

13.5 GW solar PV + 9 GW electrolyzer + 72 GWh hydrogen storage + 2.25 GW hydrogen turbine or fuel cell = 2.25 GW nuclear plant. Hydrogen storage round trip efficiency is 40% based on hydrogen production with electrolyzer at 80% efficiency and electric power generation with a hydrogen gas turbine or fuel cell at 50% efficiency. Calculating back from output electrical energy to be supplied by storage to find size of additional PV generation: The 36 GWh electric output from 50% efficient hydrogen gas turbine or fuel cell requires 72 GWh of hydrogen storage. The 72 GWh of hydrogen storage requires 90 GWh PV electric energy into 80% efficient electrolyzer. 90 GWh PV production in

> 8 hours per day requires 11.25 GW solar PV generator. The overall solar PV plant with hydrogen storage for PV generation of 2.25 GW at 90% CF utilizing hydrogen storage on an ideal 24-hour cycle would require the following: Total of 13.5 GW (2.25 GW + 11.25 GW) of PV solar generation capacity with a 9 GWh electrolyzer and a 2.25 GW gas turbines or fuel cells."

14.4.2 Cost of Vogtle 3 and 4 Nuclear is Less than PV with Storage for 90% Capacity Factor

The calculated costs in the report below (DelGatto 9-1-23) for 2 GW with 90% CF from PV solar generation with battery storage are $68.4 billion for a two-day scenario. In contrast, the cost for Vogtle Units 1 and 2 was $8.87 billion in 1989 ($17.1 billion in 2021 dollars). The present total cost for the new Vogtle Units 3 and 4 is $35 billion. Each Vogtle unit is near 1.2 GW with over 90% CF. Upon completion of Units 3 and 4 in 2023, Vogtle became the largest nuclear power station in the United States.

> "A two-day PV generation with storage scenario was estimated for 2 GW nominal 90% CF based on 1 day with 8 hours of full sun followed by a fully clouded day. This scenario would require 12 GW of PV generation and 80 GWh of battery storage since only 8 hours of charging time is available in this 48-hour period. (A total of 16 GWh of energy from 2 GW PV for 8 hours + 80 GWh from batteries over 40 hours + 10 GW PV to charge the 80 GWh batteries during the 8 hours of sun in this two-day scenario.)
>
> PV and battery installed system costs were based on NREL 2022 cost benchmarks (US DOE. 2022. NREL -Solar. . .and Storage Costs). The estimated cost of continuous 2 GW from PV with 16-hour battery storage for the 8-hour full sun day is $29.4 billion, ((6 GW PV × $1.7 billion/GW PV = $10.2 billion) + (32 GWh battery × $0.6 billion/GWh battery = $19.2 billion)). The estimated cost of continuous 2 GW with 24-hour battery storage for each consecutive cloudy day is $39 billion, ((6 GW PV × $1.7 billion/GW PV = $10.2 billion) + (48 GWh battery × $0.6 billion/GWh battery = $28.8 billion)).
>
> The total cost for 2 GW with 90% CF in this 2-day scenario is $ 29.4 billion PV generation + $39 billion battery storage = $68.4 billion.

Conclusions: High levels of VRE integration will require substantial additional storage costs to maintain a reliable electric grid. The costs associated with energy storage for variable renewable energy (VRE) are very substantial. Nuclear power could be a more cost-effective carbon-free transition source of electric energy compared to solar PV with battery or hydrogen storage. Furthermore, an extensive investment in solar with storage could limit funds available for future transitions to fusion power (Nicholas et al. 2021).

14.5 Conversion: The Combustion Process

As noted earlier, hydrogen is an *energy carrier* that is produced using energy from feedstocks such as natural gas (particularly methane), oil, coal, and biomass. Electrochemical production of hydrogen from carbon-free nuclear power plant electricity could become a major supply of hydrogen in the future. As such, the hydrogen could be considered a storage medium to be converted back into electricity as needed. This hydrogen storage cycle would act somewhat like a pumped hydro storage system. To be useful, the energy carried by hydrogen must be converted into a different form, such as heat or electricity. This conversion can be accomplished through combustion using turbines or reciprocating engines, or through an electrochemical process using a fuel cell. There are also a number of opportunities to design hybrid energy systems including high- or low-temperature stationary fuel cells integrated with gas turbines in large-scale combined-cycle hybrid systems that employ both conventional and fuel cell energy conversion technologies. Figure 14.3 shows how the Hydrogen Economy and the Electric Grid could form a very positive and substantial synergistic relationship.

Hydrogen can be combusted in the same manner as natural gas, synthetic gas, diesel, gasoline, and other common fuels. The benefit of using hydrogen is that no carbon dioxide is produced, and water is the only major byproduct. The use of hydrogen was successfully demonstrated a half-century ago by both the National Aeronautics and Space Administration, which has used hydrogen in the space shuttle's main engines, and by the Department of Defense, which has used hydrogen in unmanned rocket engines.

The advantages of power generation using hydrogen combustion include:

1) Fuel flexibility, through the ability to burn hydrogen and blends of fossil fuels
2) Fuel security through integration with hydrogen storage
3) The ability to meet large demands for electricity
4) The flexibility to follow loads from variable generation (Goldmeer 2019).

Growing attention is on hydrogen turbines, particularly for large-scale generation. The industry has developed materials and systems to increase the concentration of hydrogen that can be combusted, and these advances have allowed hydrogen to be fired at concentrations over 90% in simple-cycle turbines or aero-derivative machines, and at concentrations of up to 50% in large frame combined-cycle turbines. Interestingly, large-frame turbines capable of firing hydrogen/natural gas blends of up to 30% hydrogen, and aero-derivative turbines capable of firing over 90% hydrogen are already commercially available today.

In a hydrogen turbine power plant, the oxidizer can be air, as in a typical gas turbine, or pure oxygen, with the advantage to eliminate NO_x emissions (Theodore 2008). The combustion of hydrogen with pure oxygen requires careful burner design, due to the high temperature and combustion characteristics of this flame option. The DOE National Energy Technology Laboratory (NETL) has conducted research on this technology. Several companies are also engaged in development, including Kawasaki Heavy Industries, General Electric and Tascosa Services.

However, there are concerns. Studies are needed to address issues such as auto-ignition, flashback, thermo-acoustics, mixing requirements, aerothermal heat transfer, material issues, turndown/combustion dynamics, NO_x emissions, and other combustion-related phenomena. In addition, when hydrogen concentration exceeds 75%, there is a significant change in combustion behavior, requiring new combustor designs, different sensor locations, and new control schemes. Higher flame temperatures and increased water content could also reduce the lifetime of metal and ceramic parts exposed to hot gases, thereby increasing the need for new materials and thermal barrier coatings as well as improved cooling schemes when combusting hydrogen.

Hydrogen can also fuel combustion-based power generators such as ICE and gas turbines. Hydrogen and hydrogen-based liquid ammonia are strong candidates for carbon-free transportation fuels for trucking, aviation, and shipping as well. The reaction energy of hydrogen combustion is similar to that of hydrocarbon fuels. Hydrogen internal combustion engine generators are available from a few industrial companies, including Siemens, Mann, and General Electric. The maximum power output of the currently available systems is approximately 1 MW. Hydrogen turbine generators can have much larger power capacities. Several large turbine manufacturers, including Siemens, Mitsubishi, and Kawasaki, are in various stages of development and operational testing of hydrogen-powered turbines for electric grid use. The largest project of this kind is currently in construction at the Intermountain Power Project in Utah, where wind, solar, and hydro energy will be converted to hydrogen to be burned in a modified gas turbine (Green Hydrogen Coalition 2023).

14.6 Conversion: The Fuel Cell Process

Fuel cells can be more efficient than ICE because the electrochemical reactions in a fuel cell generate electricity *directly* – while combustion converts the energy in the fuel first into mechanical energy and then into electrical energy. Fuel cell efficiencies of over 60% have been demonstrated (Lohse-Busch et al. 2018), and over 80% efficiency is possible when fuel cells are used in combined heat and power applications (U.S. DOE 2011; Lohse-Busch et al. 2018).

The reader is referred to Part I, Chapter 6 for additional detail on fuel cells. The fuel cell is a device that uses a hydrogen-rich fuel and oxygen to produce electrical energy by means of an electrochemical reaction. The cell consists of two electrodes – an anode (negative), and a cathode (positive) – sandwiched around an electrolyte. Hydrogen is introduced to the anode and oxygen to the cathode. The electrolyte in the cell causes each proton and electron in the hydrogen atom to separate and take different paths to the cathode. The electron goes through an external circuit, creating an electrical charge, i.e., electricity. The proton migrates through the electrolyte to the cathode, where it reunites with an electron and reacts with the oxygen to produce water and energy in the form of heat.

Fuel cells are similar to batteries in that they are composed of positive and negative electrodes separated by an electrolyte or membrane, and both are highly efficient. Fuel cells, however, do not need to be recharged the way batteries do, so they may run for extended periods as long as fuel and air are provided. In fuel cells, power and energy are decoupled and can be tuned independently, i.e., for a fixed fuel cell stack, more hydrogen allows higher energy capacity without changing the fuel cell size or power. Like batteries, fuel cells have advantages over combustion engines: they have no moving parts, are quiet, and require no oil changes and minimal maintenance. Fuel cells are also easily scalable, as individual cells can be stacked together to provide a wide range of power. They can range in size from less than a watt for portable power, to many megawatts for large-scale stationary power.

There are several types of fuel cells, all of which have particular advantages that make them well-suited for various applications. The key features that distinguish the different types of fuel cells include the type of electrolyte they employ, their operating temperature, and the level of hydrogen purity required. A brief reintroduction to three relatively new fuel cell types is provided below.

1) *Polymer electrolyte membrane fuel cells* (PEMFCs). These fuel cells typically operate at about 80°C and can respond quickly to changing loads, making them suitable for transportation applications as well as stationary, backup, or portable power applications that require fast start-up times or must react to variable loads.
2) *Solid oxide fuel cells* (SOFCs). These fuel cells operate at much higher temperatures (typically 800–1,000°C) and are more suitable for use in modular and utility-scale stationary power systems, since the high temperatures make rapid start-up challenging.

3) *Intermediate temperature fuel cells*. This category of fuel cells includes *molten carbonate* (600–700°C) and *phosphoric acid fuel cells* including *polymer-phosphoric acid-based systems* (150–200°C), as well as other low-temperature fuel cells *like alkaline fuel cells* and emerging *alkaline exchange membrane fuel cells* (≤80°C).

Application-specific targets are developed by each DOE office to guide activities and ensure that PEMFC and SOFC technologies are on the pathway to competitiveness in terms of cost, performance, durability, and reliability.

Finally, fuel cells can use a wide range of fuels and feedstocks and can provide power for a number of applications across multiple sectors. In addition to these applications, DOE is exploring the use of fuel cells for tri-generation, which can use fuels such as coal syngas, biogas, or natural gas to co-produce power, heat, and hydrogen. Efforts are also focused on low- and high-temperature reversible fuel cells that can operate in two modes, to produce either hydrogen in electrolysis mode or electricity in fuel cell mode. Reversible fuel cell systems would offer the ability to provide easily dispatchable power in a single unit, using only water as a feedstock. For example, a coal gasification facility can generate hydrogen for use in an SOFC to produce electricity, while during periods of low electricity demand, the SOFC can operate reversibly to produce hydrogen for storage and/or chemical production. Alternatively, reversible fuel cells can be integrated with intermittent renewables to generate hydrogen from electricity that would otherwise be curtailed during periods of low demand. For all these applications, there is a common need to reduce cost and improve durability, while maintaining efficiency. Depending on the type of fuel cell, a key contributor to cost is the catalyst, typically based on platinum group metals, which rely on foreign imports. Other key components needing improvements to meet cost and durability targets are: membranes or electrolytes; bipolar plates, which serve multiple functions including water removal and the collection of the electric current produced; and other components such as gas diffusion layers. Reliable and low-cost balance-of-plant components such as compressors, blowers, and power electronics are also required (U.S. DOE 2011, 2020).

References

DelGatto, V.J. (2023). "Costs Multiply to Achieve 90% Capacity Factor for Large Solar PV", Unpublished technical report, presented to IEEE-USA Energy Policy Committee on Jan. 13, 2023.

DelGatto, V.J. (2023) "Costs of Vogtle 3 and 4 Nuclear is Less than PV with Storage for 90% Capacity Factor", Unpublished technical report, presented to IEEE-USA Energy Policy Committee on Sept.1, 2023

Goldmeer, J. (2019). GE Vernova Report – Power to gas: hydrogen for power generation. Fuel flexible gas turbines as enablers for a low or reduced carbon energy ecosystem. GEA33861. Presented at Electrify Europe 2018. Vienna, Austria.

Green Hydrogen Coalition (2023). Conversion of intermountain power project to green hydrogen. https://www.ghcoalition.org/green-hydrogen at-scale (accessed 16 December 2023).

Lohse-Busch, H., Duoba, M., Stutenberg, K., Iliev, S., Kern, M., Richards, B., Christenson, M. and Loiselle-Lapointe, A. (2018). *Technology Assessment of a Fuel Cell Vehicle: 2017 Toyota Mirari*. ANL/ESD-18/12. Argonne, IL: Energy Systems Division, Argonne National Laboratory, U.S. Department of Energy. https://publications.anl.gov/anlpubs/2018/06/144774.pdf.

Nicholas, T.E.G., Davis, T.P., Federici, F., Leland, J.E., Patel, B.S., Vincent, C., and Ward, S.H. (2021). Re-examining the role of nuclear fusion in a renewables-based energy mix. arXiv:2101.05727v1 [physics.soc-ph].

Theodore, L. (2008). *Air Pollution Control Equipment Calculations*. Hoboken, NJ: John Wiley & Sons.

Theodore, L. (2014). *Chemical Engineering: The Essential Reference*. New York: McGraw-Hill.

Theodore, L., Ricci, F., and VanVliet, T. (2009). *Thermodynamics for the Practicing Engineer*. Hoboken, NJ: John Wiley & Sons.

U.S. Department of Energy (2011). *Hydrogen and Fuel Cells Program Record 11014. Medium-scale CHP Fuel Cell Systems Targets*. Washington, DC: Office of Energy Efficiency and Renewable Energy. Hydrogen and Fuel Cell Technologies Office. https://www.hydrogen.energy.gov/docs/hydrogenprogramlibraries/pdfs/11014_medium_scale_chp_target.pdf.

U.S. Department of Energy (2020). *Hydrogen Program Plan*. DOE/EE-2128. Washington, DC: U.S. Department of Energy. https://www.hydrogen.energy.gov/docs/hydrogenprogramlibraries/pdfs/hydrogen-program-plan-2020.pdf?Status=Master.

U.S. Depertment of Energy (2022). NREL report: "Solar Photovoltaic System and Energy Storage Cost Benchmarks, With Minimum Sustainable Price Analysis: Q1 2022." https://doi.org/10.2172/1891204.

University of Michigan Center for Sustainable Systems (2024). Hydrogen factsheet.

World Nuclear Association (2021). Hydrogen production and uses. https://world-nuclear.org/information-library/energy-and-the-environment/hydrogen-production-and-uses.aspx.

15

Hydrogen Uses

15.1 Introduction

Hydrogen has many end-uses, including utilities, transportation, industrial and manufacturing, heating, and as a medium for storing energy. While traditional uses for hydrogen involved industrial applications, other hydrogen applications are increasingly being explored across different forms in the utility (power) industry and in the transportation sector.

Today, hydrogen can be used to power vehicles, run turbines or fuel cells to produce electricity, and/or generate heat and electricity for buildings. Hydrogen is used as a chemical in the petrochemical, electronics, and food industries. Several industrial and manufacturing processes typically require large volumes of hydrogen, including oil refining, ammonia production, treating metals, producing fertilizer, processing foods, and to lower the sulfur content of some fuels. The National Aeronautics and Space Administration (NASA) began using liquid hydrogen in the 1950s as a rocket fuel, and NASA was one of the first to use hydrogen fuel cells to power the electrical systems on spacecraft. Interest in using hydrogen as a power plant fuel continues to grow. Hydrogen is also considered an alternative vehicle fuel under the Energy Policy Act of 1992. The interest in hydrogen as an alternative transportation fuel stems from its ability to power fuel cells in zero-emission vehicles, its potential for domestic production, and the fuel cell's potential for high efficiency. It has been reported that a fuel cell may be two to three times more efficient than an internal combustion engine running on gasoline. Hydrogen can also fuel internal combustion engines, but burning hydrogen can result in nitrogen oxide emissions and is less efficient than use in fuel cells.

This chapter is concerned with hydrogen uses and applications. Here is what Webster has to say about both terms:

1) *Use.* "... to put into action or service ... to carry out a purpose or action by means of ... the act or practice of employing something . . . a method or manner of employing or applying something ... (Merriam-Webster 2023a)"
2) *Application.* ". . . an act of putting something to use . . . a use to which something is put ... (Merriam-Webster 2023b)"

Hydrogen applications include power generation (utilities), transportation, industrial processes, etc. However, hydrogen has the potential for use in diverse applications across multiple sectors, where it can provide substantial environmental and economic benefits, as well as improved energy

Hydrogen Energy: Principles and Applications, First Edition. Vincent J. DelGatto, Louis Theodore, R. Ryan Dupont, and Matthew C. Ogwu.

security and resiliency. Large amounts of hydrogen can be used in the power generation, transportation, industrial, and manufacturing sectors. Integrated energy systems, which can span sectors, offer additional opportunities by using hydrogen as an energy carrier to improve the economics of existing and emerging electric power generation systems.

A brief introduction to five of the key sectors is presented below:

1) *Power generation*. This sector includes electricity-generating facilities and power plants. Hydrogen application in the power sector is limited today, but there is potential for increasing its role in the future. Hydrogen-powered gas turbines could be a source of flexibility in the electricity system. Also, Solid Oxide Fuel Cells (SOFCs) can offer the highest conversion efficiency.
2) *Transportation*. This sector includes cars, trucks, trains, airplanes, and ships. Hydrogen applications in the transportation sector are possible with the use of internal combustion engines (ICE) or turbines, and with the use of fuel cells. While ICE technology is more mature, fuel cell technology has received faster market pull thanks to a higher efficiency of energy conversion.
3) *Industry*. This sector includes manufacturing, construction, mining, farming, fishing, and forestry. At present, hydrogen is used in large-scale industrial processes like oil refining, petrochemicals, ammonia production, methanol production, and steel production. In oil *refineries*, hydrogen is used to remove sulfur from crude oil and produce lighter oil products from heavy oil. (But electrification of transport means that this market will shrink.) Hydrogen with a source of carbon will probably replace natural gas and to a smaller degree oil as the basis for production of synthetic gases.
4) *Heating*. This application includes residential and commercial buildings. Commercial buildings include office buildings, hospitals, stores, restaurants, and schools. Residential and commercial energy use are often lumped together because homes and businesses use energy in the same way – for heating, air conditioning, water heating, lighting, and operating appliances. Hydrogen can play an important role in decarbonization of buildings. It can be applied in buildings for power and heating appliances. The potential of hydrogen in this sector is currently a debatable topic. Technologies such as heat pumps are efficient and economical options. While the electrification of the building sector on the back of renewable energy sources is a growing trend, the question is whether a 100% electrified system is feasible given the intermittence of most renewable energy sources in combination with large-scale electric heating and electric mobility.
5) *Medium for Storing Energy*. Hydrogen may become a long-term option to balance seasonal variation in electricity demand.

Section numbers and their accompanying titles for this chapter concerned with hydrogen uses and applications are listed below.

15.2 Power Generation

Today? Hydrogen plays a negligible role in the power sector today; it accounts for less than 1% of electricity generation. This is linked mostly to the use of gases from the steel industry, petrochemical plants, and refineries. But there is potential for this to change in the future. Co-firing of ammonia could reduce the carbon intensity of existing conventional coal power plants, and hydrogen-fired gas turbines and combined-cycle gas turbines could be a source of flexibility in electricity systems with increasing shares of variable renewables (IEA 2019).

Tomorrow? Hydrogen can be used in a broad range of stationary power-generation applications, including large-scale power generation, distributed power, combined heat and power (CHP), and backup power. As noted in a previous chapter, hydrogen can provide power through electrochemical conversion using fuel cells or through combustion of hydrogen using turbines in simple- or combined-cycle generation.

Fuel cells can efficiently convert hydrogen into power with low emissions, and the inherent modularity of fuel cell systems makes them ideally suited for a broad range of stationary-power applications ranging from less than a kilowatt up to the multi-megawatt scale. These fuel cells are commercially deployed worldwide, providing primary and backup power for industrial facilities, businesses, homes, telecommunications towers, data centers, and more (U.S. DOE 2020).

Combustion in simple- or combined-cycle power generation is also a viable approach for using hydrogen or hydrogen-rich blends (e.g., blended with natural gas) in a number of stationary applications. Combustion turbines can generate electric power while also providing heat for residential, commercial, and industrial applications. While combustion blends, including renewable hydrogen, offer the benefits of reduced carbon dioxide emissions, their use in existing distribution infrastructure and combustion equipment poses a number of challenges related to material compatibility and combustion characteristics. Progress has been made in the modification of natural gas burners in commercially available combustion turbines (U.S. DOE 2020).

Hydrogen also offers a number of opportunities to provide value to the electric power sector through its integration into hybrid energy systems. Broadly defined, a hybrid energy system (HES) combines electricity generation, energy storage, and/or energy conversion technologies that are integrated together through an overarching control framework to achieve enhanced capabilities, value, and/or cost savings compared to the standalone alternatives. Hydrogen technologies integrated in an HES offer unique benefits in both on-grid and off-grid electric-power applications. Examples include mid- to long-duration/seasonal energy storage (Hunter et al. 2020); grid leveling and stabilization services that leverage the fast-acting dynamic response of electrolyzers (Kurtz et al. 2017); and the ability to co-produce (along with electricity) hydrogen or other hydrogen-based fuels, chemicals, or products for use in diverse markets, potentially at higher value than electricity. Multiple hybrid scenarios are being explored that could optimally leverage the benefits of integrated energy systems. Three such examples are described below.

1) *Grid integration and renewable hybrid systems.* As the electric grid evolves with higher penetrations of variable renewable energy sources, grid-integrated electrolyzers can provide energy storage and other grid services to improve reliability and resiliency. Hydrogen produced via electrolysis can be used as a means of bi-directional energy storage, where it is converted on-demand back to electricity through low-emissions power generation technologies such as fuel cells or combustion turbines. It can also be used in one-way chemical energy applications, for example, through injection into the natural gas pipeline infrastructure or through onsite co-production of value-added commodities such as ammonia or methanol (U.S. DOE 2020).

2) *Fossil-energy hybrid systems.* Hybrid energy systems integrating natural gas or coal conversion with hydrogen technologies can provide significant value for industrial applications. Pilot-scale plants have been deployed that integrate systems for steam methane reforming of natural gas with vacuum-swing adsorption to co-produce hydrogen for petroleum refining along with concentrated carbon dioxide for use in enhanced oil recovery. Large-scale gasification facilities that co-fire coal, biomass, and waste plastics can be integrated with thermal storage, hydrogen production and utilization technologies, and carbon capture to achieve low-emissions power generation.
3) *Nuclear hybrid systems.* There is growing interest in integrating hydrogen production at nuclear power plants as a means to enhance load-following capabilities, utilize unused energy, and provide an additional revenue stream. For context, a 1-GW nuclear power plant can produce about 41,000 tons of hydrogen in a year, assuming the plant produces electricity for 70% of the time and hydrogen for 26% of the time (with 4% downtime for maintenance). Leveraging nuclear-generated electricity and heat, hydrogen can be efficiently produced using low- or high-temperature electrolyzer technologies, and utilized onsite to service hydrogen needs of the nuclear plant (e.g., in turbine generator cooling), or exported/monetized for other end-uses. While nuclear hybrid systems are not yet ready for full-scale commercial deployment, pilot plant projects currently underway at existing nuclear plants are expected to rapidly resolve many remaining uncertainties, accelerating their availability for full-scale implementation (U.S. DOE 2020).

Although pure hydrogen is not currently being employed as a fuel, hydrogen and hydrogen-based fuels such as ammonia and synthetic natural gas can be fuels for power generation in the future. Ammonia can be co-fired in coal-fired power plants to reduce coal usage and reduce the carbon footprint of these plants. Hydrogen and ammonia can also be used as fuels in gas turbines, combined-cycle gas turbines, or fuel cells, thus providing a flexible and potentially low-carbon generation option.

Using ammonia as fuel raises concerns about an increase in NO_x emissions, but earlier work (IEA 2019) demonstrated an ability to keep emissions within the usual limits and to avoid any ammonia slip into exhaust gas. Higher blending shares of up to 20% ammonia in energy terms might be feasible with only minor adjustments to a coal power plant. In smaller furnaces with a capacity of 10 MW thermal, blending shares of 20% ammonia have been achieved without problems, and in particular without any slippage of ammonia into exhaust gas (IEA 2019).

Hydrogen can also be used as a fuel in gas turbines. Most existing gas turbine designs can already handle a hydrogen share of 3–5%, and some can handle shares of 30% or higher. The industry is confident that it will be able to provide standard turbines that are able to run entirely on hydrogen by 2030 (IEA 2019). Ammonia is another potential fuel for gas turbines. The direct use of ammonia has been successfully demonstrated in micro gas turbines with a power capacity of up to 300 kW (IEA 2019).

15.3 Transportation

Transportation accounts for nearly 40% of U.S. carbon dioxide emissions and can be a key contributor to localized air pollution. The combustion of hydrogen and the use of fuel cells are an important part of a portfolio of options to reduce transportation-related emissions because they can be used in specific applications that are hard-to-decarbonize, such as long-haul heavy-duty trucks. Additional examples include other medium- and heavy-duty vehicles that require longer driving ranges, involve heavy loads, or demand faster refueling times than may be available with battery electric vehicles alone. With an increased urgency to reduce emissions and energy-related expenses, significant opportunities exist, as the medium- and heavy-duty sector accounts for 25% of annual vehicle fuel use even though it is only 4% of the vehicle fleet (ORNL 2017).

In addition to its use in fuel cells, hydrogen can also be combined with carbon dioxide to produce synthetic fuels, offering even more ways to meet the needs of various transportation applications. These synthetic fuels could allow certain applications or regions to continue using ICE and the vast existing liquid-fuel infrastructure for hard-to-decarbonize end-uses such as long-distance commercial aircraft. Similar to these pathways that utilize hydrogen and carbon dioxide, hydrogen and nitrogen can be used to produce ammonia for use in multiple applications (U.S. DOE 2020).

There is no question that there is a heightened interest in the use of hydrogen in the road transportation sector, and in particular in the heavy-duty vehicle market segment. Hydrogen fuel cell electric vehicles (FCEVs), which are considered zero-emission vehicles, are attractive replacements for internal combustion engine vehicles because they can offer performance similar to that of conventional vehicles, along with several additional advantages. These advantages include improved environmental performance, quiet operation, rapid acceleration from a standstill, and lower maintenance requirements. Furthermore, these FCEVs can potentially perform functions for which conventional vehicles are poorly suited, such as providing remote electrical power and acting as distributed electricity generators when parked and connected to a fuel supply (K&L Gates 2020).

Light-duty FCEVs are now available in limited quantities to the consumer market. The market is also developing for fleet vehicles, material handling equipment, ground support equipment, medium- and heavy-duty vehicles, and stationary applications. Thus, hydrogen FCEVs offer a low-carbon alternative to fossil fuels and are the only vehicles alongside battery electric vehicles (BEVs) that offer no exhaust emissions. BEVs may be more competitive in light-duty and lower utilization applications, while FCEVs will have an advantage in heavier-duty and higher utilization applications. Any road transport can use hydrogen either by a fuel cell or by hydrogen-based fuels.

For cars, the priority is to bring down the cost of fuel cells and onboard hydrogen storage. This could make them competitive with BEVs at driving ranges of 400–500 km and make them potentially attractive for consumers who prioritize range. For trucks, the priority is to reduce the delivered price of hydrogen. In the early stages of deployment, building hydrogen stations that serve captive fleets on hub-and-spoke missions could help to secure high refueling station utilization and thus could be a way to get infrastructure construction off the ground (IEA 2019).

Together with BEVs, FCEVs are the only vehicles with no exhaust emissions and thus offer the potential to drastically reduce local air pollution, especially in cities. They can also dramatically reduce CO_2 emissions when low-carbon hydrogen is used. The driving range and pattern of refueling for FCEVs are similar to internal combustion engine vehicles. Furthermore, hydrogen has some attractive attributes compared to biofuels, as it does not generally face resource constraints or competition for land use. FCEVs have nevertheless been slow to take off. Technical challenges and high prices have delayed their market introduction. The theoretical potential for future use of hydrogen in road transport is very large. Any road transport mode can technically be powered using hydrogen, either directly using fuel cells or via hydrogen-based fuels in ICE. It should also be noted that the installation of hydrogen refueling infrastructure, while relatively limited to date, continues to pick up momentum.

Regarding rail traffic, rail is the mainstay of transport in many countries. Electrified rail is currently used on less than 1% of U.S. railroad tracks, compared to more than one-third of the electric energy that powers trains globally. Because rail is already among the lowest greenhouse gas emitting modes of transportation, the massive overhaul required to electrify railroad systems may provide only incremental benefits and not justify the costs. On the other hand, hydrogen-powered trains could play a role in decarbonizing rail systems without incurring the high cost of electrifying tracks. Hydrogen-powered trains are less expensive, do not require massive track overhauls, and can be created by retrofitting existing diesel trains (K&L Gates 2020).

Hydrogen-fueled trains pose a host of benefits and disadvantages. Like electric trains, they are much quieter than their diesel counterparts. Hydrogen-powered trains have additional benefits, such as the ability to switch to fuel cells when electricity lines are down. Although the percentage share of electrified tracks is still expanding in most countries, further electrification of rail networks is likely to come up against diminishing returns on investment, since highly utilized lines are the first to be electrified. Plans involving hydrogen trains already exist in a number of countries, with several companies working to supply them. Under optimistic assumptions about fuel cell cost reductions, hydrogen trains could become competitive against other passenger service options.

Oil fuel – as the source of energy – currently dominates the shipping sector, and the use of hydrogen-based fuels in shipping is accordingly very limited. Ships do not currently use ammonia as fuel. Satisfactory combustion of ammonia in existing engines would generally require ignition promotors (to overcome its lower ignition energy) and engine modifications. Hydrogen use is currently limited to demonstration projects for small ships and onboard power supply in larger vessels. Air pollution regulations and greenhouse gas targets could promote hydrogen-based fuels for shipping in the near future. Ammonia and hydrogen have the potential to address environmental targets in shipping, but their cost of production is high relative to oil-based fuels.

Aviation applications are currently limited to small demonstration projects and feasibility studies. However, aviation is the fastest-growing passenger transport mode. Efficiency improvements should reduce energy consumption and slow the increase in energy demand, but alternative fuels will eventually be needed to avoid increases in emissions from the sector. Advanced biofuels and hydrogen-based fuels are leading options. These hydrogen-based liquid fuels provide a potentially attractive option for aviation at the expense of higher energy consumption and potentially higher costs. Government support in the form of low-carbon targets or other approaches will likely be critical to their prospects for wider use in the aviation sector.

15.4 Industry Feedstock

Today, the primary demand for hydrogen is as a chemical feedstock in petroleum refining and ammonia production, with smaller amounts being used in other industrial applications such as methanol production. There is also interest in using hydrogen for the production of synthetic fuels. Approximately 10 million metric tons (MMT) of hydrogen are currently produced in the U.S. each year for these end-uses, mostly from natural gas. In the future, hydrogen's versatility as both a chemical feedstock and an energy carrier will be harnessed to serve expanded end-uses. Emerging market opportunities include the use of hydrogen for multiple transportation applications (e.g., in FCEVs – particularly heavy duty applications – as a feedstock for synthetic fuels and to upgrade petroleum and biofuels); as a feedstock for industry (e.g., in steel and cement manufacturing); for power generation (for large-scale power, off-grid distributed power, and back-up or emergency power); for heat in industrial systems and buildings; and energy storage. Hybrid energy systems, which integrate energy generation, storage, and/or conversion technologies to optimize the overall value of the energy generated, are another promising market opportunity. For example, the integration of hydrogen production technologies with utility-scale power generation plants is a concept receiving increased interest, due to its potential to improve profitability of these plants while supporting grid resiliency. A relevant example is a wind farm or nuclear power plant that produces hydrogen from surplus electricity (via electrolysis) and then either sells or uses the hydrogen for other purposes or reconverts it to electricity (via a fuel cell or turbine) at times of higher demand.

The ultimate goal is for hydrogen to be affordably produced and delivered utilizing several feedstocks, processing methods, and delivery options at a variety of scales ranging from large central production to small local production, depending on what is most practical from an economic and logistical perspective for a given location and level of market demand.

Steelmaking, in particular, is receiving increasing attention as a source of demand for hydrogen. Steel is the most commonly used metal product worldwide, and the conventional way to produce it involves using coal in blast furnaces to reduce iron ore to iron. Between 7% and 9% of global greenhouse gas emissions are due to steel manufacturing (World Steel Association 2020), and by using hydrogen as the primary reducing agent, those emissions can be dramatically reduced. A number of demonstrations of the use of hydrogen in steelmaking are currently underway, including operational facilities in Austria and Sweden.

Additional emerging industrial and chemical uses of hydrogen include cement production, which is an energy-intensive process responsible for about 8% of global carbon dioxide emissions (Lehne and Preston 2018), where the use of hydrogen in place of coal could reduce both carbon dioxide and NOx emissions; and synfuel (or "e-fuel") production, which involves reacting carbon dioxide with clean hydrogen, offering an option for versatile net-zero-carbon fuels.

One can conclude from the above analysis that most hydrogen today is used in three industrial sectors: oil refining, chemicals, and iron and steel. Production of hydrogen to meet the needs of these sectors is at a commercial scale and is almost entirely from natural gas, coal, and oil today, with associated environmental impacts. However, technologies are currently available to reduce emissions in these sectors, as discussed below.

1) *Oil refining*. Oil refining – turning crude oil into various end-user products such as transport fuels and petrochemical feedstock – is one of the largest users of hydrogen today. Much is consumed by refineries as feedstock, reagent, and as an energy source. Approximately two-thirds of this hydrogen is produced in dedicated facilities at refineries or acquired from merchant suppliers. *Hydrotreatment* and *hydrocracking* are the main hydrogen-consuming processes in the refinery. Hydrotreatment is used to remove impurities, especially sulfur (it is often simply referred to as desulfurization), and accounts for a large share of refinery hydrogen use globally. Today refineries remove around 70% of naturally incurring sulfur from crude oils. With concerns about air quality increasing, there is growing regulatory pressure to further lower the sulfur content in final products. Hydrocracking is a process that uses hydrogen to upgrade heavy residual oils into higher-value oil products. Demand for light and middle distillate products is growing and demand for heavy residual oil is declining, leading to an increase in the use of hydrocracking. In addition to hydrotreatment and hydrocracking, some hydrogen that is used or produced by refineries cannot be economically recovered and is burned as fuel as part of a mixture of waste gases. Hydrogen is also used for upgrading oil sands and hydrotreating biofuels. For oil sands, the amount of hydrogen needed to remove sulfur from the raw bitumen varies considerably depending on the upgrading technology and the quality of the synthetic crude oil produced. For biofuels, hydrotreatment removes oxygen and improves the fuel quality of vegetable oils and animal fats processed into diesel substitutes. In recent decades, refinery hydrogen demand has grown substantially as a result of growing refining activity and rising

requirements for hydrotreating and hydrocracking. This trend is set to continue as fuel specifications globally further reduce acceptable levels of sulfur content.

2) *The chemical sector.* Hydrogen is part of the molecular structure of almost all industrial chemicals, but only some primary chemicals require large quantities of dedicated hydrogen production for use as feedstock, notably ammonia and methanol, the second- and third-largest sources of demand for hydrogen today, respectively. Lesser hydrogen demand is associated with the production of ethylene, propylene, benzene, toluene, and mixed xylenes. It is also a large producer of by-product hydrogen, which is both consumed within the sector itself and distributed for use elsewhere. The vast majority of the hydrogen that the chemical sector consumes is produced using fossil fuels, and this generates considerable quantities of greenhouse gas emissions. Reducing the level of emissions represents an important challenge for the sustainability of the sector's energy use, and a significant opportunity to make use of low-carbon hydrogen (IEA 2019). The chemical sector produces a complex array of outputs, from plastics and fertilizers to solvents and explosives. The seven primary chemicals listed above account for around two-thirds of the chemical sector's energy consumption, and the vast majority of its demand for energy products as raw material inputs. Fossil fuels have long been a convenient and cost-effective source of both hydrogen and carbon for ammonia and methanol production. Ammonia is mostly used in the manufacture of fertilizers such as urea and ammonium nitrate (around 80%). The remainder is used for industrial applications such as explosives, synthetic fibers, and other specialty materials, which are an increasingly important source of demand, Methanol is used for a diverse range of industrial applications, including the manufacture of formaldehyde, methyl methacrylate, and various solvents. Methanol is also used in the production of several other industrial chemicals, and for the methanal-to-gasoline process that produces gasoline from both natural gas and coal, which has proven attractive in regions with abundant coal or gas reserves but with little or no domestic oil production. The development of methanol-to-olefins and methanol-to-aromatics technology has opened up an indirect route from methanol to high-value chemicals (HVCs), and thus to plastics. Unlike ammonia and methanol, HVCs – the precursors of most plastics – are produced mostly from oil products such as ethane, liquefied petroleum gas, and naphtha. HVCs produced directly from oil products do not require hydrogen feedstock, but their production generates by-product hydrogen that can be used in oil refining and other chemical sector operations, such as the upgrading of other cracker by-products (IEA 2019).

3) *The iron and steel sector.* Like the chemical sector, the iron and steel sector produces a large quantity of hydrogen mixed with other gases as a by-product (e.g., coke oven gas), some of which is consumed within the sector and some of which is distributed for use elsewhere. Virtually all of this hydrogen is generated from coal and other fossil fuels. To reduce emissions, efforts are underway to test steel production using hydrogen as the key reduction agent (as opposed to carbon monoxide derived from fossil fuels). In the meantime, low-carbon hydrogen could be blended into existing processes that are currently based on natural gas and coal to lower their overall CO_2 intensity (IEA 2019). More than three-quarters of global steel demand today is met using primary production methods that convert iron ore to steel, as opposed to the secondary production route, which utilizes limited supplies of recycled scrap steel. The two main primary production routes already involve some production and consumption of hydrogen.

4) The *blast furnace-basic oxygen furnace* (BF-BOF) route accounts for about 90% of primary steel production globally. It produces hydrogen as a by-product of coal use. This hydrogen, contained in so-called "works-arising gases" (WAG), is produced in a mixture with other gases such as carbon monoxide. WAG is used for various purposes on site, but also transferred for use in other sectors including power generation and methanol production.

15.5 Hydrogen-Containing Feedstock Chemicals

Hydrogen is an important gaseous raw material for the petroleum, chemical, and manufacturing industries. It reacts with many compounds. It combines with nitrogen in the presence of a catalyst to form ammonia, with sulfur to form hydrogen sulfide, with chlorine to form hydrogen chloride, and with oxygen to form water. Hydrogen also combines with some metals, such as sodium and lithium, to form hydrides. It acts as a reducing agent on metallic oxides, such as copper oxide, removing the oxygen and leaving the metal in a free state. Finally, hydrogen reacts with unsaturated organic compounds to form corresponding saturated ones. This section has been included to provide the reader with information regarding four hydrogen-containing feedstock chemicals: ammonia, hydrogen chloride, hydrogen peroxide, along with hydrocarbons.

1) *Ammonia.* Ammonia is a colorless, pungent gas, NH_3, highly soluble in water. A saturated aqueous (water) solution of ammonia contains 45% ammonia by weight at 0°C (32°F) and 30% at ordinary room temperatures. On solution in water, ammonia becomes ammonium hydroxide, NH_4OH, which is strongly basic and similar in chemical behavior to the hydroxides of the alkali metals. Ammonia melts at −77.7°C (−107.9°F), boils at −33.35°C (−28.03°F), and has a specific gravity of 0.68 at its boiling point and 1 atm of pressure. Ammonia is one of the truly fundamental raw materials utilized in modern civilization. The total inorganic chemical tonnage in the country is approximately 12 million tons per year. The largest consumer of anhydrous ammonia is the fertilizer industry, which utilizes it in the production of calcium and sodium nitrate; ammonium sulfate, nitrate, and phosphate; ammoniated superphosphates; urea; and aqueous ammonia. Ammonia synthesis gas is prepared by high-pressure catalytic reforming of hydrocarbon feed gas (e.g., natural gas) in a primary reformer with superheated steam, and in a secondary reformer with air to furnish nitrogen. The amount of air is adjusted eventually to supply a H_2/N_2 molar ratio of 3:1. CO is converted by catalytic iron oxide shift conversion with H_2O (from the reformer plus steam) to CO_2. The CO_2 is removed by a reactivated hot K_2CO_3 solution wash (with reboiler regeneration yielding CO_2. This is followed by methanation.
2) *Hydrogen chloride.* Hydrogen chloride is a colorless, nonflammable gas, formula HCl, with a characteristic penetrating, suffocating odor. It melts at −114.2°C (−173.6°F), boils at −85.1°C (−121.2°F). Hydrogen chloride dissolves readily in water. One volume of water at 20°C (68°F) absorbs 142 volumes of hydrogen chloride gas at atmospheric pressure. The resulting solution – hydrochloric acid – contains 40.3% hydrogen chloride by mass and has a specific gravity of 1.20. This solution fumes strongly in moist air, but dilution stops the fuming. Hydrogen chloride becomes less soluble in water as the water temperature rises, and it is less soluble in alcohol, ether, and other organic liquids. In solution in water, the molecules of hydrogen chloride ionize, becoming positively charged hydrogen ions and negatively charged chloride ions. Because it ionizes easily, hydrochloric acid is a good conductor of electricity. The hydrogen ions give hydrochloric acid its acidic properties, so that all solutions of hydrogen chloride and water have a sour taste; corrode active metals, forming metal chlorides and hydrogen; turn litmus red; neutralize alkalies; and react with salts of weak acids, forming chlorides and weak acids. Hydrogen chloride is produced industrially as a by-product of the reaction of chlorine with hydrocarbons to produce organic chlorides. Hydrochloric acid may be made by the reaction of sodium chloride with sulfuric acid or by combining hydrogen and chloride. Crude industrial hydrochloric acid is called muriatic acid. It is used in large quantities in the preparation of chlorides, for cleaning metals, and in industrial processes such as the preparation of corn syrup and

glucose from cornstarch. Small amounts of hydrochloric acid are secreted by cells in the lining of the stomach to aid in food digestion. Hydrogen chloride is typical of the hydrogen compounds formed by all the elements of the halogen family, which includes fluorine, chlorine, bromine, and iodine.

3) *Hydrogen peroxide*. Hydrogen peroxide is a chemical compound of hydrogen and oxygen with the formula H_2O_2. Pure, anhydrous hydrogen peroxide is a colorless, syrupy liquid with a specific gravity of 1.44. It blisters the skin and has a metallic taste. The liquid solidifies at −0.41°C (31.3°F). Concentrated solutions are unstable, and the pure liquid may explode violently if heated to a temperature above 100°C (212°F). It is soluble in water in all proportions, and the usual commercial forms are a 3 and a 30% aqueous solution. To retard the decomposition of the peroxide into water and oxygen, organic substances, such as acetanilide, are added to the solutions, and they are kept in dark bottles at low temperature. Hydrogen peroxide is manufactured in large amounts by the electrolysis of aqueous solutions of sulfuric acid, potassium bisulfate, or ammonium bisulfate. It is also prepared by the action of acid on other peroxides, such as those of sodium and barium.
4) *Hydrocarbons*. Hydrogenation is a chemical reaction involving the combination of hydrogen with saturated organic compounds, i.e., hydrocarbons. When an unsaturated compound such as ethylene (C_2H_4), is treated with hydrogen at a suitable temperature and in the presence of a catalyst, such as finely divided nickel, platinum, or palladium, the double bond between the carbon atoms is broken and a hydrogen atom attaches itself to each carbon atom; in the case of ethylene, ethane (C_2H_6) is formed. This reaction also takes place with more complicated molecules, yielding a great variety of synthetic products that are of importance in the laboratory and for industrial purposes. The hydrogenation reaction is applied on a large industrial scale to a number of processes, the most important of which in the United States is the hydrogenation of vegetable oils to produce many edible fats, such as margarine. Oils are esters of fatty acids containing one or more double-bond linkages between carbon atoms; solid fats are saturated compounds. To convert the oils, which often have an unpleasant taste and odor, to fats with odor and taste sufficiently innocuous so that they can be used for cooking purposes, the oils are hydrogenated. The reaction is performed at a temperature of about 200°C (392°F) in the presence of finely divided nickel, and under a pressure of 3–4 atmospheres of hydrogen. Large volumes of cheap and plentiful vegetable oils, such as soybean and cottonseed oil, are hydrogenated sufficiently to yield creamy solids that resemble lard at room temperature. Such products are widely used in foods. Low-grade oils, such as fish oils, are hydrogenated and used in the manufacture of soap and wax. The hydrogenation process is also applied to the production of synthetic gasoline.

15.6 Heating

Hydrogen can be readily used for space heating and, because it burns without noxious exhaust products, can be used in unvented appliances without safety and/or environmental concerns. Heating systems still rely significantly on fossil-based fuels. The IEA estimates that nearly 28% of global energy-related carbon dioxide emissions result from energy use in buildings. Localized emissions are produced from residential heating systems and can degrade indoor air quality in homes, which poses a danger to human health. These greenhouse emissions can be eliminated by replacing fossil fuel energy sources with clean electricity. The electrons from the dissociation of hydrogen in a hydrogen fuel cell are clean, which can be employed to drive electric heaters in residential areas. Powering home heating systems with hydrogen-generated electricity can effectively mitigate localized emissions and ensure the preservation of indoor air quality. Hydrogen-generated electricity for heating can reduce the risk of lung cancer, asthma, and other respiratory sicknesses resulting from conventional energy generation gas emissions.

Hydrogen could reduce these heating-related emissions if green hydrogen is blended with natural gas to reduce the carbon intensity of the feedstock. Trials have used blends of up to 20% hydrogen, but the production costs of low-carbon hydrogen for residential heating are very high (Cassarino and Barrett 2022), and although decreasing, are likely to be an initial barrier to wider adoption of this low-carbon option. Overall, however, the decreasing costs of hydrogen and related technologies will be the most important factors in the widespread adoption of hydrogen for heating uses (K&L Gates 2020).

Most process plants require process heating. Many plants utilize industrial boilers, some with fairly large capacities to generate steam for space heating and process heating needs. In addition, plants may include standby boilers for emergency use as well as emergency generators for emergency electrical power. Some plants may also utilize cogeneration facilities for power and heating needs. These boilers and process heaters represent an opportunity to combust hydrogen while simultaneously recovering its heating value. Environmental applications include combusting a pollutant-laden air stream as a source of combustion air. A fairly high-fuel-value vent gas stream is required for boilers or process heaters to function without added fuel (Santoleri, Reynolds, and Theodore 2000; Theodore 2012).

The building sector accounts for 30% of global energy use, nearly three-quarters of which is used for space heating, hot water production, and cooking. Including the traditional use of solid biomass in developing countries, nearly half of this was produced directly from fossil fuels. Most of the rest came from conventional electric equipment (e.g., electric resistance radiators and cookstoves) and commercial heat (e.g., district heating). Replacing this heat demand with low-carbon alternatives and reducing heat demand through improving buildings is challenging. Decision-making for energy use in buildings is both complex and site-specific, and depends on building type, location, ownership, customer preferences, equipment costs, energy prices, and overall convenience, among other factors. This host of variables means that various energy sources and technologies are likely to co-exist in the future, from natural gas boilers to electric heat pumps, district heating, and solar thermal heating. Hydrogen has the potential to contribute to this energy transition (e.g., through blending or methane production). In doing so, it can make use of existing building and energy network infrastructure to provide both flexibility and continuity. However, hydrogen is only occasionally used as a source of energy in the global buildings sector today, although various potential uses are now being considered. There are numerous demonstration projects examining hydrogen blending in the gas grid. In the final analysis, hydrogen will not make sense for all building applications, and numerous factors will influence eventual hydrogen demand in buildings, including existing natural gas infrastructure, heat densities, other building energy needs, and safety considerations. There are also barriers related to cost and consumer acceptance, and a variety of policy design challenges that will need to be overcome to expand hydrogen's use in this application.

15.7 Energy Storage

This abbreviated section discusses hydrogen energy storage options. Once hydrogen is produced, for example, through electrolysis, it can be stored as a compressed gas, cryogenic liquid, or a wide variety of loosely bonded hydride compounds for later use. Unlike batteries, which suffer from storage degradation and can store a limited amount of energy, hydrogen fuel can be stored for long periods of time and in quantities only limited by the size of the storage facility. Hydrogen compares well to other long-duration storage technologies. Hydrogen definitely offers the potential to provide energy-storage solutions for off-grid electricity systems and to balance electric grids (K&L Gates 2020).

As noted, hydrogen and hydrogen-based fuels (such as methane, liquid organic hydrogen carriers [LOHCs], and ammonia produced from electricity via electrolysis) are potential options for long-term and large-scale storage of energy. Salt caverns are the best choice for the underground storage of pure hydrogen because of their tightness and low risk of contamination. Alternative underground hydrogen storage options, such as pore storage and storage in depleted oil and gas fields, are also being investigated. Converting electricity into methane via power-to-gas is a further long-term storage option, and one that could take advantage of the existing transport and storage infrastructure for natural gas. Approximately 70 power-to-gas projects to produce methane are in operation today, most of them in Europe. Storing electricity in the form of ammonia is another long-term and large-scale storage option. Large steel tanks are already commonly used in the fertilizer industry for storing ammonia. However, hydrogen-based storage options suffer from low round-trip efficiency; in the process of converting electricity through electrolysis into hydrogen and then hydrogen back into electricity, around 60% of the original electricity is lost, whereas in a lithium-ion battery the losses of a storage cycle are around 17%. Pumped-hydro storage facilities offer one alternative. They have been used for more than a century to store electricity for relatively long periods. Batteries offer another alternative, although they are unlikely to be used for long-term and large-scale storage because they suffer from self-discharge and because of the immense number of batteries that would be needed for large-scale storage (IEA 2019).

All the alternatives have advantages and disadvantages. For shorter time periods, e.g., below a few hours, hydrogen and ammonia are much more expensive than pumped-hydro storage or battery storage. With longer discharge durations, compressed hydrogen and ammonia become more attractive, benefiting from their relatively low capital costs for energy storage volumes (the investment costs to develop underground salt caverns or storage tanks). Among the different storage technologies considered here, compressed hydrogen becomes the most economical option for discharge durations beyond approximately 1 day.

References

Cassarino, T.G. and Barrett, M. (2022). Meeting UK heat demands in zero emission renewable energy systems using storage and interconnectors. *Applied Energy* 306, 118051. https://doi.org/10.1016/j.apenergy.2021.118051.

Hunter, C., Reznicek, E., Penev, M., Eichman, J., and Baldwin, S. (2020). *Energy Storage Analysis*. Golden, CO: National Renewable Energy Laboratory, U.S. DOE. https://www.hydrogen.energy.gov/pdfs/review20/sa173_hunter_2020_o.pdf.

IEA (2019). *The Future of Hydrogen: Seizing Today's Opportunities*. Paris, France: OECD. https://doi.org/10.1787/1e0514c4-en.

K&L Gates (2020). *The H2 Handbook. Legal, Regulatory, Policy, and Commercial Issues Impacting the Future of Hydrogen*. Washington, DC: K&L Gates LLC. https://marketingstorageragrs.blob.core.windows.net/webfiles/Hydrogen-Handbook-UNITEDSTATES.pdf.

Kurtz, J., Harrison, K., Hovsapian, R., and Mohanpurkar, M. (2017). *Dynamic Modeling and Validation of Electrolyzers in Real Time Grid Simulation – TV031*. Golden, CO: National Renewable Energy

Laboratory, U.S. DOE. https://www.hydrogen.energy.gov/docs/hydrogenprogramlibraries/pdfs/review18/tv031_hovsapian_2018_o.pdf.

Lehne, J. and Preston, F. (2018). Making concrete change. Innovation in low-carbon cement and concrete. Chatham House Report. London, England: Energy, Environment and Resources Department, Chatham House, The Royal Institute of International Affairs. https://www.chathamhouse.org/sites/default/files/publications/2018-06-13-making-concrete-change-cement-lehne-preston-final.pdf.

Merriam-Webster (2023a). Use. Merriam-Webster.com. https://www.merriam-webster.com/dictionary/use (accessed 16 December 2023).

Merriam-Webster (2023b). Application. Merriam-Webster.com. https://www.merriam-webster.com/dictionary/application (accessed 16 December 2023).

Oak Ridge National Laboratory (2017). *Transportation Energy Data Book: Edition 36*. ORNL/TM-2017/513. Oak Ridge, TN: Vehicle Technologies Office, Office of Energy Efficiency and Renewable Energy, U.S. Department of Energy. https://info.ornl.gov/sites/publications/Files/Pub104063.pdf.

Santoleri, J., Reynolds, J., and Theodore, L. (2000). *Introduction to Hazardous Waste Incineration*, 2nd Edition. Hoboken, NJ: John Wiley & Sons.

Theodore, L. (2012). *Chemical Reactor Analysis and Applications for the Practicing Engineer*. Hoboken, NJ: John Wiley & Sons.

U.S. Department of Energy (2020). *Hydrogen Program Plan*. DOE/EE-2128. Washington, DC: U.S. Department of Energy. https://www.hydrogen.energy.gov/docs/hydrogenprogramlibraries/pdfs/hydrogen-program-plan-2020.pdf?Status=Master.

World Steel Association (2020). *Steel's Contribution to a Low Carbon Future and Climate Resilient Societies – Worldsteel Position Paper*. Brussels, Belgium: World Steel Association. https://www.acero.org.ar/wp-/uploads/2020/02/Position_paper_climate_2020_vfinal.pdf.

16

The Quintessential Hydrogen Byproduct: Potable Water

16.1 Introduction

Quintessential? Here is Webster's definition (Merriam-Webster 2023) ". . . perfectly typical or representative of a particular kind of person or thing . . . something that is a typical part or pure example." No help here. But this chapter is about water, a byproduct of the combustion of hydrogen in either oxygen or air . . . but not just water . . . *potable water*.

There is no question that of all of the compounds on the Earth, water possesses unique properties including:

1) High dielectric constant
2) Low density
3) High melting (fusion) temperature
4) High boiling temperature
5) High heat capacity

Another characteristic of water also needs to be noted: lifeforms on Earth consist primarily of water.

Potable drinking water should not be taken for granted since it is the key to sustainable human existence. It has therefore become the most sought-after and prized iconic resource on planet Earth, especially since less than 3% of all water on Earth is directly potable. The hydrologic cycle provides many times the world's annual water needs, but fresh water supplies vary widely not only over the Earth's surface but also over time in a given region, and climate change pressures have exacerbated the sustainable water supply challenge (Theodore and Dupont 2020).

Both the technical community and responsible government officials have to come to grips with the fact that potable water is not a given, particularly in the coming years. There will come a time – perhaps much sooner than later – that there either may be no potable water or only wealthy individuals may be able to afford it. The recent (August 2021) severity of the drought gripping parts of the United States lends credence to this argument.

There are many chemical and physical threats to drinking water supplies. Chemical threats include contamination from improper chemical handling and disposal, animal wastes, pesticides, human wastes, and naturally occurring substances. Drinking water that is not properly treated or disinfected, or that travels through an improperly maintained distribution system, can also become contaminated and impose a health risk. Physical threats are probably the greatest threats; they include failing water supply infrastructure and threats posed by tampering or terrorist activity. These physical threats pose a major national security problem (Theodore and Dupont 2020).

Hydrogen Energy: Principles and Applications, First Edition. Vincent J. DelGatto, Louis Theodore, R. Ryan Dupont, and Matthew C. Ogwu.

An obvious way to increase potable water availability is to recover fresh water through traditional desalination methods from seawater or from large underground stores of brackish water. An alternative is to find other ways to use saline waters for agriculture and other purposes or develop non-desalination processes that produce fresh water. This chapter introduces the reader to the potable water problem and outlines another approach that may be employed to address this problem facing both this nation and the rest of the world.

Improving the security of the nation's drinking water and wastewater infrastructure has become a top priority. Significant actions, e.g., the Bioterrorism Act of 2003, are underway to assess and reduce vulnerabilities to potential terrorist attacks; to plan for and practice response to emergencies and incidents (Theodore and Dupont 2022); and develop new security technologies to detect and monitor contaminants and prevent security breaches involving potable water systems.

Section numbers and their accompanying titles for this chapter concerned with potable water as a byproduct of hydrogen combustion are listed below.

16.1 Introduction
16.2 Physical and Chemical Properties of Water
16.3 The Hydrologic Cycle
16.4 The Desalination Process
16.5 Traditional Seawater Desalination Processes
16.6 New Process Options for Potable Water Production
16.7 The Theodore Hydrogen Water Byproduct Process
References

16.2 Physical and Chemical Properties of Water

A water (H_2O) molecule consists of three atoms: two hydrogen atoms, each of which is bonded to a central oxygen atom. Water can exist in three states: solid, liquid, and gas. At room temperature and atmospheric pressure, water is a liquid, but below 0°C (32°F) it freezes and turns into ice. Water is present in the gaseous state above 100°C (212°F) and 1.0 atm. This is the boiling point of water, at which water will evaporate.

Water has a number of unusual physical properties that are a consequence of its hydrogen bonding among neighboring water molecules. The hydrogen bond is a weak bond that is the result of the dipolar nature of all water molecules. The hydrogen bonds between water molecules impart water with a larger heat capacity, heat of vaporization, and heat of fusion than that expected for a molecule of its size. Hydrogen bonds are also important biologically. Bonding between adjacent base pairs holds double-stranded DNA together. Many proteins also utilize hydrogen bonding to hold their three-dimensional shape and assist in enzymes binding to their substrate (Theodore 2021).

Hydrogen bonding also causes water to expand upon freezing, which results in solid water being less dense than liquid water. As a result, water collecting in the cracks of rocks will expand upon freezing, which is an important mechanism for mechanically breaking rocks apart. This mechanical weathering breaks rocks into smaller fragments, which increases their surface area. This, in turn, increases the breakdown of the rock by surface chemical reactions, a process known as chemical weathering.

Regarding chemical properties, water is referred to as the "universal solvent" because it is capable of dissolving many substances. This chemical property of water arises from the dipolar nature of water molecules. Water molecules effectively surround positively charged ions (cations) and negatively charged ions (anions), which serve to prevent them from precipitating as solids. This means that wherever water goes, either through the ground or through one's body, it carries with it various solutes such as dissolved minerals, nutrients, organics, and heavy metals.

Even pure water contains some amount of hydrogen ion (H^+) and hydroxide ion (OH^-) as a result of a chemical reaction known as the autoionization of water. The concentration of these ions changes as acids and bases are added to water; however, the product of their concentrations is always constant. The relative concentration of H^+ and OH^- is measured by the pH, which is defined as the negative logarithm of the hydrogen ion (H^+) concentration in mol/L. A decrease in pH indicates an increase in the H^+ concentration and a corresponding decrease in the OH^- concentration.

Pure water has a neutral pH of exactly 7.0. Values of pH less than this are considered acidic, while pH values greater than this are termed alkaline. The typical pH range of water is from 0 to 14, although values outside of this range can be attained under extreme conditions. The pH of water is an important chemical property, which controls the distribution of chemical species among various forms. For example, the pH is buffered at precise values in animal cells to maintain the functionality of specific enzymes and proteins. Likewise in the environment, the pH controls the distribution of chemicals among their various forms and also controls the rate at which many chemical reactions occur. It is therefore often necessary to precisely measure the pH of water to understand its water chemistry (Theodore, Ricci, and VanVliet 2009; Theodore and Ricci 2010; Theodore 2014).

16.3 The Hydrologic Cycle

Water is the original renewable resource. Although the total amount of water on the surface of the Earth remains fairly constant over time, individual water molecules carry with them a rich history. The water molecules contained in the fruit one ate yesterday may have fallen as rain last year in a distant place or could have been used decades, centuries, or even millennia ago by one's ancestors.

Water is always in motion, and the hydrologic cycle describes this movement from place to place. The vast majority (96.5%) of water on the surface of the Earth is contained in the oceans. Solar energy heats the water at the ocean surface, and some of it evaporates to form water vapor. Air currents move the vapor up into the atmosphere along with water transpired from plants and evaporated from the soil and water bodies. The cooler temperatures in the atmosphere cause the vapor to condense into clouds. Clouds move around the world until the moisture capacity of the clouds is exceeded and the water falls as precipitation. Most precipitation in warm climates falls back into the oceans or onto land, where the water flows over the ground as surface runoff. Runoff can enter rivers and streams, which transport the water to the oceans, accumulate and be stored as freshwater in lakes, or soak into the ground as infiltration. Some of this water may infiltrate deep into the ground and replenish aquifers, which store huge amounts of freshwater for long periods of time. In cold climates, precipitation falls as snow and can accumulate as ice caps and glaciers, which can store water for thousands of years (Theodore 2021).

Throughout this cycle, water picks up contaminants originating from both naturally occurring and anthropogenic sources. Depending upon the type and amount of contaminant present, water present in rivers, lakes, and streams or beneath the ground may become unsafe for use.

16.4 The Desalination Process

The overall desalination process is presented in Figure 16.1. Feedwater (1) is introduced to the desalination unit (2) where, following the application of some form of energy (3) it is separated into (relatively) pure water (4) and a more concentrated brine solution (5).

The various methods (to be discussed in the next section) for separating pure water and the aforementioned dissolved salts from seawater have two things in common:

1) They all involve the same theoretical (minimum possible) work energy – approximately 3.67 kWh is required to separate 1,000 gal of freshwater from seawater at 25°C – leaving a concentrated brine solution containing the same 1,000 gal of water (50% recovery).
2) They all involve some sort of physical separation process.

The above two considerations and the accompanying mode of energy delivery serve to differentiate among the various desalination processes, i.e., the type of energy employed and its accompanying mode of delivery to the solution plus the subsequent separation technique employed are the key factors. Each is briefly described below.

The various forms of energy that are available to do useful work include:

1) Fossil fuels: coal, oil, natural gas, shale oil, and tar sands
2) Solar
3) Nuclear
4) Geothermal
5) Hydroelectric
6) Wind
7) Hydrogen
8) Biomass

For purposes of desalination, the three major sources of energy are fossil fuels, solar, and nuclear, although geothermal does receive consideration later. However, since most of the aforementioned water problems exist in arid regions, it would seem that solar energy would be the most attractive option.

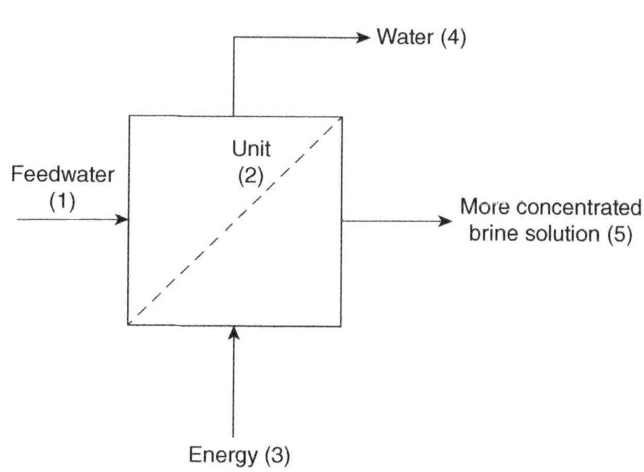

Figure 16.1 Simplified desalination process.

Any of the traditional unit operations (Theodore and Ricci 2010; Theodore 2014) are potential candidates for providing the separation technology. These can include:

1) Evaporation
2) Reverse osmosis (RO)
3) Crystallization
4) Electrodialysis
5) Solvent extraction
6) Thermal diffusion
7) Polarization
8) Ion exchange
9) Hydration

The energy requirement is proportional to the quantity of water recovered, in separation processes in which water crosses a phase boundary, and the actual work requirements are effectively independent of the salt concentration of the water processed, whether it is seawater or slightly brackish water. In processes in which the salt crosses the phase boundary, the energy requirement increases roughly in proportion to the salt concentration of the feed. Therefore, processes of the second type are generally less expensive to operate with brackish water than with seawater.

Of all the above separation processes, the only ones that are currently economically feasible are evaporation, RO, and crystallization. At present, the economic feasibility of all other processes, irrespective of the energy employed, is considered doubtful with the exception of solvent extraction (Theodore and Ricci 2010) and hydrate formation, which may have some possibility in the future.

16.5 Traditional Seawater Desalination Processes

As noted above, three prime candidate desalination processes have emerged and have been employed in the past. Each is briefly reviewed in the subsections to follow.

16.5.1 Evaporation Processes

The oldest and best-developed process for saline water conversion is the evaporation (or what some define as distillation) method. Evaporation remains a major method today for commercial production of freshwater from seawater.

In principle, traditional evaporation is the simplest method. Seawater is boiled in an evaporator by passing hot steam through a steam chest, where the steam condenses on the inside of the tubes of the chest and is usually returned to the boiler (Theodore and Ricci 2010; Theodore 2011). The vapors rising from the seawater feed are cooled in a condenser and thus converted into pure liquid water, which is collected in a storage vessel. The resulting concentrated brine solution is continuously or intermittently withdrawn from the evaporator.

The most advanced commercial evaporation processes are the following:

1) Multiple-effect evaporation
2) Flash evaporation
3) Vapor compression evaporation

In multiple-effect units, the vapors from the first evaporator condense in the second effect, where their heat of condensation is employed to boil the seawater in the second effect (chest), etc. In the flash evaporation process, seawater is first heated in tubes and then made to evaporate in chambers where the pressure is lower than in the heating tubes. While multiple-effect and flash evaporation units use an external supply of heated steam as the primary heat source, vapor compression evaporation literally uses its own steam as a heat source after it has been compressed. Here, seawater is preheated in a tubular heat exchanger by the outgoing streams of concentrated brine and freshwater and boiled in the tubes of the still. The (steam) vapors are compressed and fed back to the still to condense outside the tubes, thus providing the heat necessary for the boiling process. The two major drawbacks to the evaporation processes are corrosion and scale formation (Theodore 2011).

16.5.2 Reverse Osmosis

Electrodialysis was the membrane separation process employed for desalination a century ago. Here, the ions forming the salt are separated from the saltwater by electric forces and concentrated in different compartments. As noted earlier, the higher the salinity of the raw water, the more electric power is needed for this process. Hence, this process was applied primarily in the treatment of moderately brackish waters containing 1,000–2,000 ppm dissolved salts.

In recent years, RO has displaced electrodialysis as the membrane separation desalination process of choice, leaving the latter as the choice for medical kidney applications (Theodore and Ricci 2010). An RO system consists of an intake, a pretreatment component, a high-pressure pump, a membrane apparatus, remineralization, and pH adjustment components, as well as a disinfection step. Generally, a pressure of approximately 1.7–6.9 MPa is required to overcome the osmotic pressure of salt water (the pressure needed to stop the flux of solvent through the membrane).

This advanced separation technique may be used whenever low-molecular weight solutes such as inorganic salts or small organic molecules (e.g., glucose), are to be separated from a solvent (usually water). In normal (as opposed to reverse) osmosis, water flows from a less concentrated salt solution to a more concentrated salt solution because of an innate concentration driving force (thermodynamically referred to as the chemical potential). As a result of the migration of water, an "osmotic pressure" is created on the side of the membrane to which the water flows. In RO, the membrane is permeable to the solvent or water and relatively impermeable to the solute or salt. To make water pass through an RO membrane in the desired direction (i.e., away from a concentrated salt solution), a pressure must be applied that is higher than the osmotic pressure.

The membranes used for RO processes are characterized by a high degree of semi-permeability, high water fluxes, good mechanical strength, chemical stability, and relatively low operating and capital costs. Early RO membranes were composed of cellulose acetate, but restrictions on process stream pressure, temperature, and organic solute rejection spurred the development of noncellulosic and composite materials (membrane "sandwiches"). These membranes may be configured into certain geometries for system operation, including plate and frame, tubular, spiral wound (composite), and hollow fiber.

Obviously, the heart of the membrane process is the membrane itself. As noted above, a membrane is an ultra-thin, semi-permeable barrier separating two fluids that permits the transport of certain species through the barrier from one fluid to the other. The membrane is selective since it permits the transport of certain species while rejecting others. The term semi-permeable is frequently used to describe this selective action. New and improved membranes will probably involve nanomaterials.

16.5.3 Crystallization Processes

Crystallization is an important mass transfer operation that is often employed in the preparation of a pure product. In the process, a crystal usually separates as a substance of definite composition from a solution of varying composition. Any impurities in the liquid (often referred to as the mother liquor) are carried in the crystalline product only to the extent that they adhere to the surface or are occluded (retained) within the crystals that may have grown together during or after the crystallization operation (Theodore and Ricci 2010; Theodore 2014).

The separation of a solid from a solution onto a crystal occurs only if there is a state of imbalance involving a mass driving force, namely, a decrease in chemical potential (or concentration) between the bulk of the liquid solution and the crystal interface. This effectively means that the solution must be supersaturated.

Although crystallization is ordinarily thought of as the deposition of a solid crystalline phase from a liquid phase by cooling, evaporation, or both, the same principles apply to crystal formation by precipitation caused by the addition of a third substance. This other component may either react to form a precipitate or simply decrease the solubility of the precipitated material. It should also be noted that when crystallization begins, heat is evolved (liberated). The amount of heat evolved is defined as the heat (or enthalpy) of crystallization.

There are several different ways that crystallization can occur. The four most often encountered in practice are (Theodore and Ricci 2010; Theodore 2014):

1) Cooling
2) Evaporation
3) Cooling and evaporation (also denoted adiabatic evaporation)
4) A salting-out process

Process 1 is most commonly employed, provided the solubility of the component being crystallized decreases with decreasing temperature.

The problems concerning crystallization that are most frequently encountered by the practicing engineer are listed below.

1) Yield of a given product
2) Purity of the product
3) Energy requirements for cooling, evaporation, etc.
4) Shape of the individual crystals
5) Size of the crystals
6) Uniformity or size distribution of the crystals
7) Rate of production of the desired crystals
8) Caking

16.6 New Process Options for Potable Water Production

Two new processes were recently developed by Theodore for potable water production purposes. These include one that utilizes the flue gas following fossil fuel combustion and the other that utilizes geothermal energy. The patents associated with these two new options are briefly discussed below.

16.6.1 System and Method for Obtaining Potable Water from Fossil Fuels

This is a patented process of producing potable water, by combining a hydrocarbon containing fossil fuel with oxygen, in a combustion device, such as a home heating or utility unit to produce a flue gas of water vapor and carbon dioxide and condensing the water vapor in the flue gas to yield potable water. The combustion device can produce heat or electricity (Flynn et al. 2018). The water vapor can be condensed with one or more heat exchange devices. The source of oxygen can be air, pure oxygen, or nitrogen-reduced air. The source of oxygen can be humidified, such as with a non-potable water source or non-potable water can be added to the flue gas. The carbon dioxide and/or nitrogen in the flue gas can be reduced or removed before the condensation step(s). The pressure of the flue gas can be increased prior to condensation of the water vapor. Natural gas is a preferred fuel for this process.

What is claimed in the patent (Theodore 2022a) is:

1) A process of producing potable water, comprising: combining a fossil fuel with oxygen in a combustion device to produce a flue gas including water vapor and carbon dioxide; condensing the water vapor in the flue gas; and producing potable water as the condensed water vapor from the flue gas.
2) The process of Claim 1, wherein the combustion device is a home natural gas furnace.
3) The process of Claim 1, wherein the water vapor is condensed with a heat exchange device.
4) The process of Claim 1, wherein the heat exchange device is a multi-stage heat exchanger, a distillation device, or multiple heat exchangers.
5) The process of Claim 1, wherein the source of oxygen comprises a source of pure oxygen.
6) The process of Claim 1, wherein the source of oxygen is humidified with a source of non-potable water.
7) The process of Claim 1, wherein non-potable water is sprayed into the flue gas before it is condensed.
8) The process of Claim 1, wherein the carbon dioxide and/or nitrogen in the flue gas is reduced or removed before the condensation step.
9) The process of Claim 1, wherein the pressure of the flue gas is increased prior to condensation.
10) The process of Claim 1, wherein the hydrocarbon fuel is natural gas.
11) The process of Claim 1, wherein the combustion device is a home heating unit.
12) The process of Claim 1, wherein the flue gas is cooled to below 150°C to condense the water vapor.
13) The process of Claim 1, wherein the flue gas is cooled to below 112°F to condense the water vapor.
14) The process of Claim 1 and producing 2.500 lb of water.
15) A system for producing potable water comprising: a fuel inlet adapted to receive a flow of natural gas and an air inlet adapted to receive a flow of air containing a quantity of oxygen; a natural gas combustion device having the fuel inlet and a flue, adapted to receive the natural gas and oxygen, combust the natural gas, and produce a flue gas comprising carbon dioxide and water vapor out of the flue; a first water vapor condensation device adapted to condense the water vapor in the flue gas into a first quantity of potable water.
16) The system of Claim 15 and comprises a second condensation device adapted to condense any of the flue gas not condensed by the first condensation device and produce a second quantity of potable water.
17) The system of Claim 15, wherein the first water condensation device comprises a heat exchanger.
18) The system of Claim 16, wherein the second water condensation device comprises a heat exchanger.

16.6.2 System and Method for Obtaining Potable Water Employing Geothermal Energy

A new process and system (referred to herein as the Theogeo Process) is proposed where potable water is obtained from seawater or other sources of non-potable water at any salinity (or other aqueous solutions) by evaporative means employing geothermal energy. The energy is employed to evaporate the water in the solution into steam. Various known cooling methods can be employed to condense the resulting steam into potable water. Various evaporative cooling methods can also be used, e.g., seawater can be used as the cooling medium.

What is claimed in the patent (Theodore 2022b) is:

1) A method for obtaining purified potable water from a non-potable aqueous solution. comprising the steps of providing a passageway extending from the surface of the Earth to an underground enclosure at a temperature above the boiling point of water within the enclosure; inserting a water vapor receiving conduit through the passageway from the surface to the underground enclosure, the water vapor receiving conduit having an external diameter less than the diameter of the passageway so as to define an annular space; feeding the aqueous solution down the annular space between the conduit and the passageway so that it flows into a pool within the underground enclosure; permitting the pool of the aqueous solution in the underground enclosure to be heated to a temperature sufficient to form water vapor above the pool; drawing the water vapor into the hollow pipe from the underground enclosure to the surface: and condensing at least a portion of the water vapor traveling to the surface into potable water.
2) The method of Claim 1 and purifying the condensed water vapor.
3) The method of Claim 1 and comprising withdrawing a residue of the evaporated aqueous solution.
4) The method of Claim 1, wherein the aqueous solution is in heat exchange contact with the water vapor, is to withdraw heat from the water vapor as part of the condensing of the water vapor.
5) The method of Claim 1, wherein the aqueous solution is seawater.
6) The method of Claim 1, wherein the aqueous solution is fed down the passageway through a pipe.
7) A system for obtaining purified potable water from a non-potable aqueous solution, comprising the steps of: a passageway extending from the surface of the Earth down an underground enclosure at a temperature above the boiling point of water within the enclosure; a water vapor conduit extending down through the passageway from the surface to the underground enclosure, the pipe having an external diameter less than the diameter of the passageway, and defining an annular space therebetween; a pumping system adapted to feed the aqueous solution down the annular space between the pipe and the passageway so that it flows into a pool within the underground enclosure; the water vapor conduit adapted to permit water vapor from the aqueous solution in the pool to be drawn to the surface; and, a condensing apparatus adapted to condense at least a portion of the water vapor traveling to the surface to provide purified potable water.
8) The system of Claim 7, wherein the aqueous solution is seawater.
9) The system of Claim 7 includes purification devices to further purify the condensed water vapor.
10) The system of Claim 7, and comprising a mechanism adapted to withdraw a residue of the evaporated aqueous solution from the enclosure.

11) The system of Claim 7, wherein the condensing apparatus is constructed to place the aqueous solution in heat exchange contact with the water vapor, to withdraw heat from the water vapor as part of the condensing of the water vapor.
12) The system of Claim 11, wherein the aqueous solution is sea water.
13) The system of Claim 7, and comprising a pipe adapted to feed the aqueous solution down the passageway through a pipe.
14) The system of Claim 7, wherein the passageway is over 2 miles deep.
15) The system of Claim 7, wherein the water vapor conduit is part of a hybrid geothermal heating or electricity-producing system, and the condensing apparatus is located downstream from the geothermal heating or electricity-producing system and condenses the steam used to produce heat or electricity.
16) A system for obtaining purified potable water from a non-potable aqueous solution comprising the steps of: a passageway means for flowing water and water vapor, extending from the surface of the Earth to an underground enclosure at a temperature above the boiling point of water; a water pipeline extending to the passageway means, adapted to bring the non-potable water to the passageway means; water vapor collection means for drawing water vapor from the enclosure up the Earth's surface; water condensation means for condensing at least a portion of the water vapor into liquid water; and a water storage system adapted to store the liquid water.

16.6.3 Water Requirement of Electrolysis

A possible projection of hydrogen demand in the distant renewable future could be 2.3 Gt/year. The amount of water needed is for the predicted hydrogen economy, including the total water that is withdrawn and consumed for electrolysis. Withdrawals are water that is directly returned to the body of water from which it was extracted. Any water that is converted into another unusable form or is not returned to the original body of water will be considered consumed.

Based on the reaction stoichiometry, for every kg of hydrogen produced, 9 kg of water must be consumed. Therefore, 2.3 Gt of hydrogen requires 20.5 Gt, or 20.5 billion m^3, per year of freshwater. The only sector in which the use of hydrogen does not regenerate the entirety of the water feedstock by fuel cell or combustion is chemical synthesis, which will account for 540 Mt of hydrogen, using at most 4.8 billion m^3 of global freshwater annually (Beswick et al. 2021).

When fossil fuels are used for primary energy production and power generation, the water requirement is quite significant. In 2014, 251 billion m^3 of freshwater was withdrawn for power generation and energy production from fossil fuels such as coal, oil, and natural gas, and 31 billion m^3 were consumed as the water was used for cooling, mining, hydraulic fracturing, and refining. In comparison, even though the 20.5 billion m^3 of water withdrawn for hydrogen production by electrolysis must be consumed, this is still 33% less than the current fossil fuel energy-related uses. Furthermore, electrolysis will render fossil fuel energy sources obsolete as the energy sector is able to move more toward renewable technologies, saving 10 billion m^3 of freshwater that would have been consumed by energy-related uses of fossil fuels. Thus, it is possible that using hydrogen as a method to reach a renewable energy society will lead to drastic water savings, not expenditures, see Figure 16.2 as reproduced from Beckwith, Olivera, and Yan ACS Energy Letter 2021.

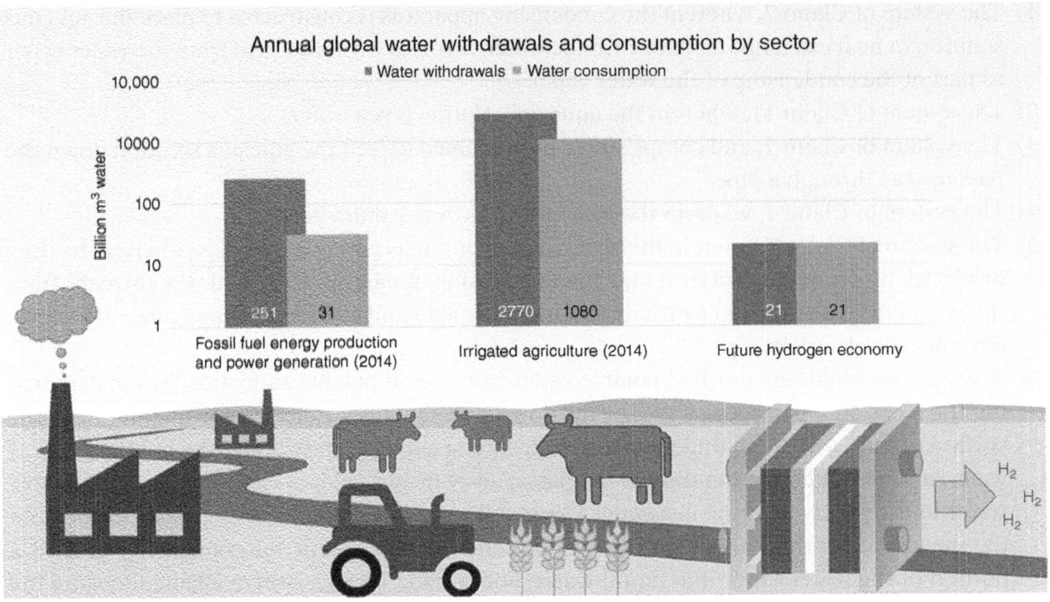

Figure 16.2 Comparison of the global freshwater withdrawal and consumption of three different sectors: fossil fuel energy production and power generation, agriculture, and the implementation of a global hydrogen economy. Note that the bar chart is on a log scale.

16.7 The Theodore Hydrogen Water Byproduct Process

The technical community has long considered hydrogen as a feedstock chemical. One feedstock option involves the conversion of hydrogen, via the combustion process, to energy in the form of heat or electricity. The byproduct of this reaction is *water*, which in many instances could serve as a source of relatively pure potable water. As shown earlier, the stoichiometric combustion reaction of hydrogen with oxygen results in Equation 16.1.

$$2H_2 + O_2 \rightarrow 2H_2O \tag{16.1}$$

with the accompanying liberation of 132,000 cal. When the reaction is conducted in air, Equation 16.1 becomes:

$$2H_2 + O_2 + \{3.76N_2\} \rightarrow 2H_2O + \{3.76N_2\} \tag{16.2}$$

with the brackets indicating that the nitrogen does not participate in the reaction. The "flue" gas produced contains 2 mol of H_2O per 3.76 mol of N_2, or 5.76 mol of total gas. This leads to a water vapor concentration, y, of:

$$y = 2.0/5.76 = 0.347 = 34.7\% \tag{16.3}$$

The corresponding water vapor temperature to this concentration is 163°F, representing the temperature at which water will begin to condense. If the temperature of the flue gas is lowered, i.e., by cooling with a heat exchanger, more water will be condensed. The process can be enhanced by increasing the pressure of the system and/or combusting with oxygen-enriched air.

Based on the above calculations, one can apply this process to not only a power plant but also to small domestic (or commercial) units for generating energy plus water that is either potable or water that can be easily further treated by traditional methods to ensure that it is potable. It has been shown that the combustion of natural gas (a clean fossil fuel) produces only trace quantities of undesirable gases, of which only trace quantities will be absorbed by the condensed water; preliminary economic calculations indicate that the cost of producing byproduct water from this combustion process is favorable. In any event, the technical community should realize that a valuable byproduct is generated following the combustion of hydrogen, and this process should be explored further, especially in water-stressed areas across the globe.

References

Beswick, R., Oliveira, A., and Yan, Y. (2021). Does the green hydrogen economy have a water problem? *ACS Energy Letters* 6(9): 3167–3169.

Flynn, A.M., Akoshige, T., and Theodore, L. (2018). *Kern's Process Heat Transfer*, 2nd Edition. Salem, MA: Scrivener-Wiley.

Merriam-Webster (2023). *Quintessential*. Merriam-Webster.com. https://www.merriam-webster.com/dictionary/quintessential (accessed 20 December 2023).

Theodore, L. (2011). *Heat Transfer Applications for the Practicing Engineer*. Hoboken, NJ: John Wiley & Sons.

Theodore, L. (2014). *Chemical Engineering: The Essential Reference*. New York: McGraw-Hill.

Theodore, M.K. (2021). *Introduction to Environmental Management*, 2nd Edition. Boca Raton, FL: CRC Press/Taylor & Francis Group.

Theodore, L. (2022a). System and process for obtaining potable water from fossil fuel. Personal patent #17/579,045, Patent #US 2022/0242728 A1.

Theodore, L. (2022b). System and process for obtaining potable water employing geothermal energy. Personal patent #17/736,239, Patent #US pending.

Theodore, L. and Dupont, R.R. (2020). *Water Resource Management Issues: Basic Principles and Application*. Boca Raton, FL: CRC Press/Taylor & Francis Group.

Theodore, L. and Dupont, R.R. (2022). *Introduction to Desalination: Principles, Processes and Calculations*. Hoboken, NJ: John Wiley & Sons.

Theodore, L. and Ricci, F. (2010). *Mass Transfer Operations for the Practicing Engineer*. Hoboken, NJ: John Wiley & Sons.

Theodore, L., Ricci, F., and VanVliet, T. (2009). *Thermodynamics for the Practicing Engineer*. Hoboken, NJ: John Wiley & Sons.

17

Safety Considerations

17.1 Introduction

No book on hydrogen would be complete without a discussion of safety concerns. Safe practices in the production, storage, distribution, and use of hydrogen are an integral part of future plans. Like most fuels, hydrogen can be handled and used safely with appropriate sensing, handling, and engineering measures.

Webster's defines safety as (Merriam-Webster 2023): ". . . the condition of being safe from undergoing or causing hurt, injury, or loss . . . to protect against failure, breakage, or accident." For hydrogen, the term safety has generally referred to concerns with employees. More specifically, the Occupational Safety and Health Administration (OSHA) and EPA regulations require employers to provide health and safety training to all employees involved in:

1) The cleanup of uncontrolled hazardous waste sites
2) Corrective actions under the Resource Conservation and Recovery Act
3) Emergency response actions involving the release of hazardous substances
4) The clean-ups at sites recognized by any government agency as uncontrolled hazardous waste sites

In determining who should receive health and safety training, employers should also consider the liability that they or their agency may incur in the event of an accident. It is imperative that anyone working in a hazardous environment receive the appropriate health and safety training as required by law.

Though hydrogen is a dangerous material, the safety concerns are essentially no different than those associated with the storage of any flammable gas, such as propane or natural gas. In fact, a number of industries have been producing, storing, and transporting hydrogen safely for decades. There are some hydrogen-specific considerations though, and if the use of hydrogen expands with new technologies to store, transport, and use hydrogen, safety standards will need to be kept current to ensure safe handling.

Increasing the use of hydrogen and exploring new processes for the production and use of hydrogen using the existing gas distribution and storage infrastructure will require the identification of and responses to several safety issues. Recent studies on market penetration of hydrogen anticipate many applications, which will require considerable additions to the transportation and storage infrastructure. Some of these will result in new safety concerns that need to be addressed. The state of existing regulations and standards is unclear or not written with new uses of hydrogen in mind.

Hydrogen Energy: Principles and Applications, First Edition. Vincent J. DelGatto, Louis Theodore, R. Ryan Dupont, and Matthew C. Ogwu.

These regulations need to be updated if hydrogen is to have the opportunity to fulfill its potential in the future. It is also important to identify the options being pursued to ensure that safety regulations exist to cover these applications, e.g., processes, infrastructure, and end uses. The current framework provides a sound basis, but gaps have been recognized, and critical needs should be addressed during the developmental phase of any new technologies.

The issue of hydrogen safety will have to be regulated to minimize risks for investors, and facilitate planning procedures, e.g., through agreement on consistent safety distances from hydrogen or blended infrastructure. The safe handling of hydrogen presents a challenge to technology to work out the methods necessary to utilize hydrogen for commercial, residential, and transportation uses. Many people remember May 6, 1937, when as the German airship *Hindenburg* was landing in Lakehurst, New Jersey, a random spark (or perhaps a saboteur) ignited a leaking hydrogen tank and the dirigible burst into a hovering ball of fire. The blaze killed 36 of the 97 passengers on board and set back the use of hydrogen for many decades. In this case, hydrogen was used as a buoyant, not as a fuel. Hydrogen has a relatively low ignition temperature, and a spark can ignite a leaking tank or gas bag such as that of the *Hindenburg*.

In open air and well-ventilated places, leaks or spills diffuse so rapidly (hydrogen being the lightest of all elements) that the risks of ignition or spreading flames are actually less than those for gasoline. As a consequence, hydrogen explosions are a rarity. For over a century before natural gas was widely distributed, cities and industries depended on so-called "town gas," manufactured from coal or coke, which contained up to 50% hydrogen, without danger to homes or industries. In general, hydrogen is hardly more hazardous than gasoline or even natural gas, although, having different characteristics, it requires different treatment. The basic requirements are adequate ventilation, leak prevention, and elimination of stray ignition sources.

Safety actions can also include *safety reviews/safety audits* (Theodore and Dupont 2012, 2023). These consist of a walk-through on-site inspection that can include interviews with plant staff to identify plant conditions or operating procedures that could lead to an "accident." While this qualitative technique is most commonly applied to operating process plants, it is also applicable to smaller systems. The walk-through on-site inspection can vary from an informal, routine function that is principally visual, with emphasis on housekeeping, to a formal one extended duration examination by a team of inspectors with appropriate backgrounds and responsibilities. The latter is sometimes referred to as a *safety review*, a *process safety review*, a *loss prevention review*, or a *process review* (Theodore and Dupont 2012, 2023). Such a program would be intended to identify hydrogen plant operating conditions/procedures that could lead to an accident and significant losses of life or property, and actively mitigate them.

H2Tools (PNNL 2023) is a national best practices resource and free, online national hydrogen safety training resource for emergency responders supporting safety practices involving hydrogen. The Hydrogen Safety Bibliographic Database is part of the H2Tools site and provides references to reports, articles, books, and other resources for information on hydrogen safety as it relates to production, storage, distribution, and use. The database includes references related to the following topics:

1) Hydrogen properties and behavior
2) Safe operating and handling procedures
3) Leaks, dispersion, and flammable vapor cloud formation
4) Embrittlement and other effects on material properties
5) Fuel cells and other energy conversion technologies
6) Sensors, tracers, and leak detection technologies
7) Accidents and incidents involving hydrogen

Section numbers and their accompanying titles for this chapter concerned with hydrogen safety considerations are listed below.

17.2 Hydrogen Details

Gaseous hydrogen is usually stored at high pressure, typically 350 bar (5,000 psi) or 700 bar (10,000 psi) to improve the energy density of the storage systems. Storage of any material at such high pressures is a safety risk that requires proper system design to overcome. Paramount among mitigation strategies are those typical for storage of gases of any kind at these pressures, such as proper selection of storage system materials and proper placement of pressure relief valves to avoid any major failures (Theodore and Dupont 2012, 2023).

Liquefied hydrogen is stored at cold temperatures (20 K or −253°C) and generally moderate pressures (<150 psi). This low storage temperature necessitates the use of boil-off ventilation systems specifically designed for hydrogen. Both oxygen and nitrogen in the air can condense and solidify in the presence of cold hydrogen gases, which can foul venting systems that are not specifically designed to handle this issue. Liquefied oxygen from venting or liquid hydrogen transfer lines can be a fire hazard as localized areas with very high oxygen concentrations and therefore potentially flammable conditions can form while the liquid evaporates. Setback distances from occupied buildings for liquid hydrogen systems should be carefully considered.

Fire safety is always a concern whenever flammable gases are being handled, and hydrogen does have some key characteristics that require special consideration relative to other gases. The flammability or explosive limit of flammable gases is defined as the minimum (lower) and maximum (upper) concentrations of the gas in a gas/oxidizer (e.g., air) mixture that will lead to flame propagation if the mixture is ignited. Hydrogen has a wider range of flammable concentrations (4–74%), as compared to other gases. For instance, the flammable range of propane is approximately 2–10% and methane is flammable at concentrations of 5–15%. The wider flammability range is often cited as a point of concern for hydrogen systems, but the lower flammability limit of hydrogen, which is the design limit for fire safety systems, is comparable to that of other flammable gases as indicated above. With proper ventilation, enough air can be introduced into a space to keep concentrations below the lower flammability limit and mitigate ignition concerns.

For hydrogen systems, leak detection devices should be placed toward the top of enclosed spaces or ventilation systems because hydrogen is less dense than air, unlike heavier gases such as propane or gasoline vapor. In outdoor systems, the high buoyancy of hydrogen reduces the associated safety hazards because any gas released disperses very quickly, quickly reaching the lower flammability limit (LFL) and lowering the chances of fire or explosions. Another important consideration is the amount of energy needed to ignite a hydrogen gas mixture or the minimum ignition energy, MIE (Theodore, Ricci, and VanVliet 2009). At the most ignitable concentration of hydrogen

(29% hydrogen in air), a minimum amount of ignition energy is needed, and a simple static discharge can light the mixture. However, at concentrations near the LFL, the MIE is comparable to that of other fuels. In cases where a hydrogen leak is ignited, the high dispersion rate and buoyancy lead to quick dissipation unlike with heavier gases that can remain concentrated leading to more concentrated damage. Although there are fire hazards associated with hydrogen, they are practically no more severe than those for other flammable gases.

17.3 Worker Safety Regulations and Requirements

The U.S. OSHA has regulations governing employee workplace safety (OSHA 2023). The Occupational Safety and Health Act of 1970 (OSH Act) was passed to prevent workers from being killed or otherwise harmed at work. The law requires employers to provide their employees with working conditions that are free of known dangers. The OSH Act created the OSHA, which sets and enforces protective workplace safety and health standards. OSHA also provides information, training, and assistance to employers and workers. The OSH Act gives workers the right to safe and healthful working conditions. It is the duty of employers to provide workplaces that are free of known dangers that could harm their employees. This law also gives workers important rights to participate in activities to ensure their protection from job hazards.

Federal OSHA covers most private sector employers and workers in 29 states, the District of Columbia, Virgin Islands, American Samoa, Guam, Northern Mariana Islands, Wake Island, Johnston Island, and the Outer Continental Shelf Lands. Private sector workers in the remaining 21 states and Puerto Rico are covered by OSHA-approved State Plans. State plans are OSHA-approved workplace safety and health programs operated by individual states instead of Federal OSHA. The OSH Act encourages states to develop and operate their own workplace safety and health programs and precludes state enforcement of OSHA standards unless the state has an approved State Plan. OSHA approves and monitors all State Plans and provides up to 50% of the funding for each program. State Plans must be at least as effective as the Federal OSHA program.

Employers have the responsibility to provide a safe and healthy workplace that is free from serious recognized hazards. This is commonly known as the General Duty Clause of the OSH Act. OSHA standards are rules that describe the methods that employers must use to protect their employees from hazards. There are four groups of OSHA standards: General Industry, Construction, Maritime, and Agriculture. (General Industry is the set that applies to the largest number of workers and worksites.) These standards are designed to protect workers from a wide range of hazards. These standards also limit the amount of hazardous chemicals, substances, or noise that workers can be exposed to; require the use of certain safe work practices and equipment; and require employers to monitor certain hazards and keep records of workplace injuries and illnesses.

Examples of OSHA standards include requirements to:

1) Provide fall protection, such as a safety harness and lifeline.
2) Prevent trenching cave-ins.
3) Ensure the safety of workers who enter confined spaces such as manholes or grain bins.
4) Prevent exposure to high levels of noise that can damage hearing.
5) Put guards on machines.
6) Prevent exposure to harmful levels of substances like asbestos and lead.
7) Provide workers with respirators and other needed safety equipment (in almost all cases, free of charge).

8) Provide healthcare workers with needles and sharp instruments that have built-in safety features to prevent skin punctures or cuts that could cause exposure to infectious diseases.
9) Train workers using a language and vocabulary they understand about hazards and how to protect themselves.

As indicated above, employers must also comply with the General Duty Clause of the OSH Act. This clause requires employers to keep their workplaces free of serious recognized hazards and is generally cited when no specific OSHA standard applies to the hazard.

The Hazard Communication standard, known as the "right-to-know" standard, was first promulgated on November 25, 1983, and is printed in 29 CFR Part 1910.120. This standard requires employers to inform and train workers about hazardous chemicals and substances in the workplace. Employers must:

1) Provide workers with effective information and training on hazardous chemicals in their work area. This training must be in a language and vocabulary that workers can understand.
2) Keep a current list of hazardous chemicals that are in the workplace.
3) Make sure that hazardous chemical containers are properly labeled with the identity of the hazardous chemical and appropriate hazard warnings.
4) Have and make readily available to workers and their representatives SDSs (formerly known as Material Safety Data Sheets or MSDSs) for each hazardous substance that provides detailed information about chemical hazards, their effects, how to prevent exposure, and emergency treatment if an exposure occurs.
5) Before their initial assignment or whenever a new hazard is introduced into their work area, inform employees of both (1) the requirements of the standard operations in their work area where hazardous chemicals are present, and (2) the location and availability of the written hazard communication program.
6) Train employees how to identify and protect themselves from chemical hazards in the work area, how to recognize the physical and health hazards of chemicals in their work area, and how to obtain and use the employer's written hazard communication program and appropriate hazard information.
7) Develop, implement, and maintain a written communication program for each workplace that describes how SDSs, labeling, and employee information and training requirements will be met. The written program must also include a list of hazardous chemicals present in the workplace, the methods that will be used to inform employees of the hazards associated with performing nonroutine tasks, and chemicals present in unlabeled pipes.
8) Employers who have multiemployer workplaces must include methods the employer will use to provide contractors at their facility with SDSs. These employers must also describe how they will inform subcontractor's employees about (1) the precautions that must be followed and (2) the labeling system used in the workplace.

Any incident represents a potentially hostile situation. Chemicals such as hydrogen that are combustible and potentially explosive, as well as chemicals that are corrosive, toxic, or reactive, along with biological and radioactive materials, can affect the general public or the environment as well as response personnel. Physical hazards may also be encountered. Workers may fall, trip, be struck by objects, or be subjected to danger from electricity and heavy equipment. Injury and illness may also occur due to the physical stress to personnel responding to accident and release events. While the response activities needed at each individual incident are unique, there are many similarities. One similarity is that all responses require protecting the health and ensuring the safety of response personnel.

To reduce the risks to personnel responding to hazardous substance incidents, an effective health and safety program must be developed and followed. At a minimum, a comprehensive worker health and safety program should address the following:

1) Safe work practices
2) Engineered safeguards
3) Medical surveillance
4) Environmental and personnel monitoring
5) Personnel protective equipment (PPE)
6) Education and training
7) Standard operating safety procedures

As part of a comprehensive program, standard operating safety procedures provide instructions on how to accomplish specific tasks in a safe manner. In concept and principle, standard operating safety procedures are universal, with specific requirements for PPE, for example, dependent on the type of specific incident to be encountered.

To safeguard the health of response personnel, a medical program must be developed, established, and maintained. This program has two essential components: routine health care and emergency treatment. Regarding the latter, the medical program must address emergency medical care and treatment of response personnel, including possible exposures to toxic substances and injuries resulting from accidents or physical hazards.

17.4 Site Safety Plans

The purpose of a site safety plan is to establish policies and procedures for protecting the health and safety of response personnel during all operations conducted at an incident. It contains information about the known or suspected hazards, routine and special safety procedures that must be followed, and other instructions for safeguarding the health of the responders.

A site safety plan is prepared and reviewed by qualified personnel for each hazardous substance response. Before operations at an incident commence, all safety aspects of site operations should be thoroughly examined. A safety plan is then written based on the anticipated hazards and expected work conditions. The site safety plan must be written to avoid misinterpretation, ambiguity, and mistakes that can result from verbal orders. The plan should be conspicuously posted or distributed to all response personnel and discussed with them. Once the safety plan is implemented, it must be periodically reviewed to keep it current and technically correct, and to reflect any changes in site work and conditions.

The plan must contain safety requirements for routine (but hazardous) response activities and also for unexpected site emergencies. The major distinction between routine and emergency site safety planning is the ability to predict, monitor, and evaluate routine activities. A site emergency is unpredictable and may occur at any time. Routine operations are all those activities that may be required in responding to an emergency in order to identify, evaluate, and control (including cleanup) an incident. These activities may involve a high degree of risk, but are standard operations generally involved in responding to a specific type of incident. Safety practices for routine operations closely parallel accepted procedures used in industrial hygiene and industrial safety. Whenever a hazardous incident progresses to the point where operations become more routine, the associated site safety plan becomes a more refined document. As a minimum, the following must be included as part of the site safety plan for routine operations:

1) Key Personnel and Alternates.
2) Known Hazards and Risks.
3) Routine or Special Training Requirements.
4) Levels of Protection Required.
5) Site-Specific Medical Requirements.
6) Environmental Surveillance Program.
7) Work Areas (exclusion zones, contamination reduction zones, support zones).
8) Site Control Procedures.
9) Decontamination.
10) Emergency Response Plan.
11) Confined Space Entry Procedures.
12) Weather-Related Problems.

Information on these site safety plan details is provided in the literature (Holms, Singh, and Theodore 2000).

17.5 Chemical Safety Data Sheets

One of the authors of this text co-authored a reference book titled *Regulatory Chemical Handbook* (Spero, DeVito, and Theodore 2000). The book provided detailed and pertinent information on three major classes of regulated chemicals: Hazardous Air Pollutants (HAPs), Priority Water Pollutants, and OSHA Chemicals. It is interesting to note that hydrogen, despite its explosive properties, was *not* listed in any of these categories. In addition, methane was also not listed.

With this introduction, this section introduces the reader to a major mechanism that serves to transmit health and hazard/safety information concerning chemicals, including hydrogen. The Safety Data Sheets (SDSs) (formerly known as Material Safety Data Sheets or MSDSs) have for more than 40 years played a key role in safety and hazard communication programs. The information contained in an SDS is largely the same as the MSDS, except now the SDSs are required to be presented in a consistent, user-friendly, 16-section format (OSHA 2012).

The SDS is a detailed information bulletin prepared by the manufacturer or importer of a chemical that describes its formulation, composition, precautions for proper use and handling, as well as the potential physical and health problems it might pose to users. The SDS details routes of exposure, emergency and first-aid procedures, and control measures in case of an uncontrolled release. Information on an SDS aids in the selection of safe products and helps prepare employers and employees to respond effectively to daily exposure situations as well as to emergency releases. It is also a source of information for identifying chemical hazards (accidents).

In line with OSHA requirements, employers must maintain a complete and accurate SDS for each hazardous chemical that is used in their facility. They are entitled to obtain this information automatically upon purchase of the material. When new and significant information becomes available concerning a product's hazards or ways to protect against the hazards, chemical manufacturers, importers, or distributors must add it to their SDS within 3 months and provide this updated information to their customers with the next shipment of the chemical. If there are multiple suppliers of the same chemical, there is no need to retain multiple SDSs for that chemical.

While SDSs are not required to be physically attached to a shipment, they must accompany or precede the shipment. When the manufacturer/supplier fails to send an SDS with a shipment labeled as hazardous, the employer must obtain one from the chemical manufacturer, importer, or distributor as soon as possible. Similarly, if the SDS is incomplete or unclear, the employer should contact the manufacturer or importer to seek clarification or obtain missing information.

When an employer is unable to obtain an SDS from a supplier or manufacturer, a written complaint, with complete background information, should be submitted to the nearest OSHA area office. OSHA will then call and send a certified letter to the supplier or manufacturer to obtain the needed information. If the supplier or manufacturer still fails to respond within a reasonable time, OSHA will inspect the supplier or manufacturer and take appropriate enforcement action against them.

It is important to note that OSHA specifies the information to be included in an SDS but does not prescribe the precise format for an SDS. A nonmandatory SDS form that meets the Hazard Communication Standard requirements can be used as is or expanded as needed.

It is also interesting to note that petroleum is considered less hazardous than hydrogen. Proponents of hydrogen argue, however, that it is no more or less dangerous than petroleum or other fossil fuels. Petroleum accidents and spillages in the past have had far from trivial environmental consequences. More research needs to be done into accidents involving hydrogen, as there have been relatively few experimental studies of hydrogen spills, fires, and explosions, and their subsequent impacts on human health and safety, and the environment.

An SDS includes information such as the properties of each chemical; the physical, health, and environmental health hazards; protective measures; and safety precautions for handling, storing, and transporting the chemical. The information contained in the SDS must be in English (although it may be in other languages as well). In addition, OSHA requires that SDS preparers provide specific minimum information as detailed in Appendix D of 29 CFR 1910.1200. The SDS preparers may also include additional information in various section(s).

Sections 1–8 contain general information about the chemical, identification, hazards, composition, safe handling practices, and emergency control measures (e.g., firefighting). This information should be helpful to those who need to get the information quickly. Sections 9–11 and 16 contain other technical and scientific information, such as physical and chemical properties, stability and reactivity information, toxicological information, exposure control information, and other information including the date of preparation or last revision. The SDS must also state that no applicable information was found when the preparer did not find relevant information for any required element.

The SDS must also contain Sections 12–15, to be consistent with the UN Globally Harmonized System of Classification and Labeling of Chemicals (GHS), but OSHA will not enforce the content of these sections because they concern matters handled by other agencies. A description of all 16 sections of the SDS, along with a brief review of their contents, is presented below (OSHA 2012).

1) *Section 1: Identification.* This section identifies the chemical on the SDS as well as the recommended uses. It also provides the essential contact information of the supplier. The required information consists of:
    a) Product identifier used on the label and any other common names or synonyms by which the substance is known.
    b) Name, address, phone number of the manufacturer, importer, or other responsible party, and emergency phone number.
    c) Recommended use of the chemical and any restrictions on use.

2) *Section 2: Hazard(s)*. Identification. This section identifies the hazards of the chemical presented on the SDS and the appropriate warning information associated with those hazards. The required information consists of:
   a) The hazard classification of the chemical.
   b) Signal word.
   c) Hazard statement(s).
   d) Pictograms of hazard symbols in black and white or a description of the name of the symbol.
   e) Precautionary statement(s).
   f) Description of any hazards not otherwise classified.

   For a mixture that contains an ingredient(s) with unknown toxicity, a statement describing the percentage of the mixture that consists of ingredient(s) with unknown acute toxicity.
3) *Section 3: Composition/Information on Ingredients*. This section identifies the ingredient(s) contained in the product indicated on the SDS, including impurities and stabilizing additives for substances, mixtures, and all chemicals where a trade secret is claimed. The required information consists of:
   a) Substances – Chemical name; Common name and synonyms; Chemical Abstracts Service (CAS) number and other unique identifiers; Impurities and stabilizing additives, which are themselves classified and which contribute to the classification of the chemical.
   b) Mixtures – Same information required for substances; The chemical name and concentration (exact percentage) of all ingredients that are classified as health hazards are: Present above their cut-off/concentration limits or present a health risk below the cut-off/concentration limits; The concentration (exact percentages) of each ingredient must be specified, except concentration ranges may be used in the following situations: A trade secret claim is made, there is batch-to-batch variation, or the SDS is used for a group of substantially similar mixtures.
   c) Chemicals where a trade secret is claimed – A statement that the specific chemical identity and/or exact percentage (concentration) of composition has been withheld as a trade secret is required.
4) *Section 4. First-Aid Measures*. This section describes the initial care that should be given by untrained responders to an individual who has been exposed to the chemical. The required information consists of:
   a) Necessary first-aid instructions by relevant routes of exposure (inhalation, skin and eye contact, and ingestion).
   b) Description of the most important symptoms or effects, and any symptoms that are acute or delayed.
   c) Recommendations for immediate medical care and special treatment are needed, when necessary.
5) *Section 5: Fire-Fighting Measures*. This section provides recommendations for fighting a fire caused by the chemical. The required information consists of:
   a) Recommendations of suitable extinguishing equipment, and information about extinguishing equipment that is not appropriate for a particular situation.
   b) Advice on specific hazards that develop from the chemical during the fire, such as any hazardous combustion products created when the chemical burns.
   c) Recommendations on special protective equipment or precautions for firefighters.
6) *Section 6: Accidental Release Measures*. This section provides recommendations on the appropriate response to spills, leaks, or releases, including containment and cleanup practices to prevent or minimize exposure to people, properties, or the environment. It may also include

recommendations distinguishing between responses for large and small spills where the spill volume has a significant impact on the hazard. The required information may consist of recommendations for:

a) Use of personal precautions (such as removal of ignition sources or providing sufficient ventilation) and protective equipment to prevent the contamination of skin, eyes, and clothing.
b) Emergency procedures, including instructions for evacuations, consulting experts when needed, and appropriate protective clothing.
c) Methods and materials used for containment (e.g., covering the drains and capping procedures).
d) Cleanup procedures (e.g., appropriate techniques for neutralization, decontamination, cleaning, or vacuuming; adsorbent materials; and/or equipment required for containment/cleanup).

7) *Section 7: Handling and Storage*. This section provides guidance on the safe handling practices and conditions for safe storage of chemicals. The required information consists of:
   a) Precautions for safe handling, including recommendations for handling incompatible chemicals, minimizing the release of the chemical into the environment, and providing advice on general hygiene practices.
   b) Recommendations on the conditions for safe storage, including any incompatibilities. Provide advice on specific storage requirements.
8) *Section 8: Exposure Controls/Personal Protection*. This section indicates the exposure limits, engineering controls, and personal protective measures that can be used to minimize worker exposure. The required information consists of:
   a) OSHA Permissible Exposure Limits (PELs), American Conference of Governmental Industrial Hygienists (ACGIH) Threshold Limit Values (TLVs), and any other exposure limit used or recommended by the chemical manufacturer, importer, or employer preparing the safety data sheet, where available.
   b) Appropriate engineering controls (e.g., use local exhaust ventilation, or use only in an enclosed system).
   c) Recommendations for personal protective measures to prevent illness or injury from exposure to chemicals, such as PPE.
   d) Any special requirements for PPE, protective clothing, or respirators (e.g., type of glove material, such as PVC or nitrile rubber gloves; and breakthrough time of the glove material).
9) *Section 9: Physical and Chemical Properties*. This section identifies physical and chemical properties associated with the substance or mixture. The minimum required information consists of:
   a) Appearance (physical state, color, etc.); Odor; Odor threshold; pH; Melting point/freezing point; Initial boiling point and boiling range; Flash point; Evaporation rate; Flammability (solid, gas).
   b) Upper/lower flammability or explosive limits; Vapor pressure; Vapor density; Relative density; Solubility(ies); *n*-octanol/water partition coefficient; Auto-ignition temperature; Decomposition temperature; and Viscosity.

   The SDS may not contain every item on the above list because information may not be relevant or is not available. When this occurs, a notation to that effect must be made for that chemical property. Manufacturers may also add other relevant properties, such as the dust deflagration index (Kst) for combustible dust, used to evaluate a dust's explosive potential.

10) *Section 10: Stability and Reactivity*. This section describes the reactivity hazards of the chemical and the chemical stability information. This section is broken into three parts: reactivity, chemical stability, and others. The required information consists of:
    a) Reactivity – Description of the specific test data for the chemical(s). This data can be for a class or family of the chemical if such data adequately represents the anticipated hazard of the chemical(s), where available.
    b) Chemical stability – Indication of whether the chemical is stable or unstable under normal ambient temperature and conditions while in storage and being handled. Description of any stabilizers that may be needed to maintain chemical stability. Indication of any safety issues that may arise should the product change in physical appearance.
    c) Other – Indication of the possibility of hazardous reactions, including a statement of whether the chemical will react or polymerize, which could release excess pressure or heat, or create other hazardous conditions. Also, a description of the conditions under which hazardous reactions may occur; List of all conditions that should be avoided (e.g., static discharge, shock, vibrations, or environmental conditions that may lead to hazardous conditions); List of all classes of incompatible materials (e.g., classes of chemicals or specific substances) with which the chemical could react to produce a hazardous situation; List of any known or anticipated hazardous decomposition products that could be produced because of use, storage, or heating. (Hazardous combustion products should also be included in *Section 5 (Fire-Fighting Measures)* of the SDS).
11) *Section 11: Toxicological Information*. This section identifies toxicological and health effects information or indicates that such data are not available. The required information consists of:
    a) Information on the likely routes of exposure (inhalation, ingestion, skin, and eye contact).
    b) Description of the delayed, immediate, or chronic effects from short- and long-term exposure.
    c) The numerical measures of toxicity (e.g., acute toxicity estimates such as the LD50 (median lethal dose)).
    d) Description of the symptoms associated with exposure to the chemical including symptoms from the lowest to the most severe exposure.
    e) Indication of whether the chemical is listed in the National Toxicology Program (NTP) Report on Carcinogens (latest edition) or has been found to be a potential carcinogen in the International Agency for Research on Cancer (IARC) Monographs (latest editions) or found to be a potential carcinogen by OSHA.
12) *Section 12: Ecological Information (nonmandatory)*. This section provides information to evaluate the environmental impact of the chemical(s) if it were released into the environment. The information may include:
    a) Data from toxicity tests performed on aquatic and/or terrestrial organisms, where available (e.g., acute or chronic aquatic toxicity data for fish, algae, crustaceans, and other plants; toxicity data on birds, bees, and plants).
    b) Whether there is a potential for the chemical to persist and degrade in the environment either through biodegradation or other processes, such as oxidation or hydrolysis.
    c) Results of tests of bioaccumulation potential, making reference to the octanol-water partition coefficient and the bioconcentration factor, were available.
    d) The potential for a substance to move from the soil to the groundwater (indicate results from adsorption studies or leaching studies).

e) Other adverse effects (e.g., environmental fate, ozone layer depletion potential, photochemical ozone creation potential, endocrine disrupting potential, and/or global warming potential).

13) *Section 13: Disposal Considerations (nonmandatory).* This section provides guidance on proper disposal practices, recycling or reclamation of the chemical(s) or its container, and safe handling practices. To minimize exposure, this section should also refer the reader to *Section 8 (Exposure Controls/Personal Protection)* of the SDS. The information may include:
   a) Description of appropriate disposal containers to use.
   b) Recommendations of appropriate disposal methods to employ.
   c) Description of the physical and chemical properties that may affect disposal activities.
   d) Language discouraging sewage disposal.
   e) Any special precautions for landfills or incineration activities.
14) *Section 14: Transport Information (nonmandatory).* This section provides guidance on classification information for shipping and transporting hazardous chemical(s) by road, air, rail, or sea. The information may include:
   a) UN number (i.e., four-figure identification number of the substance).
   b) UN proper shipping name.
   c) Transport hazard class(es).
   d) Packing group number, if applicable, based on the degree of hazard.
   e) Environmental hazards (e.g., identify if it is a marine pollutant according to the International Maritime Dangerous Goods Code (IMDG Code)).
   f) Guidance on transport in bulk (according to Annex II of MARPOL 73/783 and the International Code for the Construction and Equipment of Ships Carrying Dangerous Chemicals in Bulk (International Bulk Chemical Code (IBC Code)).
   g) Any special precautions which an employee should be aware of or needs to comply with, in connection with transport or conveyance either within or outside their premises (indicate when information is not available).
15) *Section 15: Regulatory Information (nonmandatory).* This section identifies the safety, health, and environmental regulations specific to the product that are not indicated anywhere else on the SDS. The information may include: Any national and/or regional regulatory information of the chemical or mixtures (including any OSHA, Department of Transportation, Environmental Protection Agency, or Consumer Product Safety Commission regulations).
16) *Section 16: Other Information.* This section indicates when the SDS was prepared or when the last known revision was made. The SDS may also state where the changes have been made to the previous version. Other useful information also may be included here.

Employers must ensure that the SDSs are readily accessible to employees for all hazardous chemicals in their workplace. This may be done in many ways. For example, employers may keep the SDSs in a binder or on computers as long as the employees have immediate access to the information without leaving their work area when needed and a backup is available for rapid access to the SDS in the case of a power outage or other emergency. Furthermore, employers may want to designate a person(s) responsible for obtaining and maintaining the SDSs. If the employer does not have an SDS, the employer or designated person(s) should contact the manufacturer to obtain one.

17.6 The Hydrogen SDS

This last section of this safety chapter provides a summary of the hydrogen SDS prepared by Airgas USA. More complete details on these 16 Hydrogen SDS sections can be found on the online SDS form at https://www.airgas.com/msds/001026.pdf.

1) *Section 1: Identification.* Hydrogen
   a) Product identifier – Hydrogen
   b) Name, address, and phone number of the manufacturer – Airgas USA, LLC and its affiliates, 259 North Radnor-Chester Road Suite 100, Radnor, PA 19087-5283; phone 1-610-687-5253.
   c) Recommended use – Synthetic/Analytical chemistry.
2) *Section 2: Hazard(s)* Identification. This material is considered hazardous by the OSHA Hazard Communication Standard (29 CFR 1910.1200).
   a) The hazard classification of the chemical – FLAMMABLE GASES – Category 1; GASES UNDER PRESSURE – Compressed gas
   b) Signal word – Danger
   c) Hazard statement(s) – Extremely flammable gas; contains gas under pressure; may explode if heated; may displace oxygen and cause rapid suffocation. Burns with invisible flame; may form explosive mixtures with air.
   d) Pictograms of hazard symbols – 
   e) Precautionary statement(s) – *General* – Read and follow all SDSs before use. Read the label before use. Keep out of reach of children. If medical advice is needed, have a product container or label at hand. Close the valve after each use and when empty. Use equipment rated for cylinder pressure. Do not open valve until connected to the equipment prepared for use. Use a backflow preventative device in the piping. Use only equipment of compatible materials for construction. Approach the suspected leak area with caution. *Prevention* – Keep away from heat, hot surfaces, sparks, open flames, and other ignition sources. No smoking. *Response* – Leaking gas fire: Do not extinguish unless the leak can be stopped safely. In case of leakage, eliminate all ignition sources. *Storage* – Protect from sunlight. Store in a well-ventilated place.
   f) Description of any hazards not otherwise classified – In addition to any other important health or physical hazards, this product may displace oxygen and cause rapid suffocation.
3) *Section 3: Composition/Information on Ingredients.*
   a) Substance – hydrogen; Common name and synonyms – Dihydrogen; *o*-Hydrogen; *p*-Hydrogen; Molecular hydrogen; H2; UN 1049; Chemical Abstracts Service (CAS) number – 1333-74-0.
4) *Section 4: First-Aid Measures.*
   a) Necessary first-aid instructions by relevant routes of exposure – *Eye contact* – Contact with rapidly expanding gas may cause burns or frostbite; *Inhalation* – No known significant effects or critical hazards; *Skin* – Contact with rapidly expanding gas may cause burns or frostbite, try to warm up the frozen tissues and seek medical attention; *Ingestion* – As this product is a gas, refer to the inhalation section.
   b) Description of the most important symptoms or effects – No specific data.

   c) Recommendations for immediate medical care – Treat symptomatically. Contact a poison treatment specialist immediately if large quantities have been ingested or inhaled. No action shall be taken involving any personal risk or without suitable training. It may be dangerous to the person providing aid to give mouth-to-mouth resuscitation.
5) *Section 5: Fire-Fighting Measures.*
   a) Recommendations of suitable extinguishing equipment – Use an extinguishing agent suitable for the surrounding fire. Extinguishing equipment that is not appropriate – None known.
   b) Advice on specific hazards that develop from the chemical during the fire – Contains gas under pressure. Extremely flammable gas. In a fire or if heated, a pressure increase will occur and the container may burst, with the risk of a subsequent explosion.
   c) Recommendations on special protective equipment or precautions for firefighters – Promptly isolate the scene by removing all persons from the vicinity of the incident if there is a fire. No action shall be taken involving any personal risk or without suitable training. Contact the supplier immediately for specialist advice. Move containers from the fire area if this can be done without risk. Use water spray to keep fire-exposed containers cool. If involved in fire, shut off flow immediately if it can be done without risk. If this is impossible, withdraw from the area and allow the fire to burn. Fight fire from a protected location or the maximum possible distance. Eliminate all ignition sources if safe to do so; firefighters should wear appropriate protective equipment and self-contained breathing apparatus (SCBA) with a full facepiece operated in positive pressure mode.
6) *Section 6: Accidental Release Measures.*
   a) Use of personal precautions – Accidental releases pose a serious fire or explosion hazard. No action shall be taken involving any personal risk or without suitable training. Evacuate surrounding areas. Keep unnecessary and unprotected personnel from entering. Shut off all ignition sources. No flares, smoking, or flames in the hazard area. Avoid breathing gas. Provide adequate ventilation. Wear an appropriate respirator when ventilation is inadequate. Put on appropriate personal protective equipment.
   b) Emergency procedures – If specialized clothing is required to deal with the spillage, take note of any information in Section 8 on suitable and unsuitable materials. See also the information in "N."
   c) Methods and materials used for containment – *Small spill* – Immediately contact emergency personnel. Stop leak if without risk. Use spark-proof tools and explosion-proof equipment; *Large spill* – Immediately contact emergency personnel. Stop leak if without risk. Use spark-proof tools and explosion-proof equipment.
   d) Cleanup procedures – See Section 1 for emergency contact information and Section 13 for waste disposal.
7) *Section 7: Handling and Storage.*
   a) Precautions for safe handling – Put on appropriate personal protective equipment (see Section 8). Contains gas under pressure. Avoid breathing gas. Use only with adequate ventilation. Wear an appropriate respirator when ventilation is inadequate. Do not enter storage areas and confined spaces unless adequately ventilated. Do not puncture or incinerate the container. Use equipment rated for cylinder pressure. Close the valve after each use and when empty. Protect cylinders from physical damage; do not drag, roll, slide, or drop. Use a suitable hand truck for cylinder movement. Use only nonsparking tools. Avoid contact with eyes, skin, and clothing. Empty containers retain product residue and can be hazardous. Store and use away from heat, sparks, open flame, or any other ignition source. Use explosion-proof electrical (ventilating, lighting, and material handling) equipment.

b) Recommendations on the conditions for safe storage – Store in accordance with local regulations. Store in a segregated and approved area. Store away from direct sunlight in a dry, cool, and well-ventilated area, away from incompatible materials (see Section 10). Eliminate all ignition sources. Cylinders should be stored upright, with a valve protection cap in place, and firmly secured to prevent falling or being knocked over. Cylinder temperatures should not exceed 52°C (125°F). Keep the container tightly closed and sealed until ready for use. See Section 10 for incompatible materials before handling or use.

8) *Section 8: Exposure Controls/Personal Protection.*
   a) California PEL for Chemical Contaminants (United States). Oxygen Depletion [Asphyxiant]; ACGIH TLV (United States, 3/2019). Oxygen Depletion [Asphyxiant]. Explosive potential.
   b) Appropriate engineering controls – Use only with adequate ventilation. Use process enclosures, local exhaust ventilation, or other engineering controls to keep worker exposure to airborne contaminants below any recommended or statutory limits. The engineering controls also need to keep gas, vapor, or dust concentrations below any lower explosive limits. Use explosion-proof ventilation equipment. Emissions from ventilation or work process equipment should be checked to ensure they comply with the requirements of environmental protection legislation. In some cases, fume scrubbers, filters, or engineering modifications to the process equipment will be necessary to reduce emissions to acceptable levels.
   c) Recommendations for personal protective measures – *Hygiene measures* – Wash hands, forearms, and face thoroughly after handling chemical products, before eating, smoking, and using the lavatory, and at the end of the working period. Appropriate techniques should be used to remove potentially contaminated clothing. Wash contaminated clothing before reusing. Ensure that eyewash stations and safety showers are close to the workstation location; *Eye/face protection* – Safety eyewear complying with an approved standard should be used when a risk assessment indicates this is necessary to avoid exposure to liquid splashes, mists, gases, or dust. If contact is possible, the following protection should be worn, unless the assessment indicates a higher degree of protection: Safety glasses with side shields; *Skin protection* – Chemical-resistant, impervious gloves complying with an approved standard should be worn at all times when handling chemical products if a risk assessment indicates this is necessary. Considering the parameters specified by the glove manufacturer, check during use that the gloves are still retaining their protective properties. It should be noted that the time to breakthrough for any glove material may differ for different manufacturers. In the case of mixtures, consisting of several substances, the protection time of the gloves cannot be accurately estimated. Personal protective equipment for the body should be selected based on the task being performed and the risks involved and should be approved by a specialist before handling this product. When there is a risk of ignition from static electricity, wear anti-static protective clothing. For the greatest protection from static discharges, clothing should include anti-static overalls, boots, and gloves. Appropriate footwear and any additional skin protection measures should be selected based on the task being performed and the risks involved and should be approved by a specialist before handling this product; *Respiratory protection* – Based on the hazard and potential for exposure, select a respirator that meets the appropriate standard or certification. Respirators must be used according to a respiratory protection program to ensure proper fitting, training, and other important aspects of use. Respirator selection must be based on known or anticipated exposure levels, the hazards of the product, and the safe working limits of the selected respirator.

9) *Section 9: Physical and Chemical Properties.*
   a) Appearance – gas, colorless; Odor – odorless; Odor threshold – NA; pH – NA; Melting point – −259.15°C (−434.5°F); Boiling point – −253°C (−423.4°F); Critical temperature – −240.15°C (−400.3°F); Flash point – NA; Evaporation rate – NA; Flammability – Extremely flammable in the presence of the following materials or conditions: oxidizing materials.
   b) Upper/lower flammability limits – Lower: 4% Upper: 76%; Vapor pressure – NA; Vapor density – 0.07 (Air = 1), Liquid Density at BP = 4.43 lb/ft^3 (70.96 kg/m^3); Relative density – NA; Solubility – NA; *n*-octanol/water partition coefficient – NA; Auto-ignition temperature – 500 to 571°C (932–1059.8°F); Decomposition temperature – NA; and Viscosity – NA; Molecular weight = 2.02 g/gmol; Heat of combustion – 116,486,080 J/kg.
10) *Section 10: Stability and Reactivity.*
    a) Reactivity – No specific test data related to reactivity is available for this product or its ingredients.
    b) Chemical stability – The product is stable.
    c) Other – Under normal conditions of storage and use, hazardous reactions will not occur; Avoid all possible sources of ignition (spark or flame). Do not pressurize, cut, weld, braze, solder, drill, grind, or expose containers to heat or sources of ignition; Incompatible materials include oxidizers; Under normal conditions of storage and use, hazardous polymerization will not occur.
11) *Section 11: Toxicological Information.*
    a) Information on the likely routes of exposure – Not available.
    b) Effects from short- and long-term exposure – Contact with rapidly expanding gas may cause burns or frostbite. No known significant chronic effects or critical hazards.
    c) Numerical measures of toxicity – Not available.
    d) Description of the symptoms associated with exposure – No specific data.
    e) Not listed as carcinogenic.
12) *Section 12: Ecological Information (nonmandatory).*
    a) Data from toxicity tests – Not available.
    b) Persistence and degradability – Not available.
    c) Bioaccumulation potential – Not available.
    d) Mobility in soil – Not available.
    e) Other adverse effects – No known significant effects or critical hazards.
13) *Section 13: Disposal Considerations (nonmandatory).* The generation of waste should be avoided or minimized wherever possible. Disposal of this product, solutions, and any by-products should at all times comply with the requirements of environmental protection and waste disposal legislation and any regional local authority requirements. Dispose of surplus and nonrecyclable products via a licensed waste disposal contractor. Waste should not be disposed of untreated to the sewer unless fully compliant with the requirements of all authorities with jurisdiction. Empty Airgas-owned pressure vessels should be returned to Airgas. Waste packaging should be recycled. Incineration or landfilling should only be considered when recycling is not feasible. This material and its container must be disposed of in a safe way. Empty containers or liners may retain some product residues. Do not puncture or incinerate containers.
14) *Section 14: Transport Information (nonmandatory).*
    a) UN number – 1049
    b) UN proper shipping name – HYDROGEN, COMPRESSED
    c) Transport hazard class – 2.1, Flammable Gas.

d) Packing group number – NA.
e) Environmental hazards – No.
f) Guidance on transport in bulk – Passenger aircraft/rail: Forbidden. Cargo aircraft: 150 kg. Passenger-carrying vessel: Forbidden.
g) Any special precautions – Always transport in closed containers that are upright and secure. Ensure that persons transporting the product know what to do in the event of an accident or spillage.

15) *Section 15: Regulatory Information (nonmandatory).* Exempted under TSCA, regulated flammable substance under the Clean Air Act. Non-HAP. Listed in MA, NJ, PA. Not determined in Japan, Thailand, and Turkey. Listed or exempted in New Zealand, the Philippines, the Republic of Korea, Taiwan, the United States, and Vietnam.
16) *Section 16: Other Information.* Hazardous Material Information System rating (0 to 4): Health 1; Flammability 4: Physical Hazards 3. National Fire Protection Association rating (0 to 4): Health 0; Flammability 4: Instability/Reactivity 0.

References

Holmes, G., Singh, B., and Theodore, L. (2000). *Handbook of Environmental Management and Technology*, 2nd Edition. Hoboken, NJ: John Wiley & Sons.

Merriam-Webster (2023). Safety. Merriam-Webster.com. https://www.merriam-webster.com/dictionary/safety (accessed 20 December 2023).

Occupational Safety and Health Administration (2012). *OSHA Brief: Hazard Communication Standard: Safety Data Sheets.* DSG BR-3514 2/2012. Washington, DC: OSHA, U.S. Department of Labor. https://www.osha.gov/sites/default/files/publications/OSHA3514.pdf.

Occupational Safety and Health Administration (2023). *Workers' Rights.* OSHA 3021-02R 2023. Washington, DC: OSHA, U.S. Department of Labor. https://www.osha.gov/sites/default/files/publications/osha3021.pdf.

Pacific Northwest National Laboratory (2023). *H2Tools.* Richland, WA: Hydrogen and Fuel Cell Technologies Office, Office of Energy Efficiency and Renewable Energy, U.S. Department of Energy. https://h2tools.org.

Spero, J., DeVito, B., and Theodore, L. (2000). *Regulatory Chemicals Handbook.* Boca Raton, FL: CRC Press/Taylor & Francis Group.

Theodore, L. and Dupont, R.R. (2012). *Environmental Health and Hazard Risk Assessment: Principles and Calculations.* Boca Raton, FL: CRC Press/Taylor & Francis Group.

Theodore, L. and Dupont, R.R. (2023). *Chemical Process Industries: Environmental and Health Risk Calculations.* Boca Raton, FL: CRC Press/Taylor & Francis Group.

Theodore, L., Ricci, F., and VanVliet, T. (2009). *Thermodynamics for the Practicing Engineer.* Hoboken, NJ: John Wiley & Sons.

Part III

Technical Engineering Issues

Part III consists of six chapters that one might describe as engineering oriented. Chapter numbers and their accompanying titles are listed below. The last chapter contains many applications in the form of illustrative examples, in question /answer format, that are concerned with hydrogen energy implementation, environmental impacts, and economic and ethical implications.

Chapters

18) Environmental Health and Hazard Risk Assessment
19) Energy–Environmental Interactions
20) Ethical Consideration
21) Economic Consideration
22) Optimization Considerations
23) Illustrative Examples

Hydrogen Energy: Principles and Applications, First Edition. Vincent J. DelGatto, Louis Theodore, R. Ryan Dupont, and Matthew C. Ogwu.

18

Environmental Health and Hazard Risk Assessment

18.1 Introduction

The term risk is not only used to describe the likelihood of an adverse response to a chemical or physical agent, but it has also been used to describe the likelihood of an unwanted event. This subject is treated in more detail later in this chapter. These include risks such as skiing, skydiving, flying, and bungee jumping; diseases; death due to natural causes; and many others.

Risk and hazard risk are topics that must be included in any management plan or study, particularly as it applies to the environment or a new process. It should also be noted that risk assessment and risk management are two different processes but are intertwined. Risk assessment and risk management give a framework not only for setting regulatory priorities but also for making decisions that cut across different environmental areas. Risk management refers to a decision-making process that involves such considerations as risk assessment, technology, feasibility, financial information about costs and benefits, statutory requirements, public concerns, and other factors. Therefore, risk assessment supports risk management in that the choices on whether and how much to control future exposure to the suspected hazards may be determined.

Part II, Chapter 17 Safety, introduced the reader to the role safety plays in a hydrogen energy future. A companion topic that is naturally associated with safety is not just risk but also risk assessment. Both risk and risk assessment plays an important role in addressing safety issues in the hydrogen industry. This chapter covers the general subject of risk assessment as it applies to human and environmental health and hazard risks concerning hydrogen energy. It provides background information for understanding scientific data and methods currently used to assess health and hazard risks. The content builds on two related topics health risk and hazard risk. The former topic is addressed in the next two sections to follow, while the latter is discussed in Sections 18.4 and 18.5. The chapter concludes with a discussion of future trends in the risk field as it pertains to hydrogen energy.

The origin of risk assessment can be dated back to the earliest years of human evolution. The first evidence of human health risk assessment was discovered around 3,200 B.C. Humans realized that animal meat represented a food source in lieu of edible plants. They had to balance the risk-benefit relationship of being killed by the animal versus starvation. Evidence leading to modern quantitative risk analysis can also be traced to early religious ideas regarding the probability of an afterlife. The similarities that exist between the practices and procedures of modern risk analysts and those of our ancestors underscore the point that people have been dealing with problems of risk for a long time.

Hydrogen Energy: Principles and Applications, First Edition. Vincent J. DelGatto, Louis Theodore, R. Ryan Dupont, and Matthew C. Ogwu.

Part of the basis for the concern in risk and its assessment developed between the 16th and 18th centuries. The following are some of the problems that were discovered during the period:

1) Adverse health effects due to various mining and metallurgical practices
2) The use of tobacco snuff with cancer of the nasal passage
3) Outbreaks of cholera due to contaminated water pumps
4) Cancer due to medicinal exposure to lead

Although these and many other problems were observed, very little headway was made to combat them. This was mainly due to the lack of instrumentation and the lack of rigorous observation and testing.

The early to mid-1900s spearheaded advances in occupational disease recognition, toxicology studies, and low-level dose concerns. Finally, in the 20 years between 1950 and 1970, safety levels for acceptable daily intakes were established. New approaches for setting acceptable daily intake levels were continually investigated. Most recently, public health officials, toxicologists, statisticians, and risk assessors began to question so-called dose–response models to estimate the incidence of tumors in exposed human populations. The latest approach has been to incorporate available biological data into the understanding of the mechanisms of cancer to quantitatively estimate delivered doses and responses.

In recent years, the need to quantify the risk associated with exposure to environmental and occupational toxicants has led to what is called risk assessment. Risk assessment provides:

1) Characterization of the types of effects expected.
2) An estimate of the probability of occurrence of these effects.
3) An estimate of the number of cases with these effects.

Chemical risk assessment involves the analysis of exposures that have taken place in the past and the adverse effects of which may or may not have already occurred. It also involves the prediction of the probable repercussions of exposure to the future. This chapter provides background information for understanding scientific data and methods currently used to assess health and hazard risks but is by no means a complete survey of the complex subject of risk assessment. Many of the principles discussed are widely accepted in the scientific community. The chapter has attempted to describe various points of view and has also identified the viewpoint that has been extensively adopted by public health and regulatory officials.

Section numbers and their accompanying titles for this chapter concerned with health and hazard risk assessment are provided below:

18.1 Introduction
18.2 The Health Risk Assessment Process
18.3 The Health Risk Assessment Process Components
18.4 Hazards Risk Assessment Process
18.5 The Hazard Risk Assessment Process Components
18.6 Future Trends
References

18.2 The Health Risk Assessment Process

Health risk assessments provide an orderly, explicit, and consistent way to deal with scientific issues in evaluating whether a health hazard exists and what the magnitude of the hazard may be. This evaluation typically involves significant uncertainties because the available scientific data are limited, and the mechanisms for adverse health impacts or environmental damage are only imperfectly understood. When one examines risk, how does one decide how safe is safe, or how clean is clean? To begin with, one has to look at both sides of the risk equation – that is, both the toxicity of a pollutant and the extent of public exposure. Information is required at both the current and potential exposure levels, considering all possible exposure pathways. In addition to human health risks, one needs to look at potential ecological or other environmental effects. In conducting a comprehensive risk assessment, one should remember that there are always uncertainties, and these must be included in the analysis (Paustenbach 1989).

In recent years, several guidelines and handbooks have been produced to help explain the approaches for doing health risk assessments. As discussed by a special National Academy of Sciences committee in 1983, most human or environmental health problems can be evaluated by dissecting the analysis into four parts: health problem identification, dose–response assessment or toxicity assessment, exposure assessment, and risk characterization (see Figure 18.1). For some perceived health problems, the risk assessment might stop with the first step, health problem identification, if no adverse effect is identified or if an agency elects to take regulatory action without further analysis (Burke, Singh, and Theodore 2000). Regarding health problem identification, it is defined as a toxic agent or a set of conditions that has the potential to cause adverse effects on human health or the environment. This identification involves an evaluation of various forms of information to identify the different health risks. Dose–response or toxicity assessment is required in an overall assessment; responses/effects can vary widely since all chemicals and contaminants vary in their capacity to cause adverse effects. This step frequently requires that assumptions be made to relate experimental data for animals to humans. Exposure assessment is the determination

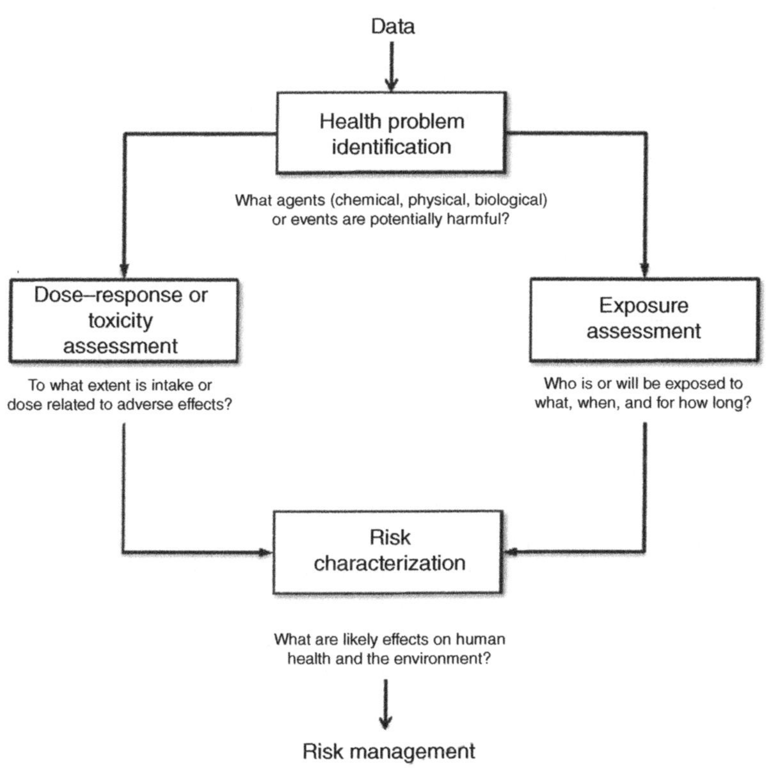

Figure 18.1 Health risk assessment process.

of the magnitude, frequency, duration, and routes of exposure of human populations and ecosystems. Finally, in risk characterization, toxicology and exposure data/information are combined to obtain a qualitative or quantitative expression of risk.

18.3 The Health Risk Assessment Process Components

Health risk assessment can be separated into four phases: problem identification, dose–response assessment, exposure assessment, and risk characterization. The four phases commonly employed in an overall risk assessment are provided in Figure 18.1. Each of the four phases is discussed in greater detail in the following sections.

18.3.1 Health Problem Identification

Health problem identification is the most easily recognized of the actions of regulatory agencies. It is defined as the process of determining whether human exposure to an agent could cause an increase in the incidence of a health condition (cancer, birth defect, etc.) or whether exposure to a nonhuman receptor, e.g., fish, birds, or other wildlife, might adversely be affected. It involves characterizing the nature and strength of the evidence of causation. Although the question of whether a substance causes cancer or other adverse health effects in humans is theoretically a yes –no question, there are few chemicals or physical agents for which the human data are definitive. Therefore, the question is often restated in terms of effects in laboratory animals or other test systems, i.e., "Does the agent induce cancer in test animals?" Positive answers to such questions are typically taken as evidence that an agent may pose a cancer risk to any exposed human. Information for short-term in vitro tests and structural similarity to known chemical hazards may, in certain circumstances, also be considered adequate information for identifying a health problem (Burke, Singh, and Theodore 2000; Theodore, M.K. and Theodore, L. 2021).

A health problem identification effort for a chemical plant or industrial application can include information about (AIChE 1985; Theodore and Dupont 2012):

1) Chemical identities.
2) The location of facilities that use, produce, process, or store hazardous materials.
3) The type and design of chemical containers or vessels.
4) The quantity of material that could be involved in an airborne release.
5) The nature of the hazard (e.g., airborne toxic vapors or mist, fire, explosion, large quantities stored or processed, and handling conditions) is most likely to accompany hazardous material spills or releases.

An important aspect of identification is a description of the pervasiveness of the problem. For example, most environmental assessments require knowledge of the concentration of the material in the environment, weighted in some way to account for the geographical magnitude of the site affected; that is, a 1-acre or 300-acre site, a 1,000–1,000,000 gal/min stream, etc. All too often, environmental incidents regarding chemical emissions have been described by statements like "concentrations as high as 150 ppm" of a chemical were measured at a 1,000-acre waste site. However, following closer examination, one may find that only 1 of 200 samples collected on a 20-acre portion of a 1,000-acre site showed this concentration and that 2 ppm was the geometric mean level of contamination in the 200 samples.

An appropriate sampling program is critical in the conduct of a health risk assessment. This topic could arguably be part of the exposure assessment, but it has been placed within health problem identification because, if the degree of contamination is small, no further work may be necessary. Not only is it essential that samples be collected in a random and representative manner, but the number of samples must be sufficient to conduct a statistically valid analysis. The number needed to ensure statistical validity will be dictated by the variability among the results. The larger the variance, the higher the number of samples needed to define the problem (Paustenbach 1989).

Toxicologists commonly refer to two general types of health risks. These are classified as:

1) *Acute*. Exposures that occur for relatively short periods, generally from minutes up to 1 or 2 days. Concentrations of (toxic) air contaminants are usually high relative to their protection criteria. In addition to inhalation, airborne substances might directly contact the skin, or liquids and sludges may be splashed onto the skin or into the eyes, leading to adverse effects. This subject area falls, in a general sense, into the domain of hazard risk assessment (HZRA), a topic treated later in this chapter.
2) *Chronic*. Continuous exposure occurs over long periods, usually several months to years. Concentrations of inhaled (toxic) contaminants are usually relatively low. This subject area falls in the general domain of health risk assessment (HRA), and it is this subject that is addressed in this section. Thus, in contrast to the acute (short-term) exposures that predominate in hazard risk assessment, chronic (long-term) exposures are the major concern in HRA.

The means of identifying health problems are complicated. Different methods are used to collect and evaluate toxic properties of chemicals (those properties that indicate the potential to cause biological injury, disease, or death under certain exposure conditions). One method is the use of epidemiological studies that deal with the incidence of disease among groups of people. Epidemiological studies attempt to correlate the incidence of adverse health effects from an emission by an evaluation of people with a particular effect and people without the effect. Another method, long-term animal exposure studies, are the most common method of hazard determination through the evaluation of disease development in laboratory animals (primarily mice and rats) exposed to increasingly higher concentrations or doses of a suspected health hazard. Responses or effects can vary widely from no observable effect to temporary and reversible effects, to permanent injury to organs, to chronic functional impairment, and ultimately, death. Increased tumor incidence in laboratory animals is the primary health effect considered in long-term animal exposure studies evaluating carcinogens. Exposure testing for a major portion of an animal's lifetime (2–3 years for rats and mice) provides information on disease and susceptibility, primarily for carcinogenicity (the development of cancer).

The understanding of how a substance is handled in the body, transported, transformed, and excreted, and of the response of both animals and humans to toxic chemical exposure, has advanced remarkably. There are many questions concerning these animal tests as to what information they provide, which kinds of studies are the best, and how animal data compare to human data. In an attempt to answer these questions, epidemiological studies and animal exposure studies are compared to determine if a particular chemical is likely to pose a health problem to humans based on animal data. Many assumptions are made in health risk assessments. For example, it is assumed that the chemical administered in an exposure study is in a form similar to that present in the environment. Another assumption is that animal carcinogens are also human carcinogens, i.e., that there is a similarity between animal and human metabolisms, and so on. With these and other assumptions, and by analyzing identification procedures, lists of problematic chemicals have been developed (Burke, Singh, and Theodore 2000; Theodore, M.K. and Theodore, L. 2021).

18.3.2 Dose–Response Assessment

Dangers are not necessarily defined solely by the presence of a particular chemical, but rather by the amount of that substance one is exposed to, also known as the dose. A dose is usually expressed in milligrams of chemical received per kilogram of body weight per day. For toxic substances other than carcinogens, a threshold dose must be exceeded before a health effect will occur, and for many noncarcinogenic substances, there is a dosage below which there is no harm. A health effect will occur or at least be detected at the threshold. For carcinogens, it is assumed that there is no threshold, and, therefore, any substance that produces cancer is assumed to produce cancer at any concentration. It is vital to establish the link to cancer and to determine if that risk is acceptable. Analyses of cancer risks are much more complicated than noncancer risks (Burke, Singh, and Theodore 2000; Theodore, M.K. and Theodore, L. 2021).

Dose–response assessment is the process of characterizing the relationship between the dose of an agent administered or received and the incidence of an adverse effect in exposed populations – estimating the incident impact as a function of exposure to an agent. This process considers such important factors as the intensity of exposure as well as other variables that might affect the response of an exposed individual, such as sex, lifestyle, and other modifying factors. A dose–response assessment usually requires extrapolation from high to low doses and extrapolation from animals to humans, or one laboratory animal species to a wildlife species. A dose–response assessment should describe and justify the methods of extrapolation used to predict incidence, and it should characterize the statistical and biological uncertainties in these methods. When possible, the uncertainties should be described numerically rather than qualitatively.

Toxicologists tend to focus their attention primarily on extrapolations from cancer studies. However, there is also a need to evaluate the risks of lower doses in affecting the various organs and systems in the body. Many scientific papers have focused on the use of a safety factor or uncertainty factor approach since all adverse effects other than cancer, and mutation-based developmental effects are believed to have a threshold – a dose below which no adverse effect should occur. Several researchers have discussed various approaches to setting acceptable daily intakes or exposure limits for developmental and reproductive toxicants. It is thought that an acceptable limit of exposure could be determined using cancer models, but today they are considered inappropriate because of these thresholds (Burke, Singh, and Theodore 2000; Theodore, M.K. and Theodore, L. 2021).

For a variety of reasons, it is difficult to precisely evaluate toxic responses caused by acute exposures to hazardous materials. First, humans experience a wide range of acute adverse health effects, including irritation, narcosis, asphyxiation, sensitization, blindness, organ system damage, and death. In addition, the severity of many of these effects varies with the intensity and duration of exposure. Second, there is a high degree of variation in response among individuals in a typical population. Third, for the overwhelming majority of substances encountered in industry, there are not enough data on the toxic responses of humans to permit an accurate or precise assessment of the substance's hazard potential. Fourth, many releases involve multi-components. There are presently no rules on how these types of releases should be evaluated. Fifth, there are no toxicology testing protocols that exist for studying episodic releases on animals. In general, this has been a neglected area of toxicology research. There are many measures available to use as benchmarks for predicting the likelihood that a release event will result in severe injury or death. Several references (Crowl and Louvar 1990; Theodore and Dupont 2023) review various toxic effects and discuss the use of various established toxicological criteria.

18.3.3 Exposure Assessment

Exposure assessment is the process of measuring or estimating the intensity, frequency, and duration of human or animal exposure to an agent currently present in the environment or estimating hypothetical exposures that might arise from the release of new chemicals into the environment. In its most complete form, an exposure assessment should describe the magnitude, duration, schedule, and route of exposure; the size, nature, and classes of the human, animal, aquatic, or wildlife populations exposed; and the uncertainties in all estimates. The exposure assessment can often be used to identify feasible prospective control options and to predict the effects of available control technologies for controlling or limiting exposure and subsequent adverse effects (Paustenbach 1989).

Much of the attention focused on exposure assessment has come recently. This is because many of the risk assessments performed in the past used too many conservative assumptions, which caused an overestimation of the actual exposure and subsequent risks. Without exposures, there are no risks. To experience adverse effects, one must first come into contact with the toxic agent(s). Exposure to chemicals can be via inhalation of air (breathing), ingestion of water and food (eating and drinking), or absorption or penetration through the skin. These are all pathways to the human body.

Generally, the main pathways of exposure considered in this step are atmospheric transport, surface, and groundwater transport, ingestion of toxic materials that have passed through the aquatic and terrestrial food chain, and dermal absorption. Once an exposure assessment determines the quantity of a chemical with which human populations may come in contact, the information can be combined with toxicity data (from the health problem identification and dose–response process) to estimate potential health risks (Paustenbach 1989). The primary purpose of an exposure assessment is to determine the concentration levels over time and space in each environmental media where humans and other environmental receptors may come into contact with chemicals of concern. There are four components of an exposure assessment: potential sources, significant exposure pathways, populations potentially at risk, and exposure estimates (Paustenbach 1989).

The two primary methods of determining the concentration of a pollutant to which target populations are exposed are direct measurement and dispersion modeling. Measurement of the pollutant concentration in the environment is used for determining the risk associated with an existing discharge source. Receptors are placed at regular intervals from the source, and the concentration of the pollutant is measured over a certain period (usually several months or a year). The results are then related to the size of the local population. This kind of monitoring, however, is expensive and time-consuming. Many measurements must be taken because exposure levels can vary under different atmospheric conditions or at different times of the year. Dispersion modeling can predict environmental concentrations of pollutants under a variety of release scenarios (strength, duration, etc.), atmospheric conditions, receptor conditions, etc., and can provide estimates of probabilities of exposure under variable conditions felt feasible for a given source and community receptor situation. This method is primarily used for assessing risk from a proposed facility or discharge. Sophisticated techniques are employed to relate reported or measured emissions to atmospheric, climatological, demographic, geographic, and other data in order to predict a population's potential exposure to a given chemical (Burke, Singh, and Theodore 2000; Theodore, M.K. and Theodore, L. 2021).

18.3.4 Risk Characterization

Not all contaminants or chemicals are created equal in their capacity to cause adverse effects. Thus, cleanup standards or action levels are based in part on the compounds' toxicological properties. Risk characterization is the process of estimating the incidence of a health effect under the various conditions of human or animal exposure described in the exposure assessment. It is performed by combining exposure and dose–response assessments. The summary effects of the uncertainties in the preceding steps should be described in this step. The quantitative estimate of a risk is the principal interest of the regulatory agency or risk manager making decisions. The risk manager must consider the results of the risk characterization when evaluating the economics, societal aspects, and various benefits of the risk assessment. Factors such as societal pressure, technical uncertainties, and severity of the potential hazard influence how decision-makers respond to the risk assessment. There is room for improvement in this step of the risk assessment (Paustenbach 1989).

A risk estimate indicates the likelihood of the occurrence of different types of health or environmental effects in exposed populations. Risk assessments should include both human health and environmental impacts (i.e., effects on ecosystems). Ecological impacts include actual or potential effects on plants and animals (other than domesticated species). The number produced from a risk characterization, representing the probability of adverse health effects being caused, must be evaluated. This is done because certain agencies will only look at risks of specific levels of concern (i.e., 1:10,000 risk and higher) and act on them (U.S. EPA 1987; Crowl and Louvar 1990; Theodore and Dupont 2023).

18.4 Hazard Risk Assessment Process

Risk evaluation of accidents serves as a dual process. It estimates the probability that an accident will occur and also assesses the severity of the consequences of an accident. Consequences may include damage to the surrounding environment, financial loss, or injury to life. This section is primarily concerned with the methods used to identify hazards and the causes and consequences of accidents. Issues dealing with health risks have already been explored. Risk assessment of accidents provides an effective way to help ensure either that a mishap does not occur or reduces the likelihood of an accident. The result of the risk assessment allows concerned parties to take precautions to prevent an accident before it happens.

Regarding definitions, the first thing an individual needs to know is what precisely an accident is. An accident is an unexpected event that has undesirable consequences (AIChE 1985). The causes of accidents have to be identified to help prevent accidents from occurring. Any situation if characteristic has to be identified to help prevent accidents from occurring. Any situation or characteristic of a system, plant, or process that has the potential to cause damage to life, property, or the environment is considered a hazard. A hazard can also be defined as any characteristic that has the potential to cause an accident. The severity of a hazard plays a large part in the potential amount of damage a hazard can cause if it occurs. Risk is the probability that personal injury, damage to property, damage to the environment, or financial loss will occur. An acceptable risk can also be defined as an accident that has a high probability of occurring with negligible consequences. Risks can be ranked qualitatively using the labels high, medium, and low. Risk can also be ranked quantitatively as an annual number of fatalities per million affected individuals. This is generally denoted as a number time one-millionth, i.e., 3×10^{-6}; this representation indicates that on average, for every million individuals, three individuals will die every year. Another quantitative approach that has become popular in the industry is the fatal accident rate (FAR) concept. The lifetime of a worker is defined as 10^5 hours, which is based on a 40-hour work week for 50 years. A reasonable FAR for a chemical plant is 3.0, with 4.0 usually taken as a maximum. The FAR for an individual at home is approximately 3.0. A FAR of 3.0 means that there are three deaths for every 1,000 workers over a 50-year period.

There are several steps in evaluating the risk of an accident (see Figure 18.2). These are detailed below if the system in question is a chemical plant.

1) A brief description of the equipment and chemicals used in the plant is needed.
2) Any hazard in the system has to be identified. Hazards that may occur in a chemical plant include fire, explosions, toxic vapor release, rupture of a pressurized vessel, slippage, runaway reactions, and corrosion.
3) The event or series of events that will initiate an accident need to be identified. An event could be a failure to follow correct safety procedures, improperly repaired equipment, or failure of a safety mechanism.

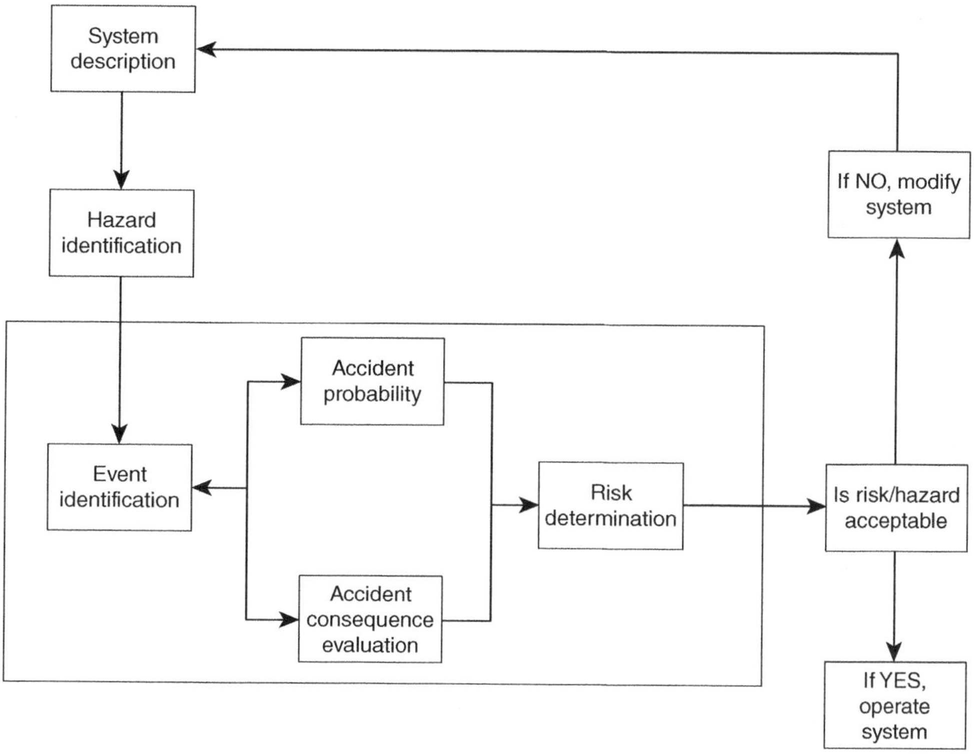

Figure 18.2 Hazard risk assessment process.

4) The probability that an accident will occur has to be determined. For example, if a chemical plant has a 10-year life, what is the probability that the temperature in a reactor will exceed the specified temperature range? The probability can be ranked from low to high. A low probability means that it is unlikely for the event to occur in the life of the plant. A medium probability suggests that there is a possibility that the event will occur. A high probability means that the event will probably occur during the life of the plant.
5) The severity of the consequences of the accident must be determined. This will be described later in detail.
6) If the probability of the accident and the severity of its consequences are low, then the risk is usually deemed acceptable, and the plant should be allowed to operate. If the probability of occurrence is too high or the damage to the surroundings is considerable, then the risk is usually unacceptable, and the system needs to be modified to minimize these effects.

18.5 The Hazard Risk Assessment Process Components

The heart of the hazard risk assessment methodology is enclosed in the dashed box of Figure 18.2. This methodology allows for a reevaluation of the process if the risk is deemed unacceptable (the assessment is repeated, starting with either the System Description or Hazard Identification steps). Details of the methodology are provided in the following sections.

18.5.1 Hazard Identification

Hazard or event identification provides information on situations or chemicals and their releases that can potentially harm the environment, life, or property. Information that is required to identify hazards includes chemical identities, quantities, and location of chemicals in question, and chemical properties such as boiling points, ignition temperatures, and toxicity to humans. There are several methods used to identify hazards. The methods that will be discussed here are the process checklist and the hazard and operability study (HAZOP).

A process checklist evaluates equipment, materials, and safety procedures (AIChE 1985; Theodore and Dupont 2012; Theodore, M.K. and Theodore, L. 2021). A checklist is composed of a series of questions prepared by engineers who know the process being evaluated. It compares what is in the actual plant to a set of safety and company standards. Some questions that may be on a typical checklist are:

1) Was the equipment designed with a safety factor?
2) Does the spacing of the equipment allow for ease of maintenance?
3) Are there pressure relief valves on the equipment in question?
4) How toxic are the materials that are being used in the process, and is there adequate ventilation to mitigate toxic gas buildup?
5) Will any of the materials cause corrosion to the pipe(s)/reactor(s)/system?
6) What precautions are necessary for flammable materials?
7) Is there an alternate exit in case of fire?
8) If there is a power failure, what fail-safe procedure(s) does the process contain?
9) What hazard is created if any piece of equipment malfunctions?

These questions and others are answered and analyzed. Changes are then made to reduce the risk of an accident. Process checklists are updated and audited at regular intervals.

A HAZOP study is a systematic approach to recognizing and identifying possible hazards that may cause a failure of a piece of equipment (Theodore and Dupont 2023). This method utilizes a team with diverse professional backgrounds to detect and minimize hazards in a plant. The process in question is divided into smaller processes (subprocesses). Guide words are used to relay the degree of deviation from the subprocesses' intended operation. The guide words can be found in Table 18.1. The causes and consequences of a deviation from the process are determined. If there are any recommendations for revisions, they are recorded, and a report is made outlining the required corrective action(s).

Table 18.1 Guide words used to relay the degree of deviation from the intended subprocess operation.

Guide Word	Meaning
No	No part of the intended function is accomplished
Less	Quantitative decrease in intended activity
More	Quantitative increase in intended activity
Part of	The intention is achieved to a certain percent
As well as	The intention is accomplished along with side effects
Reverse	The opposite of the intention is achieved
Other than	A different activity replaces the intended activity

A summary of the basic steps of a HAZOP study is (Theodore and Dupont 2023):

1) Define the objectives.
2) Define the plant limits.
3) Appoint and train a team.
4) Obtain the complete preparative work (i.e., flow diagrams and sequence of events).
5) Conduct examination meetings that select subprocesses, agree on the intention of subprocesses, state and record intentions, use guide words to find deviations from the intended purpose, determine the causes and consequences of deviation, and recommend revisions.
6) Issue the meeting reports.
7) Follow up on revisions.

There are other methods of hazard identification. A "what-if" analysis presents specific questions about a particular hazard and then tries to find the possible consequences of that hazard. Human error analysis identifies potential human errors that will lead to an accident. These other methods can be used in conjunction with the two previously described methods to develop robust hazard identification efforts.

18.5.2 Hazard/Accident Probability

There are a host of reasons why accidents occur in industry. The primary causes are mechanical failure, operational error (human error), process upsets, and design errors. Figure 18.3 provides the relative number of accidents that have occurred in the petrochemical energy field (Crowl and Louvar 1990). There are three steps that normally lead to an accident:

1) Initiation
2) Propagation
3) Termination

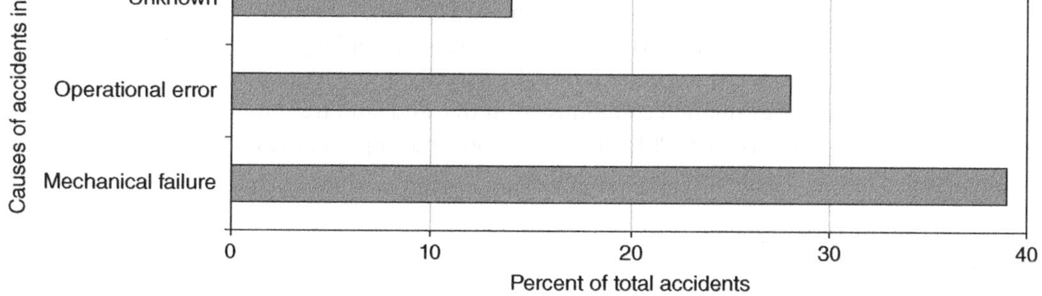

Figure 18.3 Causes of accidents in the petrochemical energy field (*Source:* Adapted from Crowl and Louvar (1990)).

The path that an accident takes through the three steps can be determined utilizing a *fault tree analysis* (AIChE 1985; Theodore and Dupont 2012). A fault tree is a diagram that shows the path that a specific accident takes. The first thing needed to construct a fault tree is the definition of the initial event. The first event is a hazard or action that will cause the process to deviate from normal operation. The next step is to define the existing conditions needed to be present for the accident to occur. The propagation event (e.g., the mechanical failure of equipment related to the accident) is discussed. Any other equipment or components that need to be studied have to be defined. This includes safety equipment that will bring about the termination of the accident. Finally, the normal state of the system in question is determined. The termination of an accident is the event that brings the system back to its normal operation. An example of an accident would be the failure of a thermometer in a reactor. The temperature in the reactor could rise and a runaway reaction might take place. Stopping the flow to the reactor and/or cooling the contents of the reactor could terminate the accident.

Event trees are diagrams that evaluate the consequences of a specific hazard (AIChE 1985; Theodore and Dupont 2012). The safety measures and the procedures designed to deal with the event are presented. The consequences of each specific event that led to the accident are also presented. An event tree is drawn (a sequence of events that led up to the accident). The accident is described. This allows the path of the accident to be traced. It shows what could be done along the way to prevent the accident. It also shows other possible outcomes that could arise had a single event in the sequence been changed.

18.5.3 Accident Consequence Evaluation

The consequences of accidents can be classified qualitatively by the degree of severity. A quantitative assessment is beyond the scope of the text; however, information is available in the literature. Factors that help to determine the degree of severity are the concentration of the hazard that is released, the length of time that a person or the environment is exposed to a hazard, and the toxicity of the hazard. The worst-case consequence or scenario is defined as a conservatively high estimate of the most severe accident identified (AIChE 1985; Theodore and Dupont 2012; Theodore, M.K. and Theodore, L. 2021). On this basis, one can rank the consequences of accidents into low, medium, and high degrees of severity (U.S. EPA 1987). A low degree of severity means that the hazard is nearly negligible, and the injury to a person, property, or the environment is observed only after an extended period. The degree of severity is considered to be medium when the accident is serious, but not catastrophic, the toxicity of the chemical released is great, or the concentration of a less toxic chemical is large enough to cause injury or death to persons and damage to the environment unless immediate action is taken. There is a high degree of severity when the accident is catastrophic or the concentrations and toxicity of a hazard are large enough to cause injury or death to many persons, and there is long-term damage to the surrounding environment. Figure 18.4 provides a graphical qualitative representation of the severity of consequences (U.S. EPA 1987).

Cause–consequence risk evaluation combines event tree and fault tree analyses to relate specific accident consequences to causes (AIChE 1985; Theodore and Dupont 2012). The process of cause–consequence evaluation usually proceeds as follows:

1) Select an event to be evaluated.
2) Describe the safety system(s)/procedure(s)/factor(s) that interfere with the path of the accident.
3) Perform an event tree analysis to find the path(s) an accident may follow.
4) Perform a fault tree analysis to determine the safety function that failed.
5) Rank the results based on the severity of the consequences.

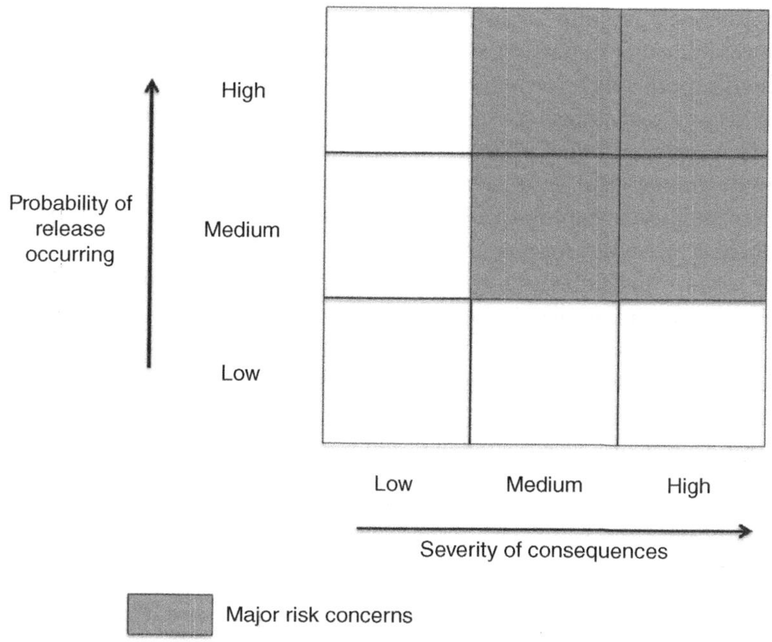

Figure 18.4 Qualitative probability-consequence analysis (U.S. EPA 1987).

As its name implies, cause–consequence analysis allows one to see how the possible causes of an accident and the possible consequences that result from that event interact with each other. Details on all of the above calculations are available in the literature (Paustenbach 1989; Theodore and Dupont 2012; Theodore 2016).

18.6 Future Trends

Today's sophisticated equipment and technologies require equally sophisticated means of accident prevention. Unfortunately, the existing methods of detection and prevention are often assumed to be adequate until proven otherwise. This approach to determining a technology's effectiveness is sometimes costly and often leads to loss of life. Chemical manufacturers and power plants are businesses and thus are not as likely to "unnecessarily" update their present controls.

Before the advent of technology, there was still a need for safety features and warnings; yet these did not exist. Many accidents occur because of a lack of knowledge of the system, process, or substance being dealt with. Many of the pioneers of modern science were sent to an early grave by their experiments. Karl Wilhelm Scheele, the Swedish chemist who discovered many chemical elements and compounds, often sniffed or tasted his finds. He died of mercury poisoning. Marie Curie died of leukemia contracted from overexposure to radioactive elements. Had either of these brilliant scientists had an accurate idea of the properties of their materials, their methods certainly would have been significantly different. In those days, safety precautions often were devised by trial and error; if inhaling a certain gas was found to make someone sick, the prescribed precaution was not to smell it. Today, since the physical properties of most known compounds are readily found in handbooks, proper care can be exercised when working with these chemicals. Laboratories are equipped with exhaust hoods and fans to minimize a buildup of gases; in addition, safety glasses and eye-wash stations are required, and gloves and smocks must be worn. Many natural disasters are now accurately predicted, buying precious time in which warnings can be made and possible evacuation plans implemented. Radar equipment commonly tracks storms, and seismographs detect slight rumblings in the earth, which can provide early warning of potential earthquakes. Volcanic eruptions can be predicted by using seismic event counters and aerial scanning of anomalies detected in the infrared region. Where natural disasters often occurred unexpectedly in the past, similar occurrences today are more predictable. Thus, there is more time for preparation and less likelihood of loss of life.

The use of computers and modern instrumentation has greatly enhanced plant safety. System overloads, uncontrollable reactions, and unusual changes in temperature or pressure can be detected, with the information being relayed to a computer. The computer can then shut down the system or take the steps necessary to minimize the danger. The industry has come a long way from sniffing and tasting its way to safety.

For the most part, future trends in health risk assessment will probably be found in expanding the dose–response database. To help promote health risk prevention, companies should expand employee training programs designed to alert the technical staff and employees about the hazards they are exposed to on the job.

Future trends in hazard risk assessment will be found in accident prevention, not hazard analysis. To help promote hazard accident prevention, improved training should cover company safety policies and the proper procedures to follow in case an accident does occur. A significant avenue to reducing risk will involve source reduction of hazardous materials (i.e., hazardous material substitution with nonhazardous alternatives, development of safer synthesis pathways, and development of nonhazardous product chemicals). Risk education and communication are two other areas that need improvement (Dupont, Ganesan, and Theodore 2017).

References

American Institute of Chemical Engineers (1985). *Guidelines for Hazard Evaluation Procedures.* New York: Battelle Columbus Division for the Center for Chemical Process Safety of the American Institute of Chemical Engineers.

Burke, G., Singh, B., and Theodore, L. (2000). *Handbook of Environmental Management and Technology*, 2nd Edition. Hoboken, NJ: John Wiley & Sons.

Crowl, J. and Louvar, J. (1990). *Chemical Safety Fundamentals with Applications*. Englewood, NJ: Prentice-Hall.

Dupont, R.R., Ganesan, K., and Theodore, L. (2017). *Pollution Prevention: Sustainability, Industrial Ecology, and Green Engineering*, 2nd Edition. Boca Raton, FL: CRC Press/Taylor & Francis Group.

Paustenbach, D. (1989). *The Risk Assessment of Environmental and Human Health Hazards: A Textbook of Case Studies*. Hoboken, NJ: John Wiley & Sons.

Theodore L. (2016). *Environmental Risk Analysis*. Boca Raton, FL: CRC Press/Taylor & Francis Group.

Theodore, L. and Dupont, R.R. (2012). *Environmental Health and Hazard Risk Assessment: Principles and Calculations*. Boca Raton, FL: CRC Press/Taylor & Francis Group.

Theodore, L. and Dupont, R.R. (2023). *Chemical Process Industries: Environmental and Health Risk Calculations*. Boca Raton, FL: CRC Press/Taylor & Francis Group.

Theodore, M.K. and Theodore, L. (2021). *Introduction to Environmental Management*, 2nd Edition. Boca Raton, FL: CRC Press/Taylor & Francis Group.

U.S. Environmental Protection Agency (1987). *Technical Guidance for Hazards Analysis*. Washington, DC: EPA/FEMA/DOT.

19

Energy–Environmental Interactions

19.1 Introduction

A big part of any discussion concerning hydrogen is the environmental issues, safety, and economic assessment. Many have come to realize the interactions exist between the two topics, energy costs and environmental costs. This chapter attempts to introduce the reader to the energy-environment issues that are likely to arise as the technical community investigates a hydrogen energy economy.

In terms of the environmental implications of hydrogen, the fundamental attraction of hydrogen and the main driving force behind the current research and development effort is its environmental advantages over fossil fuels. But hydrogen is only as clean as the technologies used to produce and use it. Replacing fossil fuels with hydrogen in providing energy services could bring major environmental benefits, but primarily on the condition that the carbon dioxide and noxious gases emitted in the hydrogen production process are reduced or eliminated. The production of hydrogen can be emission-free if the energy used in the production process is derived from solar, geothermal, nuclear power, or other renewables. If it is to be based on fossil fuels, the carbon dioxide emitted would need to be captured at the point of production.

National governments would need to provide for the elements that will promote the societal goals as well as net-zero carbon goals in a hydrogen energy economy. The role of politics is to organize a country's political and economic institutions so that they are inclusive. Thereby creating a rich environment for growth, innovation, and sustainable development from economic, societal, technological, and environmental perspectives. The political system of a nation is critical to its success, and it follows that politics will have a major role in developing hydrogen energy policy. In the modern world, a nation's energy policy is proving to be critical to its ultimate success and/or failure and its role and influence internationally.

It is important to note the economic and political roles that environmental issues play. Air, water, land, and safety have been the ongoing issues. Carbon dioxide and other gaseous carbon emissions that impact global warming are now a big concern. Hydrogen-based fuel has great potential to help remedy this concern.

Sections titles for this Chapter 19 "Energy–Environmental Interactions" are:

19.1 Introduction
19.2 U.S. Hydrogen Energy Policy
19.3 U.S. Energy–Environmental Policy Issues
19.4 Individual State Energy Policies

Hydrogen Energy: Principles and Applications, First Edition. Vincent J. DelGatto, Louis Theodore, R. Ryan Dupont, and Matthew C. Ogwu.

19.2 U.S. Hydrogen Energy Policy

The U.S. Infrastructure Investment and Jobs Act of 2021 appropriates $62 billion for the U.S. Department of Energy (DOE), including $9.5 billion for clean hydrogen. In the U.S. Inflation Reduction Act of 2022 (IRA), (Public Law 117-169) allocated more funds and tax credits for hydrogen development. This legislation provides additional policies and incentives for hydrogen including a production tax credit that has further boosted the U.S. market for clean hydrogen. The "U.S. National Clean Hydrogen Strategy and Roadmap" in 2023 was informed by extensive industry and stakeholder feedback, including workshops and listening sessions, written comments from more than 50 organizations, and ongoing engagement. In addition, this roadmap sets forth an all-government approach to clean hydrogen, with contributions across multiple agencies as well as key experts in the Executive Office of the President.

This inclusive and collaborative approach is critical to the success of this expansive technology. The report is meant to be a living strategy that provides a snapshot of hydrogen production, transport, storage, and use in the United States today, as well as an assessment of the opportunity for hydrogen to contribute to national decarbonization goals across sectors over the next 30 years. Using clean hydrogen can reduce U.S. emissions by approximately 10% by 2050 relative to 2005, consistent with the U.S. Long-Term Climate Strategy. The U.S. National Clean Hydrogen Strategy and Roadmap in 2023 also summarizes many of the U.S. hydrogen efforts such as the U.S. Hydrogen Energy Earth Shot of 2021. These Federal Acts and Laws above also amend Title VIII of the Energy Policy Act of 2005 (EPACT-2005) by adding Section 814 – National Clean Hydrogen Strategy and Roadmap. Section 814 is presented below.

> Section 814 states: A) DEVELOPMENT. —(1) IN GENERAL.—In carrying out the programs established under sections 805 and 813, the Secretary, in consultation with the heads of relevant offices of the Department, shall develop a technologically and economically feasible national strategy and roadmap to facilitate widescale production, processing, delivery, storage, and use of clean hydrogen. (2) INCLUSIONS.—The national clean hydrogen strategy and roadmap developed under paragraph (1) shall focus on—(a) establishing a standard of hydrogen production that achieves the standard developed under section 822(a), including interim goals toward meeting that standard; (b) (i) clean hydrogen production and use from natural gas, coal, renewable energy sources, nuclear energy, and biomass; and (ii) identifying potential barriers, pathways, and opportunities, including Federal policy needs, to transition to a clean hydrogen economy; (c) identifying—(i) economic opportunities for the production, processing, transport, storage, and use of clean hydrogen that exist in the major shale natural gas-producing regions of the United States; (ii) economic opportunities for the production, processing, transport, storage, and use of clean hydrogen that exist for merchant nuclear power plants operating in deregulated markets; and (iii) environmental risks associated with potential deployment of clean hydrogen technologies in those regions, and ways to mitigate those risks; (d) approaches, including sub-strategies, that reflect geographic diversity across the country, to advance

clean hydrogen based on resources, industry sectors, environmental benefits, and economic impacts in regional economies; (e) identifying opportunities to use, and barriers to using, existing infrastructure, including all components of the natural gas infrastructure system, the carbon dioxide pipeline infrastructure system, end-use local distribution networks, end-use power generators, LNG terminals, and other users of natural gas, for clean hydrogen deployment; (f) identifying the needs for and barriers and pathways to developing clean hydrogen hubs (including, where appropriate, clean hydrogen hubs coupled with carbon capture, utilization, and storage hubs) that—(i) are regionally dispersed across the United States and can leverage natural gas to the maximum extent practicable; (ii) can demonstrate the efficient production, processing, delivery, and use of clean hydrogen; (iii) include transportation corridors and modes of transportation, including transportation of clean hydrogen by pipeline and rail and through ports; and (iv) where appropriate, could serve as joint clean hydrogen and carbon capture, utilization, and storage hubs; (g) prioritizing activities that improve the ability of the Department to develop tools to model, analyze, and optimize single-input, multiple-output integrated hybrid energy systems and multiple-input, multiple-output integrated hybrid energy systems that maximize efficiency in providing hydrogen, high-value heat, electricity, and chemical synthesis services; (h) identifying the appropriate points of interaction between and among Federal agencies involved in the production, processing, delivery, storage, and use of clean hydrogen and clarifying the responsibilities of those Federal agencies, and potential regulatory obstacles and recommendations for modifications, in order to support the deployment of clean hydrogen; and (i) identifying geographic zones or regions in which clean hydrogen technologies could efficiently and economically be introduced in order to transition existing infrastructure to rely on clean hydrogen, in support of decarbonizing all relevant sectors of the economy.

Pathways for clean hydrogen to decarbonize applications are informed by demand scenarios for 2030, 2040, and 2050 with strategic opportunities for 10 million metric tons (MMT) of clean hydrogen annually by 2030, 20 MMT annually by 2040, and 50 MMT annually by 2050. These values are based not only on the opportunity for clean hydrogen production in the United States, but on demand for clean hydrogen use across sectors, informed by achieving market competitiveness in specific applications.

See reference-DOE
https://liftoff.energy.gov/wp-content/uploads/2023/05/20230523-Pathways-to-Commercial-Liftoff-Clean-Hydrogen.pdf

19.3 U.S. Energy–Environmental Policy Issues

General comments on energy–environmental interactions are provided below. While this list is not comprehensive, it provides the reader with some of the more important issues on the topic.

1) The chronic worldwide energy gap challenges domestic as well as world economic stability and security, while society expects planning to appropriately address low-carbon substitutions to primary energy sources.
2) Evolving national energy plans struggle to keep current and cause uncertainty.
3) Development of an energy plan requires a careful and pragmatic look at the technical validity of underlying assumptions and their feasibility, as well as the capital costs and how they will be affected by political and monetary uncertainties.

4) Society sees the vagaries of regulation, misallocation of resources, and institutional rigidities, as problems in industry and government.
5) Increasing focus on Environmental Social Governance (ESG) impacts corporate policies.
6) A new hard look at how an energy plan such as a hydrogen economy with energy/environment policy trade-offs could be implemented to minimize damage to the economy and safeguard national priorities and societal goals is needed.
7) There is the risk that carbon reduction targets will require investments so large and institutional changes so radical that both private enterprises and the consumer may withhold approval.
8) Concern exists with optimistic assumptions in the projected supply and demand of various energy sources and the synergy between old and new energy forms.
9) A number of energy experts consider extreme zero-carbon targets will be unattainable by 2050.

Developing a hydrogen economy in the second quarter of the 21st century will require bringing many environmental issues into the planning process. Environmental degradation can rarely be avoided during the process of obtaining and using energy in developed countries. Any energy policy should be extremely sensitive to activities that may damage as well as help the environment for decades or generations to come. A balance between issues of environmental quality and issues of national energy security is paramount. Much work is needed to address many uncertainties that presently exist to transition to synthetic hydrogen fuels.

Net zero in 2050 for the United States is the energy goal rather than a specific policy. This goal embraces both conservation and greater reliance on renewable energy resources with a lower-carbon footprint. Conservation has been a major ingredient of a national energy policy. More energy-efficient products and processes continue to be developed. However, conservation efforts alone will not significantly reduce CO2 emissions, especially with a growing global economy and population.

The United States has typically addressed issues successfully over the years because it has reconciled many opposing viewpoints in a balanced and responsible manner. In the implementation of environmental laws, this has not always been easy. Success in achieving economic growth with environmental quality also depends on well-reasoned public interest decisions. Congress must act with wisdom and deliberation, and it must process contradictory information to formulate a sense of the public will. The intent is an implementable energy policy backed by cogent, logical, and enforceable rules and regulations with flexibility to accommodate innovation and changing conditions.

The U.S. Sulphur Dioxide Cap and Trade Program was established in 1990 as part of Title IV of the Clean Air Act Amendments (CAAA). The program was the world's first large-scale cap-and-trade system for pollutants. Its goal was to reduce sulfur dioxide (SO_2) emissions from power plants by half by 2010, compared to 1980 levels. The program also aimed to reduce nitrogen oxides (NO_x), which are both key components of acid rain.

The program used a market-based approach to environmental protection, allowing companies to decide how to reduce their emissions. Compared to command-and-control regulations, the cap-and-trade market-based mechanism created financial incentives to reduce emissions. Companies that reduced their emissions more than required could sell their extra allowances to other plants. The program has been widely considered to be very effective, with sulfur emissions dropping faster than expected and at a fraction of the projected cost.

In 2009, the U.S. Congress considered but ultimately failed to enact legislation aimed at limiting U.S. greenhouse gas (GHG) emissions. Hostility toward cap-and-trade in debates about U.S. climate legislation may have reflected the broader political environment of the climate debate more than the substantive merits of market-based regulation.

Energy production and environmental improvement do not need to be conflicting goals. Both can be achieved for the benefit of all people and economies under a reasonable policy. Some individuals believe environmental protection should be pursued above all other national goals. Others advocate economic growth and production as the sole objective. But either of these extremes would impose excessive societal costs. Both a clean and healthy society with a growing economy is the ultimate goal. The issues surrounding growth, environmental quality, and affordability need to align. A low-carbon economy will require a massive transition to low-carbon renewable energy and low-carbon fuels. According to the January 2022 report *The net-zero transition: What it would cost, what it could bring* by McKinsey and Company (Krishnan et al. 2022), "the global capital spending on physical assets for energy and land-use systems in the net-zero transition between 2021 and 2050 would amount to about $275 trillion, or $9.2 trillion per year on average, an annual increase of as much as $3.5 trillion from today." This author respects the 2022 McKinsey report but questions if the projected $3.5 trillion annual global increase is too low. Follow on report by McKinsey Sustainability (Krishman 2023) does suggest larger increases. Certainly, beyond net-zero to a zero-carbon economy will require much more time and money after 2050.

19.4 Individual State Energy Policies

National energy plans are normally developed by the federal government and executed by the President in coordination with various agencies such as the U.S. DOE and other agencies and various branches such as the EIA. However, individual states and local governments have developed individual policies for their constituents that translate the federal regulations and other requirements as well as adding local rules and guidance. Due to the diversity in population, climate, building codes, economics, government, industry, and other aspects that influence individual states and local communities, applicable policies can have a wide range of objectives.

In New York, to help ensure ample and affordable energy supplies, the New York State Energy Planning Board, through the adoption of the 2002 State Energy Plan and its revisions in 2015, positioned the state to maximize the use of clean and efficient energy and transportation technologies that can provide for the safe and secure operation of the state's growing demand for energy, provide for the safe and secure operation of the state's energy and transportation systems, and support the continued development of competitive energy markets. The energy policies and strategies contained in the plan were designed to keep New York at the forefront among states in providing its citizens with abundant, competitively priced, clean, and efficient energy resources. The Energy Planning Board adopted the following public policy objectives in the 2015 State Energy Plan (New York State 2015).

1) Drive system-wide savings that benefit all customers and encourage private investment in distributed clean energy solutions that help customers better manage their energy bills and reduce fuel costs.
2) Animate clean energy markets through strategies to attract private sector capital investment, and support clean transportation alternatives, enabling New York State to meet its aggressive environmental energy goals and transition to a clean energy economy.
3) Complement and further other resiliency efforts by promoting the development of clean, local energy resources that strengthen and improve the reliability of the grid.
4) Provide New York State with the ability to operate its energy system more efficiently and at a lower cost, and enable utilities to chart a vibrant, but changing, future.

5) Help communities disproportionately impacted by air pollution pursue a clean energy future.
6) Guide the development and implementation of programs that will help fund and facilitate the clean transportation system of the future while maintaining existing infrastructure. Build an integrated energy network able to harness the combined benefits of the central grid with clean, locally generated power.

The 2020 New York State Climate Leadership and Community Protection Act dramatically pushes ahead from the 2015 Energy Plan and requires New York to reduce economy-wide greenhouse gas emissions by 40% by 2030 and no less than 85% by 2050 from 1990 levels.

In 2024, twelve states will have active carbon-pricing programs and are successfully reducing emissions. Those states are California, Washington, and the eleven Northeast states – Connecticut, Delaware, Maine, Maryland, Massachusetts, New Hampshire, New Jersey, New York, Pennsylvania, Rhode Island, Vermont – that make up the Regional Greenhouse Gas Initiative (RGGI).

RGGI is the first mandatory cap-and-trade program in the United States to limit carbon dioxide emissions from the power sector. California's program was the first multi-sector cap-and-trade program in North America. Massachusetts has also implemented regulations to establish an additional cap-and-trade program for its power sector that runs in parallel with RGGI but extends out to 2050. Washington State enacted new cap-and-invest legislation that took effect in 2023. New York is preparing an economy-wide cap-and-invest program that is expected to launch in 2025.

19.5 Global Energy Policies

The preceding sections have highlighted some of the features of an energy policy for the United States and individual states in the United States. As noted, the policy objectives vary depending on the specific governmental agency involved and continually change as new administrative leadership is placed in charge of the respective legislative and executive offices. The basis for why energy policies are difficult to construct and ultimately carry out is the result of this ever-changing leadership and control landscape in addition to public perception of the best path forward. The ever-changing economic forces on energy policy are also primary drivers and significantly affect the chosen path. Historically, energy resources have over time changed drastically, especially in the past several hundred years. Based on current trends, they are likely to change even more rapidly in the coming decades (Dupont, Ganceran, and Theodore 2017)

Most developed countries have energy policy documents, rules, and established boards of experts to provide guidance and attempt to manage their energy resources, supply and demand, and related interests. Some, like France, have chosen the nuclear approach to electricity generation, while others have chosen the resources at hand to meet their energy needs.

One of the most important international forums for analyses and discussions about energy policy is the International Energy Agency (IEA). The IEA was founded in 1974 following the first great oil crisis, which hit the world economy hard. To avoid similar crises in the future, the countries in the Organization for Economic Coopreation and Development (OECD) established the IEA as a forum to cooperate on energy policies, primarily in order to build up oil supplies to avoid getting hit hard by oil supply cuts.

The IEA was originally focused on oil supply security, but as the global energy agenda developed, the IEA changed significantly. Today, the organization has a much wider area of interest and is one of the most important international, intergovernmental forums when it comes to energy policy. It is also the international energy community's dominant think tank.

The IEA conducts a series of analyses across the energy policy spectrum on an ongoing basis via an international secretariat with approximately 200 people. They look at the security of supply as well as economic, technical, and environmental issues related to the production and consumption of energy. The IEA evaluates the market situation within coal, oil, gas, renewable energy, and energy efficiency.

The IEA's flagship publication is the World Energy Outlook, which is published each year at the beginning of November and aims to provide an overview of the entire current global energy situation and perspectives for the coming 25 years. The World Energy Outlook is approximately 700 pages long and helps to set the agenda within the international energy policy community.

The IEA's technical analyses are discussed by the secretariat and the member states in technical committees. The IEA's analyses tend to be both neutral and technical. Data is based on input from the energy and political realities in the member states. The most current World Energy Outlook can be accessed at: https://www.iea.org/reports/world-energy-outlook-2022. This is an excellent report to reference, including excellent graphs trending to 2050.

The European Union adopted a "Green Deal Policy Plan" in 2020. By 2050, Europe aims to become the world's first climate-neutral continent. To pave the way to achieve this ambitious target, the European Commission (EC) pledged to reach the following more detailed targets by 2030: Minimum 55% cuts in greenhouse gas emissions; above 32% share of renewable energy; at least 32.5% improvement in energy efficiency. In order to be able to reach these three targets on time, the EC created a dedicated action plan known as the EU Green Deal, which covers a wide range of policy areas.

To get to the Net-Zero goal, the Green Deal calls for a number of parallel and complementary lines of action: clean energy, circular economy, building efficiency, agriculture, biodiversity protection, pollution control, and social justice. Each of these areas is outlined in the 2019 document, and all are intended to be translated into more detailed plans in the years to come. This is why some measures (such as energy efficiency in buildings and limiting the use of combustion engines for cars) have been at the center of political debate in recent months. The time is coming when the broad commitments outlined in the Green Plan must be translated into legislative choices. Europe's largest focus is now on clean energy since energy production and use in all sectors accounts for 75% of Europe's greenhouse gas emissions. The cornerstone principles of the Green Deal are the rapid abandonment of coal, the massive development of renewable sources, the promotion of energy efficiency, and the digitalization and interconnection of the European energy market, to ensure equal access to energy for all. In particular, the Fit for 55 package sets a target of producing 40% of Europe's energy from renewable sources by 2030.

The United States chose to cloak its recent carbon-lowering initiatives inside its 2022 Inflation Reduction Act. In the European elections on June 9, 2024, the 2020 Green Deal lost significant support by voters fearing more the economy than the environment. Elections in the United States in 2024 may also show waning support for strong net-zero regulations as the potentially high costs of rapidly transitioning to a net-zero carbon economy become more obvious.

Shipping transportation on the world's oceans is now subject to rules limiting ship carbon intensity. International emission standards may be more widespread for ships because the International Maritime Organization (IMO) is less tied to national politics. The IMO is a special agency of the United Nations. The strong IMO 2023 Greenhouse Gas (GHG) Strategy reduces emissions from international shipping to reach net-zero emissions by 2050. The IMO GHG strategy includes the following targets: 2030 – Reduce emissions by 20–30% compared to 2008 levels, and increase the use of zero or near-zero GHG fuels to at least 5–10% of international shipping's energy, 2040 – Reduce emissions by 70–80% compared to 2008 levels, and 2050 – Reach net-zero emissions.

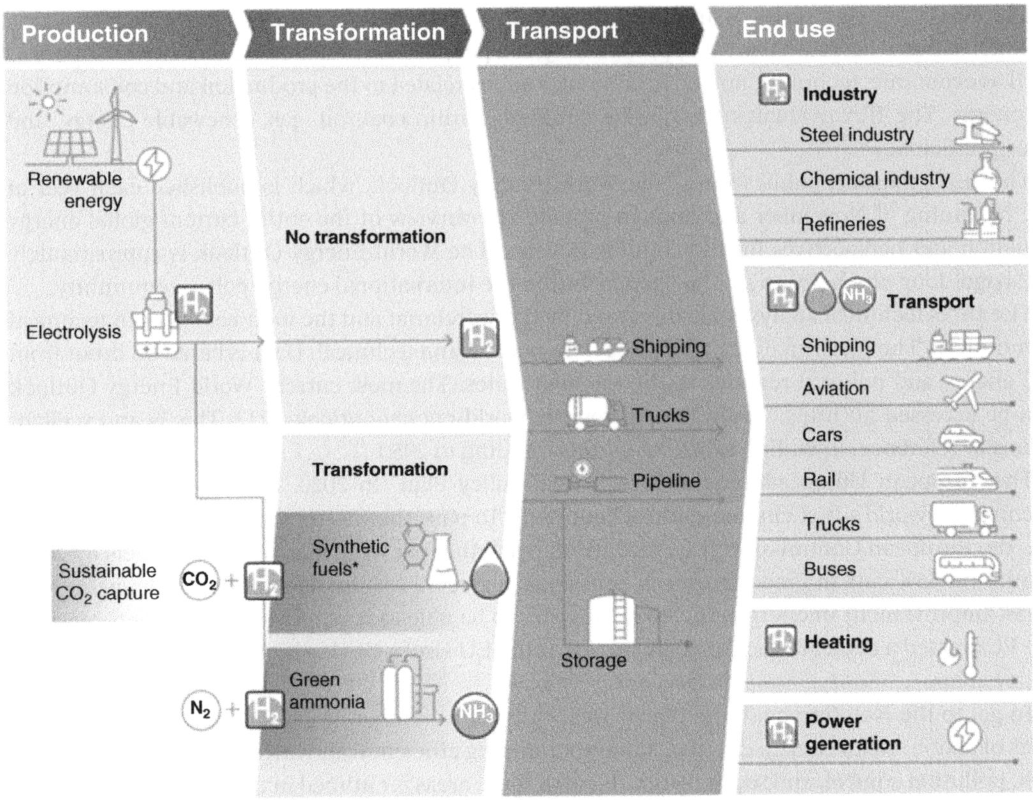

Figure 19.1 Green hydrogen production, conversion, and end uses across the energy system. *Source:* IRENA (2020).

Transportation on land, sea, and air all require a highly energy-concentrated fuel supply. Compressed hydrogen gas, liquid ammonia, and other net-zero hydrogen-based synthetic fuels are expected to be the majority of transport fuels in 2050. Figure 19.1 shows the projected green hydrogen production, conversion, and end uses across the energy system. The IMO expects that these synthetic fuels to initially meet the zero or near-zero GHG fuel 5%–10% requirement for international shipping in 2030.

The U.S. Joint Institute for Strategic Analysis (JISEA p/o NREL) predicts hybrid energy solutions will be needed for 2050 goals. The JISEA report "Scenarios of a Sustainable Energy Transition: Perspectives from the U.S" shows hybrid energy solutions will be needed for the production of hydrogen with electrolyzers powered from both thermal and intermittent renewable electric generation. This JISEA report also suggests that a major portion of industrial and transportation markets in 2050 will be served by new low-carbon or zero-carbon fuels derived from hydrogen. Over the past several decades there has been a simple quiet transformation of mixing 10% ethanol derived from corn with most gasoline and heating oil. This seamless ethanol gasoline transformation gives some confidence that hydrogen-based new lower-carbon alternative fuels for multiple uses will become a reality in the not-too-distant future.

The United Nations conducts the climate change conferences "Meetings of the Conference of the Parties (COP/CMP/CMA)": To review the implementation of the Convention, the Kyoto Protocol, and the Paris Agreement, respectively; and to adopt decisions to further develop and implement these three instruments. The conduct of the meetings and brokering of agreements within the collective decision-making framework of the COP therefore involves negotiation and compromise. Global energy policies follow from these meetings.

During COP28 in 2023, 22 nations promised to triple their nuclear power by 2050 based on the consensus that nuclear power is now considered to be a carbon-free renewable energy. Nuclear electric power plants are expected to be a major energy source for the production of hydrogen by electrolysis. One plan for nuclear-powered baseload electric generation is to produce hydrogen when there is a surplus of intermittent wind and solar electric generation. However, intermittent production of hydrogen based on periods of electric generation surplus would likely increase hydrogen production costs due to the low utilization of expensive electrolyzers.

The International Renewable Energy Agency (IRENA) 2022 report titled Energy-Transition/Tecnology/Hydrogen Report reiterates the potential huge growth of electrolytic hydrogen-rated generation from 0.7 GW in 2021 to 5 TW by 2050 (7,143 × multiple). In contrast, global total electric generation capacity is expected to grow from approximately 10 TW now to 30 TW (3× multiple) by 2050. Certainly, hydrogen has great worldwide attention but much innovation is needed to overcome conversion losses and achieve exponential growth.

The International Renewable Energy Agency (IRENA) 2018 report titled "Global Energy Transformation: A Roadmap to 2050" predicts global electricity consumption to double by 2050 to 50,000 TWh and correspondingly electric generation capacity to triple from 10 TW to 30 TW by 2050. Nonrenewable fossil fuel electric generation capacity could decline from approximately 9 TW rated power now to 3 TW in 2050. Total renewable generation capacity will increase from 1 TW now to 27 TW in 2050. If low capacity factor intermittent wind and solar are the majority of renewables in 2050, then this 27 TW of intermittent generation capacity requirement could double to 54 TW or more to meet demands for storage recharging. It is important to note that globally 71% of all renewable energy generation (not counting nuclear) in 2023 was from long established base-loaded hydroelectric generating stations. In the United States the percentage of all renewable energy generation (not counting nuclear) from hydropower electric generation has declined from approximately 97% in 1983 to 26% in 2023 (Table 19.1). This dramatic percentage change has occurred because U.S. wind and solar-generated electricity have dramatically grown due to incentives and favorable regulations.

19.6 Environmental Concerns: A Technological Mandate

The public could become hesitant to accept price increases and inconveniences of conservation, fuel substitution, and changes in transportation modes that are likely for a meaningful national energy program to achieve net-zero carbon emissions in 2050. If new net-zero carbon policies occur with higher unemployment and a higher cost of living, the public may attribute additional inconvenience and living costs to any new policy superimposed upon already voluminous environmental statutes. Congress may find environmental laws to be incompatible with a rapid increase in energy production from domestic sources, causing a regulatory impasse. Congress may then have a difficult time deciding on the appropriate path forward. This was evident by a National Renewable Energy Laboratory report comparing 12 different National Energy Plans (NEPs) dating back to 2009 (Logan and James 2009).

Major interruptions of any energy supply often cause the government to seek a temporary accommodation from environmental regulations that may not be sufficiently flexible to meet the energy priorities. For example: "variances" from existing regulations in areas of relatively mild air pollution.

Table 19.1 U.S. utility-scale electricity generation by source, amount, and share of total in 2023.

U.S. Energy Source of Electricity in 2023	Billion kWh	Share of total (%)
Total – all sources	4,178	
Fossil fuels (total)	2,505	60.0
Natural gas	1,802	43.1
Coal	675	16.2
Petroleum (total)	16	0.4
Petroleum liquids	12	0.3
Petroleum coke	5	0.1
Other gases	11	0.3
Nuclear	775	18.6
Renewables (total)	894	21.4
Wind	425	10.2
Hydropower	240	5.7
Solar (total)	165	3.9
Photovoltaic	162	3.9
Solar thermal	3	0.1
Biomass (total)	47	1.1
Wood	31	0.8
Landfill gas	8	0.2
Municipal solid waste (biogenic)	6	0.1
Other biomass waste	2	0.1
Geothermal	16	0.4
Pumped storage hydropower	−6	−0.1
Other sources	10	0.2

Data source: U.S. Energy Information Administration, *Electric Power Monthly*, February 2024; preliminary.

Legislative amendments must also consider quantifiable trade-offs between the environment, economics, power reliability, and energy shortages. It may become increasingly difficult in the future to balance energy needs and environmental management. A rational compromise to resolve this quandary will require ongoing policy adjustments even after the enactment of state and federal energy policies. However, industry planning suffers when government guidelines, policies, and timing change. For example, in New York State, the combination of transmission congestion and new regulations giving priority to new intermittent wind and solar dramatically reduced electric generation from existing upstate hydropower and nuclear baseload generators. This also caused an increase use of downstate fossil fuel generation and reduced the valuation of many older upstate generators.

Energy industry management (usually specialized in a specific sector) and government agencies (often preoccupied with pressing problems of the day) may be challenged to cope with the interdisciplinary linkage and assessment of environmental concerns needed for problem-solving under emergency conditions of any resource scarcity. The verification and critique of facts and figures, models, and simulations should help energy planners analyze the cost-benefits and policy trade-offs when investigating future energy policy.

Many consulting firms, think tanks, and policy groups have taken deep dives into analyzing the potential benefits and problems of reducing carbon emissions with plans that include alternative fuels with hydrogen feedstock. A very well-researched report titled "The net-zero transition, What will it cost, what could it bring" by McKinsey Global Institute, January 2022 provides much insight into the high costs of rapid transition. Independent technical engineering societies like the Institute of Electrical and Electronic Engineers and the American Society of Chemical Engineers (ASCHE) also provide expert technical advice to legislatures and government leaders.

19.7 Net Energy Concepts

Any decisions on energy–environmental policy should include a net energy analysis (NEA), a topic briefly introduced earlier in Chapter 2. Energy analysis (EA), in a general sense, represents a broad field of study dealing with the development and use of all aspects of energy in human society and its environment, while NEA is a more constrained field of study. NEA deals with the analysis of the energy made available to society by an energy technology *after* the deduction of energy lost to society because of the total energy to find, extract, process, deliver, and upgrade that energy to a socially useful form. This subject can also include the EA of materials production – how much energy must be invested in the total system needed for the production of a unit of material. Materials production and utilization systems should be examined for energy reduction potentials consistent with those discussed here.

There are three typical major energy questions or issues to which the subject of NEA could apply. All three major issues include the concept of cradle-to-grave energy and associated carbon expenditures.

1) How much energy is required from the industrial component of society to drive or to establish and operate an energy production process, relative to the energy yield of the process?
2) In any "extracting" process and the process of converting and moving that energy to end users, what final yields are obtained relative to losses if the total energy in the recovered resources and in the energy needed to establish and operate the energy production system(s)?
3) For a given output of energy for end use, what total amounts of the gross resources and industrial energies are necessary to establish and operate the system?

There is also the concern of defining the boundary conditions that would lead to analytical findings relevant to these major issues. Because an NEA is a relatively new technique, various individuals have defined and employed different boundaries. Some have included the human energy of employees; others have included the energy in ecosystems that are disturbed by industrial processes (Skipka and Theodore 2014).

Another method of analysis popular for electric generation is Levelized Cost of Energy (LCOE). This method is a simple measure of the average net present cost of electricity generation for a generator over its lifetime. It is used for investment planning and to compare different methods of electricity generation on a consistent basis. However, LCOE does not directly account for the costs of low capacity factor that is characteristic of intermittent renewable wind or solar. These intermittent sources need backup power generation and storage. In a report to the IEEE-USA Energy Policy Committee presented by VJ DelGatto on the estimated construction cost of 2 GW solar power with 30% capacity factor and additional battery or hydrogen storage for 48-hour operation. The report calculates the construction cost for 2 GW solar with 40 hours of storage (80 GWh) would exceed the 37 billion dollar cost of recent 2 GW nuclear units in the United States with 90% capacity factor. Please refer back to Chapter 14.4 for full details.

While LCOE can show intermittent solar to have the lowest LCOE, net energy analysis (NEA) can show solar to have the highest net energy cost because of the capacity factor. If we consider the extreme example of a 100% full solar economy with full 24/7 storage providing 95% capacity factor: NEA may likely determine that just the build portion of this extreme example of 100% full global solar economy with 100% battery backup could raise temperatures beyond the 1.5 degree mark even before construction is completed.

Neither NEA nor LCOE should be used in isolation as the primary decision factor. Dynamic System Modeling techniques should also be applied. The following topics are among the dynamic factors which generally carry more weight and should be considered:

1) Economics
2) Environment
3) National security
4) Energy mix, end-use efficiencies, and sustainability
5) Lead times
6) Transportation capacities
7) Institutional restraints such as governmental regulations and incentives
8) Availability of needed materials
9) Available water
10) Local attitudes
11) Socioeconomic impacts
12) Employment needs
13) Needs for energy
14) Comparative risks and rewards
15) Weather and climate issues
16) Infrastructure flexibility to shift to alternative fuels/technologies
17) Ability to remain open to incorporate newer game-changing technologies in the future.

Application of all three analysis techniques above would help to better refine energy policy. It would probably show that the public interest may not be well served if the power industry were required to install excessive intermittent renewable generation with full 24/7 energy storage systems. Likewise, these analysis techniques could help determine a better-integrated plan for the transition to a hydrogen economy. Extending carbon goals past 2050 would provide more transition time for an efficient development of the needed extensive green hydrogen fuel infrastructure which is projected to ultimately provide 5 TW of global power capacity. Flexibility should also remain for other technologies as they could rapidly innovate in the future. The famous energy policy author Dr. Daniel Yurgin metaphorically concluded that the energy economy needs to run on all cylinders.

A much more positive view of transitioning to a low-carbon economy is provided in the excellent progressive forward-thinking report titled "The Breakthrough Effect: How to Trigger a Cascade of Tipping Points to Accelerate the Net Zero Transition" by Systemiq, Univerisity of Exeter and Bezos Earth Fund in January 2023. The report examines where positive socioeconomic tipping points may exist for zero-emission solutions and what conditions can activate these tipping points. Setting the conditions to trigger tipping points and reinforcing feedback loops becomes a dominant market driver. The adoption of zero-carbon solutions then scale up an S-curve. This very positive report suggests transition to a low-carbon economy has the potential to create a more prosperous and just global economy on multiple fronts. Hydrogen is also one of the cornerstones of this transition.

19.8 Interaction with Other Goals

Unprecedented changes in the world economy are causing difficulties that go beyond deficits in the balance of payments or tensions within the international community. The specter of violent disruptions in world materials markets and prices (through political upheaval, international incidents, terrorism, pandemics, or curtailment of freedom on the high seas and elsewhere) could severely challenge U.S. security. These violent disruptions may occur spontaneously or are influenced by the centrally planned economies whose economic role and military leverage are growing, or due to adversary relationships inherent in U.S. foreign policy.

A cautionary detailed analysis of the zero-carbon race is extremely well presented in Lawrence Papers for Global Security No. 12 – "Examining The Energy Transition Through The Lens of Great Power Competition" by Asmeret D. Asghedom at Lawrence Livermore Labs in August 2023. Issues of limited raw materials, concentrated manufacturing locations, supply chain challenges, technical feasibility, costs, land use, pollution, energy reliability, and energy security are cogently presented as reasons why the net-zero race to 2050 is fraught with socioeconomic and geopolitical risk.

The extraneous impacts on the supplies of natural resources that affected the economic recovery from the 2008 economic downturn or the recovery from the 2019 COVID-19 pandemic are real and have caused dislocations that go beyond the scope of conventional planning and decision-making in industry(Engel-Cox 2021). Economic questions and issues that will influence the outlook for transitioning to hydrogen fuels, as well as other lower-carbon technologies, are:

1) How will the growing discoveries, production, demand, and exports of the centrally controlled economies affect U.S. oil and other energy resource imports/exports and foreign trade posture?
2) Could a monetary system in disarray compound economic uncertainty, threatening defaults on foreign loans of banks and their ability to raise the capital needed to implement transitioning to lower-carbon economies?
3) How will governments implement a carbon tax or other support measures for alternative low-carbon technologies and infrastructure?
4) What are supply chain and cost problems when new low-carbon infrastructure construction requires increasing imports of high-level manufactured goods.?
5) Can the United States develop a workable policy to induce increased investments (by industries) in sectors of national priority?
6) Capital to finance the increase in domestic and global supply of low-carbon energy as well as other national priorities is declining for the following reasons:
   a) Capital formation has been reduced after severe and protracted economic slowdowns beginning in 2008 and continues with national debt pressures and now rising cost of living and higher interest rates.
   b) International uncertainty (i.e. – the 2019 pandemic, war in eastern Europe and the Middle East, localized raw materials, and new innovations in low-carbon technologies) investors are hesitant to risk their capital into questionable ventures.

7) The danger of a new worldwide economic turmoil heightened by the high costs of transitioning to a zero-carbon economy not only threatens the rational and balanced supply of energy and industrial materials but also brings instability and discouragement to the financial markets and causes losses of revenue to the government (Acemoglu and Robinson 2012).
8) Regulatory constraint that favors solar and wind generation threatens the economic viability of nuclear and other baseload electric generation required for reliable 24/7 grid operation.
9) The danger of a full build-out of intermittent renewable generation with storage not being able to accommodate future innovation and limiting adoption of future breakthroughs in nuclear and fusion power.
10) An increasing portion of United States expenditures go into mandatory compliance with environmental health, safety, and other regulations with a legal liability rather than into increased productivity that invigorated the economy in the past.

The literature that has appeared over the last decade on the influence of carbon on global warming has been extensive. Most authors strongly believe that man is unalterably changing the world's climate by continuing to release carbon dioxide into the atmosphere, carbon that was previously bound for millennia in fossil fuels. These concerns have fostered new initiatives and concepts such as the Kyoto Protocol, carbon cap-and-trade, carbon sequestration, carbon footprint analyses, and restrictions on coal-fired utility boilers. For some, the fact that the summer of 2023 is the hottest ever recorded in human history is evidence that global warming is a reality and leading to occurrences of more wildfires, heat-related deaths, and ecological upsets.

Fewer authors believe there are lesser or insignificant impacts of man's influence, citing other phenomena such as sunspot activity, methane releases, and cosmic radiation, as examples of climate-forcing events. The difficulty with any opinion on climate change is that the subject is so broad and complex, it is difficult for any one individual to be truly knowledgeable of all competing factors and analyses used to be able to reach conclusions on the impact of each individual process and their ultimate effect on climate. However, it is extremely difficult to argue that climate change is not occurring, regardless of the causes (Skipka and Theodore 2014).

Reports from the International Panel on Climate Change (IPCC) often show a range of possible climate risks the world may face. The IPCC calls these scenarios "representative concentration Pathways" (RCPs). Pathways range from relatively mild RCP2.6 scenario all the way to RCP8.5, whose risks would be calamitous. The Deputy Director of the MIT Joint Program on the Science and Policy of Global Change, Adam Schlosser, stated in *Ask MIT Climate* October 20, 2023 "that even with no further progress on climate change a less dire RCP6.0 scenario would now look to be closest to reality." Again, the less dire RCP 6.0 scenario is projected by MIT even with no further progress on climate change. The accuracy and probability of various RCP's are challenged with a Monte-Carlo set of outcomes and complex dynamic system interactions. The more probable less dire lower numbered RCP's suggest that slower transitioning to net-zero carbon may be optimum lest economic downturn, energy shortages and geopolitical unrest caused by an accelerated transition produce an even worse scenario of war. The ongoing primary existential threat of nuclear war should always be carefully considered when planning the massive disruptive transition to a low-carbon global economy.

References

Acemoglu, D. and Robinson, J. (2012). *Why Nations Fail: The Origins of Power, Prosperity, and Poverty*, 1st Edition. New York: Currency, Crown Publishing Group, Penguin Random House LLC.

Asghedom, A.D. (2023). *Examining the energy transition through the lens of great power competition livermore papers on global security no.12*. Lawrence Livermore National Laboratory Center for Global Security Research.

Dupont, R.R., Ganesan, K., and Theodore, L. (2017). *Pollution Prevention: Sustainability, Industrial Ecology, and Green Engineering*, 2nd Edition. Boca Raton, FL: CRC Press/Taylor & Francis Group.

Engel-Cox, J. (2021). *Scenarios of a sustainable energy transition: perspectives from the U.S. Presented at 2021 International Institute for Carbon-Neutral Energy Research (I2CNER) Annual Symposium* (26 January 2021), Joint Institute for Strategic Energy Analysis, Colorado, USA.

IRENA (2018). *Global Energy Transformation: A Roadmap to 2050*. Abu Dhabi: International Renewable Energy Agency.

IRENA (2020). *Green Hydrogen: A Guide to Policy Making*. Abu Dhabi: International Renewable Energy Agency.

IRENA (2022). *Energy-Transition/Technology/Hydrogen Report*. Abu Dhabi: International Renewable Energy Agency.

Krishnan, M., Samandari, H., Woetzel, J., Smit, S., Pacthod, D., Pinner, D., Nauclér, T., Tai, H., Farr, A., Wu, W., and Imperato, D. (2022). *The net-zero transition: What it will cost, what it could bring*, Report by McKinsey Global Institute, January 2022. https://www.mckinsey.com/~/media/mckinsey/business%20functions/sustainability/our%20insights/the%20net%20zero%20transition%20what%20it%20would%20cost%20what%20it%20could%20bring/the-net-zero-transition-what-it-would-cost-and-what-it-could-bring-final.pdf

Krishman, M., Tai, H., Pacthod, D., Smit, S., Nauclér, T., Houghton, B., Noffsinger, J., and Simon, D., (2023). *An affordable, reliable, competitive path to net zero*. Report by McKinsey Sustainability, November 2023 https://www.mckinsey.com/capabilities/sustainability/our-insights/an-affordable-reliable-competitive-path-to-net-zero

Logan, J. and James, T. (2009). *A Comparative Review of a Dozen National Energy Plans: Focus on Renewables and Energy Efficiency*. NREL/TP-6A2-45046. Golden, CO: Department of Energy, National Renewable Energy Lab.

New York State (2020). 2020 New York State Energy Plan. https://energyplan.ny.gov/.aspx (accessed 15 August 2023).

Schlosser, A. (2023). Ask MIT Climate Oct 20, 2023. *Will Climate Change Drive Humans Extinct or Destroy Civilization*. MIT Joint Program on the Science and Policy of Global Change.

Skipka, K. and Theodore, L. (2014). *Energy Resources: Availability, Management, and Environmental Impacts*. Boca Raton, FL: CRC Press/Taylor & Francis Group.

Systemiq, Univ of Exeter, Bezos Earth Fund (2023). *The breakthrough effect: how to trigger a cascadeof tipping points to acceleratethe net zero transition*.

U.S. Department of Energy, References:

1 *Pathway to Commercial Liftoff: Clean Hydrogen*, March 2023. https://liftoff.energy.gov/wp-content/uploads/2023/05/20230523-Pathways-to-Commercial-Liftoff-Clean-Hydrogen.pdf

2 Satyapal, S. (2022). *2022 AMR Plenary Session*. Washington, DC: U.S. Dept of Energy. https://www.hydrogen.energy.gov/pdfs/review22/plenary4_satyapal_2022_o.pdf.

3 U.S. Department of Energy Hydrogen Program (2021). *Hydrogen Shot*. Washington, DC: U.S. Department of Energy. https://www.energy.gov/eere/fuelcells/hydrogen-shot.

4 U.S. Energy Information Administration (February 2024). *Electric Power Monthly*; preliminary data.

20

Ethical Considerations

20.1 Introduction

This chapter addresses another peripheral energy topic, ethics. It is a topic that was neglected during the latter part of the last century. However, the situation has changed. Ethics has today become an integral part of engineering, science, technology, and business education and, one might suppose, needs to be introduced and addressed in any discussion involving a hydrogen energy economy. It should be noted that this chapter provides a generic introduction to ethics. Three case studies presented at the end of the chapter are related to hydrogen energy issues.

The primary responsibility of an engineering professional is to protect public health and safety. However, engineering professionals also have a responsibility to their employers or clients, to their families, to themselves, to the environment, and for Civil and Environmental Engineers, to ". . . adhere to the principles of sustainable development; treat all persons with respect, dignity, and fairness, and reject all forms of discrimination and harassment; acknowledge the diverse historical, social, and cultural needs of the community, and incorporate these considerations in their work" among other ethical practices in the performance of their professional duties (ASCE 2020). Meeting all of these responsibilities will challenge the practicing engineer to draw upon a system of ethical values and ethical professional practice.

Well, what about ethics? Ethics means "doing the right thing" as opposed to "what you have the right to do." But doing the right thing is not always obvious or easy. In fact, ethical decisions are often difficult and may involve a certain amount of self-sacrifice. Doing the right thing for a practicing engineer can be especially challenging. Furthermore, the corporate and government world has confused this concept by developing ethics programs that emphasize only what you have the right to do. An organization, for example, may have a list – often called a Code of Conduct – of what an employee can and cannot get away with. Employees are required to sign an acknowledgment that they have read and understood the list. The company unfortunately calls this "ethics training" (Tabak 2002).

One difficulty in some situations is recognizing when a question of ethics is involved. Frequently, in the area of energy production and management, a breach of ethics involves a practice that endangers public health and safety or covers up a violation of a rule or regulation. Occasionally, however, a breach may involve a case of the exact opposite. This might seem an unlikely scenario. How can someone be too honest, too caring, or too professional?

One example of this might be lying to save a life. Suppose you are standing on a street and a woman runs past you chased by two men. She screams, "They are trying to attack me!" as she dashes into the entry of a building around a corner. The men ask you, "Which way did she go?"

Hydrogen Energy: Principles and Applications, First Edition. Vincent J. DelGatto, Louis Theodore, R. Ryan Dupont, and Matthew C. Ogwu.

What do you tell them? Clearly, the right thing is to lie. In this case, the value of caring overrides the value of honesty. This situation is exaggerated to illustrate that sometimes it is appropriate to violate certain values to protect public health and safety. In doing the right thing ideally, one should not have to make snap decisions and should take the time to investigate all of the facts, for instance, in this case, whether or not the woman was a thief, and the men were police.

Sometimes one must decide how much to sacrifice to ensure public health and safety. In establishing environmental regulations, the regulating agency must decide how safe and how stringent to make the regulations. For example, in the case of drinking water regulations, 1 standard may result in ten cancer cases per 1 million people. But why isn't the standard established for one or zero cases per one million people? Who should decide on this acceptable level of risk?

The role of ethics in energy production and management can be complex and confusing in the context of the engineer's role in providing safe, reliable energy to the public. This chapter provides a discussion of ethics with illustrative examples specifically focused on hydrogen energy issues. Topics covered include the present state of ethics, ethics of energy access, the dos and don'ts, integrity, moral issues, guardianship, engineering and environmental ethics, environmental justice, and future trends in professional ethics. The applications section provides three illustrative examples related to applied ethics that have been primarily drawn from the work of Wilcox and Theodore (1998).

Section numbers and their accompanying titles for this chapter concerned with ethical considerations are listed below.

20.2 The Present State of Ethics

Since this chapter is focused on ethics, there is a need for definitions. Here are four that might help the reader:

- *Ethics* is defined as a branch of philosophy dealing with the rules of right conduct.
- *Environmental ethics* deals with the moral issues of conduct with respect to the environment.
- *Ethical theory* is an attempt to answer certain questions about standards of conduct or ethics. It also attempts to provide a framework for decisions regarding what moral principles are correct and how one should treat one another, the environment, other species, etc.
- *Morality* is concerned with reasons for the desirability of certain kinds of actions and the undesirability of others. To say that an act is right is not to express a mere feeling or bias, but instead to assess the best moral reasons for support. Moral reason is a reason that requires individuals to respect other people. In addition, moral reasons are such that they set limits to the legitimate pursuit of self-interest. They can, for example, be used to evaluate, praise, and criticize laws.

With respect to the above, some in industry have spawned financial abuses and excesses. These individuals have turned what used to be moral and ethical questions into legal technicalities. Some industry executives are more likely to ask what one can get away with legally rather than to be concerned with what is right, honest, and/or fair. The above abuses and excesses have spread, particularly with lawyers and individuals in government. Some lawyers have become adept at creating controversy, smokescreens, and/or obstruction of the truth, and using the complexity of the law for their own aggrandizement. Some elected officials at all levels of government, on the other hand, have used their entrusted position of power to further and maintain their careers rather than to represent their constituency. As most know, nearly all elected officials are lawyers (Theodore 2003).

Consider if an environmental engineer, acting as an expert witness in a court of law, knowingly conjures up an idea that creates reasonable doubt even though it is highly unlikely. This type of conduct is clearly defined as fraud in any engineering code of ethics. Given the above, the reader should consider a lawyer's present-day conduct in and out of court (Theodore 2003).

Well, what is the answer? Is there an answer? The reality may be that the above-referenced corruption is just too widespread and amorphous – like humidity in South Florida in the summer. It seems that the only thing that can ultimately turn things around on this issue is for each individual to take a stand for basic values and virtues, and return to the likes of integrity, responsibility, and selflessness. If a grass-roots movement is to succeed, it will ultimately rest on each individual's ability and willingness to discard the practices of so many lawyers and elected officials. As Albert Schweitzer put it, "Man must cease attributing his problems to his environment and learn again to exercise his will and his personal responsibility in the realm of faith and morals" (Theodore 2004).

20.3 Dos and Don'ts

The need to make ethical decisions does not start the day one enters his or her first job. Encounters with ethical concepts begin early in life. Early on, before most have any active memory, parents almost certainly encourage their children to tell the truth. The conflict between choosing the ethical act of telling the truth and accepting the punishment that may follow, or choosing the unethical path of lying and going unpunished introduces virtually everyone to the struggle for ethical behavior that lasts a lifetime. In school, it is not long before students learn that it is unethical to copy homework and cheat on tests. The unethical nature of stealing from a classmate is another area that surely received some attention early in life. As one grows up, ethical situations often become more complex. Conflicts develop. Situations arise in which one may have to decide if it is more ethical to hide an unpleasant truth that will hurt someone or stay loyal to the truth one was taught to value so highly.

As one moves into professional life, an individual learns a long list of ethical dos and don'ts. Some examples include:

- Never falsify data
- Never offer or accept a bribe
- Judge colleagues, employees, employers, clients, and others, on their professional merits and not on any other basis such as race, gender identification, age, religion, national origin, or disability
- Never draw misleading conclusions from the data you have collected
- Avoid conflicts of interest
- Accept work only in those areas in which you are properly trained
- Obey all laws, rules, and regulations

- Never break confidentiality
- Always do the best possible work of which you are capable
- Statements to the public must always be truthful and objective
- Health and safety of the public must always be the highest priority
- Always act to protect the environment

Ethics becomes a dilemma for the professional when confronted with difficult choices or apparent conflicts between ethical imperatives. A modern term for the outcome of such dilemmas is "situational ethics," which refers to the temptation to bend the rules when a situation seems to have no satisfactory or satisfying solution. Examples of difficult choices are:

- A choice between keeping one's job and income that is needed to care for the children who depend upon you and refusing to certify substandard work as ordered by your employer (job versus family)
- A choice between whether or not to get involved when you become aware of unethical behavior on the part of one's colleagues or employer even though not personally touched by it (job versus conscience)
- Conflicts between two ethical imperative situations could include:
- The need to observe the confidentiality of a client after learning of the client's unethical behavior (protection of the client versus unethical behavior)
- The ethical imperative to faithfully serve an employer or client versus the ethical imperative to protect the public (protection of the employer versus the public). At what point does a professional stop working within the system and report a violation of regulations to authorities?

The second conflict of ethical imperatives stated above introduces issues that an energy management professional is likely to encounter. Most professionals belong to one or more organizations that commonly adopt a code of ethics for their members. Each code is designed to provide members with an ethical foundation from which to work. The ASCE (2020) Code of Professional Conduct is one such guide for practicing civil and environmental engineers.

20.4 Integrity

Scenarios are, for the most part, designed to reflect ambiguity in work situations. The ethicist hopes to get his or her hands dirty, dealing with the bottom-line motives of survival, competitiveness, and profitability as well as the mixed motives of self-interest, respect for the rights of others, and altruism. Obtaining an ethical solution to a difficult moral problem or dilemma is based on much more than choosing the correct ethical framework with its normative frame of reference. One must also be ready to examine fundamental assumptions and the values to which the assumptions give rise. Carter (1996) made this point in a discussion of "integrity."

- *Honesty in relation to integrity*. Carter explores integrity in relation to the value society places on honesty. On this subject, one of the best-known and most popular ethics books of the last five decades is Bok's (1978) *Lying: Moral Choice in Public and Private Life*. Without taking away from the merits of Lying, Carter notes: "Plainly, one cannot have integrity without being honest (although, as we shall see, the matter gets complicated), but one can certainly be honest and yet have little integrity." Honesty is far easier to practice than the tough work of figuring out what it takes to have integrity in a situation. Integrity requires a high degree of moral reflectiveness. Honesty may result in harm to another person. Furthermore, ". . . if forthrightness is not preceded by discernment, it may result in the expression of an incorrect moral judgment." The racist may be transparently honest, Carter declares, but he certainly lacks integrity because his beliefs, deeply held as they might be, are wrong. He has not engaged in the hard work of examining his fundamental assumptions, values, and beliefs.

- *Personal integrity without public responsibility?* It would appear that one cannot have integrity without responsibility since any consideration of integrity addresses the effects of our conduct on other people. In our work life and our community life, we have public responsibilities for our clients and fellow citizens.

20.5 Moral Issues

The conflict of interest between Chief Seattle (and Native Americans in general) and President Pierce (and the European-American expansion) provides a perfect example of how ethics and the resulting codes of behavior they engender can differ drastically from culture to culture, from religion to religion, and even from person to person. This enigma, too, is noted again and again by Chief Seattle (Martin and Schinzinger 1989):

> . . . I do not know. Our ways are different from your ways . . . . But perhaps it is because the red man is a savage and does not understand . . . . The air is precious to the red man, for all things share the same breath . . . the white man does not seem to notice the air he breathes . . . . I am a savage and do not understand any other way. I have seen a thousand rotting buffaloes on the prairie, left by the white man who shot them from a passing train. I am a savage and I do not understand how the smoking iron horse can be more important than the buffalo we kill only to stay alive.

Chief Seattle sarcastically uses the European word "savage" and all its connotations throughout his address. When one finishes reading the work, it becomes obvious which viewpoint (President Pierce's or his own) Chief Seattle feels is the savage one. What his culture holds dearest (the wilderness), the whites see as untamed, dangerous, and savage. What the whites hold in the highest regard (utilization of the Earth and technological advancement), the Native Americans see as irreverent to all other living things. Each culture maintains a distinct and conflicting standard for the welfare of the world. Opposing viewpoints and moralities such as these are prevalent throughout the world and have never ceased to present a challenge to international, national, state, community, and interpersonal peace.

It is generally accepted, however, that any historical ethic can be found to focus on one of four different underlying moral concepts:

1) *Utilitarianism* focuses on good consequences for all.
2) *Duties Ethics* focus on one's duties.
3) *Rights Ethics* focus on human rights.
4) *Virtue Ethics* focuses on virtuous behavior.

Utilitarians hold that the most basic reason why actions are morally right is that they lead to the greatest good for the greatest number. "Good and bad consequences are the only relevant considerations, and hence all moral principles reduce to one: 'We ought to maximize utility'. . ." (Martin and Schinzinger 1989).

Duties Ethicists concentrate on an action itself rather than the consequences of that action. To these ethicists, there are certain principles of duty such as "Do not deceive" and "Protect innocent life" that should be fulfilled even if the best does not result. The list and hierarchy of duties differ from culture to culture, religion to religion. For Judeo-Christians, the Ten Commandments provide an ordered list of duties imposed by their religion (Martin and Schinzinger 1989).

Often considered to be linked with Duties Ethics, *Rights Ethics* also assesses the act itself rather than its consequences. Rights Ethicists emphasize the rights of the people affected by an act rather than the duty of the person(s) performing the act. For example, because a person has a right to life, murder is morally wrong. Rights Ethicists propose that duties actually stem from a corresponding right. Since each person has a right to life, it is everyone's duty not to kill. It is because of this link and their common emphasis on the actions themselves that Rights Ethics and Duty Ethics are often grouped under the common heading of *Deontological Ethics* (Barbour 1993).

The display of virtuous behavior is the central principle governing *Virtue Ethics*. An action would be wrong if it expressed or developed vices, for example, bad character traits. Virtue Ethicists, therefore, focus on becoming a morally good person.

To highlight the different ways that these moral theories view the same situation, one can explore their approach to the following scenario that Martin and Schinzinger (1989) present: On a midnight shift, a botched solution of sodium cyanide, a reactant in organic synthesis, is temporarily stored in drums for reprocessing. Two weeks later, the day shift foreperson could not find the drums. Rosa, the plant manager, finds out that the batch has been illegally dumped into the sanitary sewer. She severely disciplines the night shift foreperson. Upon making discreet inquiries, she finds out that no apparent harm has resulted from the dumping. Should Rosa inform government authorities, as is required by law in this kind of situation? If a representative of each of the four different theories on ethics just mentioned were presented with this dilemma, their decision-making process would focus on different principles.

The utilitarian Rosa would assess the consequences of her options. If she told the government, her company might suffer immediately under any fines administered and later (perhaps more seriously) due to exposure to the incident by the media. If she chooses not to inform authorities, she risks heavier fines (and perhaps even worse press) in the event that someone discovers the cover-up. Consequences are the utilitarian Rosa's only consideration in her decision-making process.

The Duties Ethicist Rosa would weigh her duties, and her decision would probably be more clear-cut than her utilitarian counterpart. She is obliged foremost by her duty to obey the law and must inform the government (a course demanded by engineering ethics).

The Rights Ethicist mind-frame would lead Rosa to the same course of action as the Duties Ethicist, not necessarily because she has a duty to obey the law, but because the people in the community have the right to informed consent. Even though Rosa's inquiries informed her that no harm resulted from the spill, she knows that the public around the plant has the right to be informed of how the plant is operating.

Vices and virtues would be weighed by the Virtue Ethicist Rosa. The course of her thought process would be determined by her own subjective definition of what things are virtuous, and what things would make her a morally good person. Most likely, she would consider both honesty and obeying the law virtuous and withholding information from the government and the public as virtueless, and she would, therefore, tell the authorities.

Numerous case studies of this nature are provided by Wilcox and Theodore (1998).

20.6 Guardianship

There are two important questions that must be kept in mind when an ethical problem is being assessed:

1) Who are the guardians of the system? This question addresses the issue of who, among engineering or science professionals, is responsible for the ethical standards in the organization. If professionals point the finger at senior management, the legal department, the Environmental Protection Agency (EPA), or the Department of Justice, they have indeed misunderstood the nature of a professional calling. The first line of defense is the willingness of professionals themselves to maintain and enhance the integrity of the engineering or scientific profession through their own personal adherence to the highest standards of conduct and to assume responsibility for commitment to these standards within the companies they work for. Moreover, ethics is a positive task, not a list of dos and don'ts. To achieve excellence in one's work presumes a commitment to the client's contract, public safety, and environmental integrity, among several factors that are all too often thought of as "management" issues. They are, in reality, the ethical standards of the work itself. Thus, the ethical engineer or scientist is the one who identifies with the profession and all that is involved in the work assigned or contracted.
2) Who gives support to the guardians of the system? This second issue goes to the heart of the assessment problem but also has an impact on the first issue. Unless the organization backs those who assume positive responsibility for the ethical tenor of the group, very little will change. Why would someone risk ostracism or retaliation by confronting a person engaging in unethical behavior or illegal behavior if there is no institutional support for the one assuming responsibility?

Effective guardianship is facilitated if:

1) There are clear-cut standards of behavior and high expectations of the membership within an organization.
2) The standards are brought to the attention of the members through a well-developed training program.
3) The standards are taken seriously by the senior leadership team of the organization. They must demonstrate that seriousness by taking an active role in the training, without, at the same time, creating a chilly climate stifling discussion and participation in the training. The ethics program must be seen not as frosting on the cake or as a value-added to forestall legal problems through better compliance. The CEO needs to demonstrate a commitment to the values and principles that drive the business. Ethics training is no add-on. Ethics is what drives the organization: trust, integrity, fidelity to the client, to the public, and to the environment.
4) It is evident that the leadership "walks the talk" in all aspects of its decision-making and actions.
5) There are mechanisms in place to address the concerns of the members, such as an ombudsperson, and an ethics hotline.
6) Those who adversely affect the integrity of the business are effectively and fairly disciplined.

Another way of addressing the question of who supports the guardians is to emphasize the importance of organizational or corporate culture. A positive response to the six points above has a great impact on the culture of the organization. Unless there is what is sometimes called a "thick" culture, wherein respect for and adherence to guardianship and the tenets of integrity and trust are palpable, individually ethical persons can do very little to raise the moral climate of the organization. An organization is more than the sum total of the individuals who constitute the membership. The attitudes conveyed, values expressed, and ways of doing business in an organization profoundly affect the perceptions of the members therein and set the tone of the company. Having a positive impact on culture is a great challenge that is not easily achieved. Culture is so subtle that one often does not even realize or understand its dimensions until a significantly different culture is experienced.

An organization will not have effective guardianship of the system unless there is a concerted attempt to create, enhance, or reinforce a culture where values and ethics are clear and fully supported. There is little doubt, however, that the twin issues of guardianship and culture are much more difficult to address than the institutionalization of the ethics program itself.

20.7 Engineering Ethics

The ethical behavior of engineers is more important today than at any time in the history of the profession. The engineers' ability to direct and control the technologies they master has never been greater. In the wrong hands, the scientific advances and technologies of today's engineers could become the worst form of corruption, manipulation, and exploitation. Engineers, however, are bound by a code of ethics that carries certain obligations associated with the profession. Some of these obligations include:

1) Support one's professional society
2) Guard privileged information
3) Accept responsibility for one's actions
4) Employ proper use of authority
5) Maintain one's expertise in a state-of-the-art world
6) Build and maintain public confidence
7) Avoid improper gift exchange
8) Practice conservation of resources, pollution prevention, and sustainability
9) Avoid conflict of interest
10) Apply equal opportunity employment
11) Practice health, safety, and accident prevention
12) Maintain honesty in dealing with employers and clients
13) Practice energy and materials conservation
14) Treat all persons with respect, dignity, and fairness in a manner that fosters equitable participation without regard to personal identity

Tabak (2002) suggested that the engineer/scientist should recognize situations encountered in professional practice with conflicting interests that test one's ability to take the "right" action. Then, take each situation to a trusted colleague to determine the best course of action consistent with the above precepts and which would have the least adverse impact on all stakeholders.

There are many codes of ethics that have appeared in the literature. Excerpts from the preamble for the ASCE Code of Ethics (ASCE 2020) follow:

> Members of The American Society of Civil Engineers conduct themselves with integrity and professionalism, and above all else protect and advance the health, safety, and welfare of the public through the practice of Civil Engineering. Engineers govern their professional careers on the following fundamental principles: create safe, resilient, and sustainable infrastructure; treat all persons with respect, dignity, and fairness. . .; consider the current and anticipated needs of society; and utilize their knowledge and skills to enhance the quality of life for humanity.

Engineering ethics is tricky enough when it concerns only how engineers relate to each other and the public; it becomes trickier still when one also considers how engineers ought to react to the nonhuman environment. That is, when one asks, "What is the engineers' environmental ethic?"

Environmental ethics, to an even greater degree than ordinary ethics, is a subject without definition and without consensus. And yet, every person on this planet makes everyday decisions that relate to environmental ethics. Questions as simple as "What should I eat?" or "How should I move from place to place?" all raise environmental and ethical issues.

Environmental ethics is especially important for engineers because so much of their work affects the environment. How should the engineer balance the human gains of development against environmental damage? When should the engineer maintain client confidentiality in the face of potential environmental problems?

Now that a regulatory framework and general guidelines are in place, it is up to the individual to follow these procedures and strive to provide a safer, cleaner environment for the future of the planet. In any ethical dilemma, it will always be up to individuals, such as the engineers and scientists working on the ground, who best understand the shifting dynamic between the changing climate, the local environment, and the specific work they do to speak up if and when they see something that goes against their knowledge and ethical obligations.

As indicated earlier, ethics is the branch of philosophy dealing with the rules of right conduct. When defining environmental ethics, right conduct with respect to the environment must be evaluated. The American Society of Civil Engineers (ASCE 2020) states in the expected conduct for its first priority stakeholder, Society, that engineers should ". . . first and foremost, protect the health, safety, and welfare of the public; (and) enhance the quality of life for humanity. . ." The society's responsibilities to the second priority stakeholder, the Natural and Build Environment, are that engineers should ". . . adhere to the principles of sustainable development; . . . mitigate adverse societal, environmental, and economic effects; and use resources wisely while minimizing resource depletion. . ."

This statement of ethics can be extended to all individuals and to all aspects of engineering and technology. Whether one's particular area of responsibility is with the government, industry, or a public utility, the actions of an engineer, attorney, politician, or scientist should promote the general health and well-being of the community and the surrounding environment. If every individual makes even a small contribution to the reduction of pollution and the protection of the environment, the sum of the contributions will be great, and the goal of long-term sustainable development will be one step closer.

Regarding education, the Accreditation Board for Engineering and Technology (ABET) 2000 accreditation guidelines require academic programs to demonstrate that students are exposed to an ethics education and conduct outcome assessments. Despite indicators that reveal the value of an ethics education, few large universities require an ethics course. Ideally, a student would take an ethics course or, better yet, would be exposed to ethics issues and ethical problem-solving in a diverse set of courses throughout their engineering curriculum.

20.8 Future Trends in Professional and Environmental Ethics

Although the environmental movement has grown and matured, its development is far from stagnant. On the contrary, changes in individual behavior, corporate policy, and governmental regulations are occurring at a dizzying pace. Because of the Federal Sentencing Guidelines, the Defense Industry Initiative, as well as a move from compliance to a values-based approach in the marketplace, corporations have inaugurated company-wide ethics programs, hotlines, and senior line positions responsible for ethics training and development. The Sentencing Guidelines allow for the mitigation of penalties if a company has taken the initiative in developing ethics training programs and codes of conduct.

In the near future, these same guidelines will likely apply to infractions of environmental law. As a result, the corporate community will undoubtedly welcome ethics integration in engineering and science programs generally, but more so in those that emphasize environmental issues. Newly hired employees, particularly those in the environmental arena who have a strong background in ethics education, will allay fears concerning integrity and responsibility. Particular attention will be given to the role of public policy in the environmental arena as well as in the formation of an environmental ethic. Regulations instituted by federal, state, and local agencies have over the years become more and more stringent. The deadlines and fines associated with these regulations encourage corporate and industrial compliance of companies (the letter of the law), but it is the personal conviction of the corporate individuals that lies in the spirit of the law and the heart of a true ecological ethic. To bolster this conviction of the heart, there must be the emergence of a new *dominant social paradigm* (Barbour 1993). This is defined as "... the collection of norms, beliefs, values, habits, and survival rules that provide a framework of reference for members of a society, It is a mental image of social reality that guides behavior and expectations" (Barbour 1993). The general trend in personal ethics is steady "greener" and is being achieved at a sustainable pace with realistic goals.

It has been suggested (Gleick 1987) that "The flap of one butterfly's wings can drastically affect the weather." While this statement sounds much like one conceptualized by a romantic ecologist, it is part of a mathematical theory explored by the contemporary mathematician Gleick (1987) in his book *Chaos, Making a New Science*. The "butterfly" theory illustrates that the concept of interdependence, as Chief Seattle professed it, is emerging as more than just a purely environmental one. This embracing of the connectedness of all things joins the new respect for simplified living and the emphasis on global justice, renewable resources, and sustainable development (as opposed to unchecked technological advancement) as the new, emerging social paradigm. The concept of environmentalism is now widely held; its future is becoming *deeply* held.

Finally, one should note that an ethical analysis generally produces no absolute answers. This can be quite disconcerting because engineers expect valid, correct, and useful answers to problems. When one studies statistics, for example, one knows what the rules are, and all agree on the right answer. In ethics, however, the best an individual can do is argue that some answers are better than others, and, of course, these answers are always open to disagreement.

Engineers, as professionals, have a special responsibility to the public, and this responsibility is expressed in terms of professional ethics. Engineers invariably face situations where values become variables in the decision-making process. Indeed, the ethical aspects of a decision often prove more difficult than the technical ones. As indicated above, the engineering profession has strongly encouraged engineering schools to introduce more professional ethics into engineering curricula. Professional ethics has rightly become an integral part of engineering education.

The progress EPA has made in the environmental justice arena over the years was significantly threatened, however, by the Trump Administration's priorities for EPA as reflected in the FY 2021 EPA budget (U.S. EPA 2020). Despite statements by EPA that "In FY 2021, the Agency will enhance coordination across its programs and with federal partners to ensure alignment of mutual efforts, including consideration of vulnerable groups and communities when making decisions. EPA will leverage resources with new and existing partners to deliver services efficiently and effectively," the FY 2021 budget document indicated the Trump Administration largely abandoned environmental justice actions with the proposed reductions in environmental justice and environmental justice-related programs across the board at EPA. These cuts ranged from reducing the Environmental Programs and Management – Environmental Justice Enforcement budget by 71.4%, to completely eliminating the Hazardous Substance Superfund – Environmental Justice Enforcement program. The subsequent Biden Administration restored and, in most cases, increased funding for environmental justice, human health and environmental risk assessment, disadvantaged communities, and environmental education programs. However, with such actions and drastic policy shifts from one administration to another at the federal level, it becomes even more imperative that professional ethics drive engineering firms and engineers to meet their obligations for equity and the social science component of sustainable development in all of the projects they are associated with.

20.9 Case Studies

Three case studies complete the presentation of ethics principles highlighted in this chapter. The last two are case studies that have been adapted from Wilcox and Theodore (1998).

20.9.1 Case Study 1

Fact Pattern – Curts Inc. is a company providing consulting services to power plants. Recently, their business has taken a jump forward, and the company has started obtaining contracts faster than it can hire and train qualified individuals. John is a mechanical engineer for Curts Inc. Because of the recent lack of employees, he has been juggling three or more projects at the same time. One project that he neglected for a while was a contract with the local utility company to modify an existing burner to use hydrogen and still produce the same amount of energy (Btus).

Now, the project is near the end, and he has just started on his assignment of giving a detailed report about the fuel to be used. Initial information was gathered by a team of engineers who concluded the fuel would be easy to obtain if imported from South America and that it would produce the same amount of energy per unit volume of fuel. John finds out that the fuel, which is a very crude (tarlike) oil, is suspected of being carcinogenic when burned, according to a study done on this type of fuel. When he brings this to the attention of his peers, they accuse him of trying to destroy all of their work at the last minute.

John takes a moment to think over his options: Should he destroy the whole team's work, or forget about that one study and finish the project?

Questions for Discussion

1 What are the facts in this case?

2 What is the ethical problem John faces?

3 Does John have more than two options?

4 Which course of action would you recommend to John?

20.9.2 Case Study 2

Vincenzo has a PhD in Energy Systems Engineering, a Professional Engineer (P.E.) license, and a job as an Assistant Professor at a local university. To learn more about ISO 14000 and to earn some extra money, Vincenzo became a certified ISO 14000 auditor. During an audit of an energy facility, Vincenzo observed a significant violation of regulations related to energy production and emissions management at the facility. (Note that ISO 14000 is voluntary standard of environmental management systems, i.e., it is not a regulation and ISO 14000 audits are confidential.)

What does Vincenzo do? Does he forget about the significant violation since he is not an EPA inspector and it does not appear to violate the ISO 14000 standard he is there to audit? Does he insist to be part of the audit report? Does he call it to the attention of the facility management? Does he report it to the EPA?

Solution. The question of professional ethics presented here can be generalized as follows: does a professional have an ethical duty to report any violation of which he/she becomes aware? Most licensing organizations have codes of ethics to which their professionals must subscribe. These codes generally call upon the professional to serve their client/employer with loyalty and to protect the welfare of the public. The codes, however, are sufficiently vague as to not provide a clear-cut answer to Vincenzo's situation. In the absence of such a clear-cut code, there is no one right answer and one's answer may vary from the one presented below.

Vincenzo must first investigate further to make certain of his facts. If the violation is confirmed and has never been reported, he must insist that it be part of the findings presented to management at the exit meeting and in the report to the ISO Registrar who makes the final decisions on awarding certification. Although ISO 14000 does not require facility compliance with any regulatory requirements, knowingly failing to comply does affect the evaluation of the facility's commitment to good management system practices and continuous improvement that the standard does require.

Having done all that, does Vincenzo go the last step and report it to the U.S. EPA? One aspect of the problem that Vincenzo must evaluate is whether the violation poses an imminent threat to public health or the environment. Such a report would violate the confidentiality of the audit and might even precipitate a lawsuit. Vincenzo is also concerned that he may never be hired again as an auditor. He senses a conflict between two sets of ethics: the ethics of a professional bound to protect public health and the environment, and the ethics of a professional bound to confidentiality. Keeping silent would be easier, but ethics is not easy.

It would be easy to answer "report it!" but the real question is, what would YOU (the reader) do?

20.9.3 Case Study 3

Fact Pattern – Joanne Murray, a chemical engineer at the Pristine Water Company, is responsible for the production of potable drinking water employing a combination of geothermal and hydrogen energy. The company distributes this potable water to its municipal and industrial customers. Joanne has been reviewing the results of routine water quality monitoring conducted within the plant for process monitoring purposes, but monitoring results are not required to be reported to the local regulatory agency. Joanne notices that there is a disturbing trend in pH that, while not currently violating permit requirements, has potentially significant implications for water quality within the distribution system for the mobilization of lead and for the production of undesirable disinfection by-products during treated water chlorination.

Joanne decides to go to her manager and proposes that expensive source water and distribution system monitoring be carried out, first to attempt to determine if changes in source water quality could be the cause of these water quality changes, and secondly to ensure that undesirable conditions within the distribution system are not occurring. Her manager, concerned about quarterly profits and control of operating costs, feels that if no permit violations are occurring then no actions are warranted, so no more money or personnel time should be wasted on this voluntary monitoring program.

Ethically, Joanne feels strongly that this is wrong and feels this situation needs special attention before conditions within the distribution system become unstable and their customers are subjected to unnecessary risk if the plant's response to water quality changes is delayed. Joanne's only alternative is to go above her manager and try to convince upper management to expend the necessary funds to conduct the recommended monitoring and perhaps institute process modifications to ensure the quality of the produced water and the health and safety of their customers. However, doing this could anger her manager and also lead to Joanne's dismissal. But it might also make her appear to be confident and conscientious, could prevent a major water crisis such as that happened in Flint, Michigan, in 2014, and could also serve to demonstrate her abilities as an engineer.

Questions for Discussion

1 What are the facts in this case?

2 What issues are involved?

3 What are the choices Joanne must make?

4 What could be the consequences of each decision she could make?

5 What were the outcomes of the Flint, Michigan, water crisis, and what can be learned from this real-life event that pertains to Joanne's problem?

Solution. This is an open-ended question, and there may be a range of responses to these questions. Put your answer in the context of professional ethics and refer to ethics guidelines from ASCE (2020) for appropriate responses and actions you suggest as a Professional Engineer. Also refer to the National Resources Defense Council (2018) for a detailed summary of events and consequences of the Flint, Michigan, water system crisis.

References

American Society of Civil Engineers (2020). *Code of Ethics*. Reston, VA.: ASCE. https://www.asce.org/-/media/asce-images-and-files/career-and-growth/ethics/documents/asce-code-ethics.pdf (accessed 16 August 2023).

Barbour, I. (1993). *Ethics in the Age of Technology*. San Francisco, CA: Harper.

Bok, S. (1978). *Lying: Moral Choice in Public and Private Life*. New York: Pantheon.

Carter, S.L. (1996). The insufficiency of honesty. *Atlantic Monthly* (February), pp. 74–76.

Gleick, J. (1987). *Chaos: Making a New Science*. New York: Viking Penguin Inc.

Martin, M. and Schinzinger, R. (1989). *Ethics in Engineering*. New York: McGraw-Hill.

National Resources Defense Council (2018). Flint water crisis: everything you need to know. https://www.nrdc.org/stories/flint-water-crisis-everything-you-need-know (accessed 19 June 2020).

Tabak, H. (2002). Ethics corner: doing the right thing. *Environmental Management* 29(1): 15–16.

Theodore, L. (2003). On ethics I. *Williston Times* (November 28).

Theodore, L. (2004). On ethics II. *Williston Times* (March 26).

United States Environmental Protection Agency (2020). FY 2021 EPA Budget in Brief. EPA-190-5-20-002. Office of Chief Financial Officer. Washington, DC: U.S. EPA. https://www.epa.gov/sites/production/files/2020-02/documents/fy-2021-epa-bib.pdf.

Wilcox, J. and Theodore, L. (1998). *Engineering and Environmental Ethics: A Case Study Approach*. Hoboken, NJ: John Wiley & Sons.

21

Economic Considerations

21.1 Introduction

The development of a new energy economy, e.g., a hydrogen energy economy, requires the use of engineering principles and theories combined with a practical realization of the limits imposed by a host of other conditions. The development of new energy economics involves a wide variety of talents. However, there is one major qualification that must be added before this goal is attained – the operation must be able to proceed economically. Enter economic analysis. To a practicing engineer, the term *economics* (or, more properly, *applied economics*) refers to the analysis of all factors affecting costs and profits in any venture. The related definition of *economics*, as interpreted by economists, is "the science that investigates the conditions and laws affecting the production, distribution, and consumption of wealth."

Regardless of the type of society or government within which a business or economy is operated, it must either show a profit or will usually perish in the normal course of events. What applies to an individual business applies equally to part or the whole of an economy. Without profit, the structure is weak with profit, it *may* be strong. This conclusion is not debatable.

Because of the necessity for combining economic considerations along with a host of other considerations, the engineer or applied scientist must be aware of the many different types of costs involved in energy projects and processes. Money must be paid out for different process expenses, such as those for raw materials, labor, and equipment. In addition, many other indirect expenses are incurred, and these must be included if a complete analysis of the total cost is to be obtained. Some examples of these are addressed later in this chapter.

The methods of preparing an economic evaluation are numerous and vary with government sector, agency, business sector, company, location, etc. In some cases, shortcut or "quickie" evaluation procedures may be all that is required as a basis for justifying further expenditures on a project or process. Much more exacting information is almost always required when a final decision to commercialize a process is to be made. There are all degrees of economic analysis between these two extremes, and (as noted) it is beyond the scope of this chapter to delve into every nuance and detail of various economic analyses. Rather, this chapter aims to present several general-purpose methods with suitable references so that an individual has a sound basis for making intelligent decisions and for proceeding further with more detailed economic analyses.

Every method of economic evaluation requires cost analysis for a host of items. Two typical costs that usually appear under each item in a detailed analysis include traditional capital and operating

Hydrogen Energy: Principles and Applications, First Edition. Vincent J. DelGatto, Louis Theodore, R. Ryan Dupont, and Matthew C. Ogwu.

costs. However, the main problem with the traditional type of economic analysis is that it is difficult – in some cases impossible – to quantify some of the not-so-obvious economic costs and/or merits of a business process and/or program (Theodore 2014). A number of considerations and/or factors that need to be included in any meaningful economic analysis are listed below (Theodore 2014; Dupont, Ganesan, and Theodore 2017).

1) Long-term liabilities
2) Regulatory compliance
3) Regulatory recordkeeping
4) Dealings with the Environmental Protection Agency (EPA)
5) Dealings with the state and local regulatory bodies
6) Fines and penalties
7) Potential tax benefits
8) Customer relations
9) Stockholder support (corporate image)
10) Improved public image
11) Insurance costs and claims
12) Effect on borrowing power
13) Improved mental and physical well-being of employees
14) Reduced health maintenance costs
15) Employee morale
16) Worker safety
17) Rising costs of waste treatment and/or disposal
18) Training costs
19) Emergency response planning
20) Social consideration
21) Ethical consideration
22) Labor relations
23) Fringe analysis
24) Any combination of the above

This does not include potential contingency cost issues. A contingency factor is usually included in an estimate of capital investment to compensate for unpredictable events such as storms, floods, strikes, price changes, small design changes, errors in estimation, and other unforeseen expenses. Contingency factors usually range from 8% to 30% of the direct process costs. Many analyses have been halted in their early stages because a comprehensive economic analysis was not performed on these two dozen issues. Until the effects described above are included, the true merits of a project may be clouded by incorrect and/or incomplete economic data.

Can something be done by the industry to remedy the above problem? One approach is to use a modified version of the standard *Delphi panel*. In order to estimate these "other" economic benefits, several knowledgeable individuals within and perhaps outside an organization are asked to independently provide estimates, with explanatory details, on these economic benefits. Each individual in a panel is then allowed to independently review all responses. The cycle is then repeated until the groups' responses approach convergence. Most others refer to it as the aforementioned Delphi panel approach. The Delphi panel approach has its origin based – there are numerous versions – on the Oracle at Delphi in Greece (home to one of the author's ancestors). The city of Delphi became famous in ancient Greece because of Pythia, a priestess who muttered incomprehensible words (often with multiple meanings) that foretold the future. It is rumored that Alexander the Great violated the sacredness of Delphi by forcing Pythia to relent and provide the prophecy he desired.

To further compound any analysis, all of the above are also functions of one or more of the following:

1) Location
2) Industry
3) Local politics
4) National politics
5) International politics
6) Time

The latter concern cannot be overemphasized. Cost analysis is a *strong* function of time.

What does all the above mean? It is one thing to perform an economic analysis of, for example, the design and construction of a new ethanol plant, which can be accomplished with a considerable degree of exactness. After all, there are over a hundred already out there. But it is another thing to perform an economic analysis on something as vague as a hydrogen energy economy since there are too many intangibles to allow exact analysis, in which case engineering judgment must be applied. Thus, it is beyond the scope of this chapter to examine a hydrogen energy plant or a hydrogen energy economy in any detail, other than to provide a very rough first draft assessment.

It should be noted that this chapter was written primarily for the beginning student since it would be almost impossible at this level to cover all the various aspects of engineering economics. This chapter, therefore, introduces the philosophy of economics as a tool used in engineering and applied sciences. For the interested reader, there are many texts available that give a much broader and deeper treatment of the subject. Both the qualitative and quantitative viewpoints are emphasized in the material to follow.

Section numbers and their accompanying titles in this chapter concerned with economics are provided below:

21.2 Economic and Finance Definitions

Before proceeding to the developmental material and applications later in the chapter, it would be wise to provide the reader with certain key definitions in this field. Fourteen concepts that often come into play in an economic analysis are given below. The definitions have been drawn from Theodore and Neuser (1996).

21.2.1 Simple Interest

This term can be defined as the money paid for the use of money. It is also referred to as the value or worth of money. Two terms of concern are *simple interest* and *compound interest*. Simple interest is always computed on the original principal. The basic formula to employ in simple interest calculation is:

$$F = P(1 + ni) \tag{21.1}$$

where F = sum of interest and principal after n years, also known as the future value; P = original principal; n = time in years; and i = annual interest rate. Normally the interest period is 1 year, in which case i is referred to as the effective interest rate.

21.2.2 Compound Interest

Unlike simple interest, with compound interest, interest is added periodically to the original principal. The terms "conversion" or "compounding of interest" simply refers to the addition of interest to the principal. The interest period or conversion period in a compound interest calculation is the time interval between successive conversions of interest. It is the ratio of the stated annual rate to the number of interest periods in 1 year. Thus, if the given interest rate is 10% compounded semiannually, the interest period is 6 months, and the interest rate per interest period is 5%. Alternatively, if the given interest rate is 10% compounded quarterly, then the interest period is 3 months and the interest rate per interest period is 2.5%. One should always assume the interest is compounded annually unless otherwise stated. The basic formula to employ for compound interest is as follows:

$$F = P(1 + i)^n \tag{21.2}$$

If interest payments become due m times per year at compound interest, $(m)(n)$ payments are required in n years. The future value of this principle is based on a nominal interest rate per period, $i' = i/m$:

$$F = P(1 + i')^{mn} \tag{21.3}$$

In this case, the effective annual interest, i_{eff}, is:

$$i_{eff} = (1 + i')^m - 1 \tag{21.4}$$

In the limit (as m approaches infinity), such payments may be considered to be required at infinitesimally short intervals, in which case, the interest is said to be compounded continuously. Numerically, the difference between compounding may be significant when applied to very large sums of money.

21.2.3 Present Worth

The present worth is the current value of a sum of money due at time n and at interest rate i. This equation is the compound interest equation solved for the present worth term, P.

$$P = F(1 + i)^{1/n} \tag{21.5}$$

21.2.4 Time Value of Money

The value of a sum of money changes with time because of interest considerations: \$1,000 today, \$1,000 10 years from now, and \$1,000 10 years ago all have different values when interest is considered. The \$1,000 today would be worth more than 10 years from now because of the interest that could be accumulated in the interim. On the other hand, \$1,000 today would have been worth less 10 years ago because a smaller sum of money could have had to have been invested than to yield \$1,000 today. Therefore, one must refer to the date as well as the sum of money when considering value and when comparing costs.

Summarizing and evaluating single sums of money requires multiplying by $(1 + i)$ if the required date of evaluation is after the date associated with the obligation. If the required date of evaluation is before the date associated with the obligation, single sums of money must be multiplied by $(1+i)^{1/n}$ The term n is always the time in periods between the date associated with the obligation and the date of evaluation.

The time value of money may be applied to the evaluation of a uniform series of payments. A uniform series of equal payments is made at equal intervals. Suppose R is invested at the end of every interest period for n periods. The total value of all these payments, F, as of the date of the last payment, may be calculated from the equation.

$$F = R\frac{\left[(1+i)^n - 1\right]}{i} \tag{21.6}$$

21.2.5 Depreciation

Depreciation refers to the decrease in the value of an asset. Two approaches that can be employed are the straight-line and sinking fund methods. In the straight-line method of depreciation, the value of the asset is decreased each year by a constant amount. The annual depreciation amount, D, is given by:

$$D = \frac{(\text{Original cost} - \text{Salvage value})}{\text{Estimated life in years}} \tag{21.7}$$

In the sinking fund method of depreciation, the value of the asset is determined by first assuming that a "sinking" fund consisting of uniform annual payments had been set up to replace the asset at the end of its estimated useful life. The uniform annual payment (UAP) may be calculated from the equation:

$$UAP = (\text{Original cost} - \text{Salvage value})(\text{SFDF}) \tag{21.8}$$

where $SFDF$ is the sinking fund deposit and is given by:

$$SFDF = \frac{i}{(1+i)^n - 1} \tag{21.9}$$

The value of the asset at any time is estimated to be the difference between the original cost and the amount that would have accumulated in the sinking fund. The amount accumulated in the sinking fund is obtained by substituting the UAP into Equation 21.6 for R and calculating the future value of these regular payments at interest rate i over the corresponding number of payment periods, n.

21.2.6 Equipment Cost and Cost Indexes

Equipment cost is a function of many variables, one of the most significant of which is equipment capacity. Other important variables include operating temperature and/or pressure conditions, and the degree of equipment sophistication. Preliminary capital equipment costs are often made using simple cost–capacity relationships that are valid when the other variables are confined to narrow ranges of values. These cost–capacity relationships can be represented by approximate, linear log–log cost equations of the form:

$$C = aQ^b \tag{21.10}$$

where C = cost; Q = some measure of equipment capacity; and a, b = empirical constants that depend mainly on equipment type.

It should be emphasized that this procedure is suitable for rough estimation only; actual estimates from vendors are preferable. If more accurate values are needed and if old price data are available, the use of an indexing method is better. The method consists of adjusting the earlier cost data to present values using factors that correct for inflation using information for specific industrial sectors. A number of such indices are available; one of the most commonly used is the Chemical Engineering Plant Cost Index (CEPCI) available monthly in the *Chemical Engineering Magazine*. The CEPCI is indexed to 1957–1959 costs (Index = 100) and the January 2023 index was reported as 802.6, with a 2022 annual index of 816. The CEPCI consists of the weighted average of 41 industry and commodity indices and 12 labor cost indices, including construction labor; engineering and supervision; heat exchangers and tanks; process machinery; pipes, valves and fittings; etc., all reported by the Bureau of Labor Statistics. Another relevant cost index for public works projects is the Corps of Engineering (2018) Civil Works Construction Cost Index (CWCCI). The CWCCI is updated regularly and compiles public works project cost information on a quarterly and annual basis and provides future cost index projections and regional cost adjustment factors for 19 categories including reservoirs; dams; power plants; pumping plants; buildings, grounds and utilities; and permanent operating equipment. The CWCCI is indexed to 1967 (Index = 100) and reported a Fiscal Year 2017 composite index of 835.57 and a projected Fiscal Year 2023 index of 976.41.

All indices work in the same manner, that is, if a cost and index value for a given year are known along with the index value for a different year, the cost for that different year can be estimated using Equation 21.11:

$$Cost_{Year\,B} = Cost_{Year\,A}\frac{Index\ value_{Year\,B}}{Index\ value_{Year\,A}} \tag{21.11}$$

Generally, it is not wise to use cost data older than 5–10 years, even with the use of the cost indices. Within that period, technologies used in the processes may have changed drastically. The use of the indices could cause estimates to be much greater than actual costs. Such an error might lead to the choice of an alternative proposal other than the least costly one.

21.2.7 Capital Recovery Factor

In comparing alternative processes or different energy options for a particular process from an economic point of view, one recommended procedure to follow is that the total capital cost can be converted to an annual basis by distributing it over the projected lifetime of the facility (or the equivalent).

The sum of both the annualized capital cost (*ACC*), including installation, and the annual operating cost (*AOC*), is called the total annualized cost (*TAC*) for the project or facility. The economic merit of the proposed facility, process, or scheme can then be examined once the *TAC* is available.

The conversion of the total capital cost (*TCC*) to an *ACC* requires the determination of an economic parameter known as the capital recovery factor (*CRF*). This parameter can be found in any standard economics textbook or calculated directly from the following equation:

$$CRF = \frac{i(1+i)^n}{(1+i)^n - 1} \tag{21.12}$$

The *CRF* is a positive fractional number. Once this factor has been determined, the *ACC* can be calculated from the following equation:

$$ACC = (TCC)(CRF) \tag{21.13}$$

The *ACC* reflects the cost associated with recovering the initial capital expenditure over the depreciable life of the system, accounting for the interest accrued over the system's life.

21.2.8 Net Present Worth

There are various approaches that may be employed in the economic selection of the best of several alternatives. For each alternative in the net present worth (NPW) method of economic selection, a single sum is calculated for each alternative that accounts for all expenditures over a common period. The alternative having the least NPW is selected as the most economical. The equation to employ this method of economic evaluation is:

$$NPW = CC + PN + PWD - PWS \tag{21.14}$$

where CC = total capital costs; PN = present worth of required new investments over the comparison period; PWD = present worth of operation and maintenance costs over the comparison period; and PWS = present worth of salvage value of investment at the end of the comparison period.

If the estimated lifetimes differ for the various alternatives, one should employ a period equal to the least common multiple of the different lifetimes for comparison purposes.

21.2.9 Perpetual Life

Capitalized cost can be viewed as present worth under the assumption of perpetual life. Computing capitalized cost involves finding the present worth of an infinite series of payments. To obtain the present worth of an infinite series of payments of *R* at the end of each interest period forever, one needs simply to divide *R* by *i*, where *i* is the interest rate per interest period. Thus, to determine what sum of money, *P*, would have to be invested at 8.0% to provide payments of \$100,000 at the end of each year forever, *P*, would have to be such that the interest on it each period would be \$100,000. Withdrawal of the interest at the end of each period would leave the original sum intact to again draw \$100,000 interest at the end of the next period. For this example,

$$P = \frac{\$100{,}000}{0.08} = \$1{,}250{,}000$$

The $1,250,000 would be the present worth of an infinite series of payments of $100,000 at the end of each year forever, assuming the interest rate is 8%. To determine the present worth of an infinite series of payments of R at the end of each n period forever, first, multiply by the SFDF to convert to an equivalent single period payment and then divide by i to obtain the present worth.

21.2.10 Break-Even Point

From an economic point of view, the break-even point of a process or operation is defined as that condition when the costs (C) exactly balance the income (I). The break-even point is where the profit (Pr) is equal to zero or:

$$Pr = I - C = 0 \tag{21.15}$$

21.2.11 Approximate Rate of Return

The rate of return can be viewed as the interest that will make the present worth of net receipts equal to the investment. The approximate rate of return (ARR) may be estimated from:

$$ARR = \frac{\text{Average annual profit or earnings}}{\text{Initial total investment}} \tag{21.16}$$

To determine the average annual profit, simply divide the difference between the total income and the total expenses by the number of years in the period of the investment.

21.2.12 Exact Rate of Return

Using the ARR as a guide, one can generate the exact rate of return (ERR). This is usually obtained by trial-and-error and interpolation calculations of the rate of interest that makes the present worth of the net receipts equal to the investments. The ARR will tend to overestimate the ERR when all or a large part of the receipts occur at the end of a period of investment. The approximate rate will tend to underestimate the exact rate when the salvage value is zero and also when the salvage value is a high percentage of the investment.

21.2.13 Bonds

A bond is a written promise to pay both a certain sum of money (redemption price) at a future date (redemption date) and equal interest payments at equal intervals in the interim. The holder of a $1,000 5% bond, redeemable at 105 (bond prices are listed without the last zero) in 10 years, with interest payable semiannually, would be entitled to semiannual payments of $1,000 (0.025) or $25 for 10 years and 105% of $1,000, that is $1,050, at the end of 10 years when the bond is redeemed.

The interest payment on a bond is found by multiplying the face value of the bond by the bond interest rate per period. From the above, the face value is $1,000 and the bond interest rate per period is 0.025. Therefore, the periodic interest payment is the aforementioned $25. Redeemable at 105 means that the redemption price is 105% of the face value of the bond.

The purchase price of a bond depends on the yield rate, i.e., the actual rate of return on the investment represented by the bond purchase. Therefore, the purchase price of a bond is the present worth of the redemption price plus the present worth of future interest payments, all computed at the yield rate. The bond purchase price formula is:

$$V = \frac{C(1+i)^n + R}{i} \tag{21.17}$$

where V = purchase price; C = redemption price; i = yield rate; n = time in periods to maturity; and R = periodic interest payment.

21.2.14 Incremental Cost

By definition, the average unit increment cost is the increase in cost divided by the increase in energy production. Only those cost factors that vary with production can affect the average unit increment cost. In problems involving decisions as to whether to stay in production or (temporarily) shut down, the average unit increment cost may be compared with the unit increment cost or the unit selling price.

21.2.15 Inflation

Inflation is another important consideration when discussing capital investment and economic analysis. Money is a commodity that is universally accepted in exchange for every other commodity. The technical community has a term to describe a decrease in the value of money: it is defined as inflation. As the value of money falls due to the increase in the supply of money relative to other goods and services, the price of every other commodity must go up. Thus, inflation is a monetary phenomenon. It is also a psychological one: when the expectation of inflation becomes widespread, it becomes a self-fulfilling prophecy.

21.3 Investment and Risks

Capital investment is required for any process, and the determination of the necessary investment is an important part of an energy project. The total investment for any process consists of the fixed capital investment for the physical equipment and facilities plus the working capital for money, which must be available to pay salaries, keep raw materials and products on hand, and handle other special items requiring a direct cash outlay. Thus, in an analysis of costs in industrial processes, investment costs, factory manufacturing costs, and general expenses that include income taxes must be taken into consideration. The following sequence of steps is usually taken:

1) A process must be planned and evaluated, or even invented.
2) The equipment is designed, and the operational procedure is worked out and examined.
3) The land is acquired for the site and services are made available through roads, lines for gas, water, electricity, and sewage disposal systems, etc.
4) Requisite materials are purchased, and the buildings and all equipment are erected.
5) Funds must be made constantly available to keep the venture in operation once the plant has been put "online."

All five of the above steps require money. The total money involved is called the *capital investment*. This consists of two parts; that required for Steps (1), (2), (3), and (4), known as *fixed capital*, and that needed for Step (5) called *working capital*. In more graphic terms, the money required to bring the process from the moment of conception to the moment when the "button is pushed" and the plant is first put into operation is fixed capital. All money required for the venture thereafter (with the exception of capital improvements or expansions of the plant) is working capital. Typical items of the fixed capital variety are salaries paid to researchers and engineers for conception and design; the cost of raw materials; salaries and wages paid to production engineers, foremen, and plant workers; packaging and shipping costs; the costs of water, gas, fuel, electricity, etc., and the salaries paid to clerical personnel involved with the process (Vatavuk and Neveril 1980; Vogel and Martin 1983; Ulrich 1984).

Many processes that look good in an early stage of research and/or development have been proven to be economically unattractive after a closer and more intensive examination. The process often does not go beyond Step 2 listed above. Money invested up to this point is certainly partly lost and may be completely lost. In other cases, the process gets to the operational stage before proving to be unprofitable, in which case the loss may be significant. Certain things should be obvious. All investment of capital in energy projects involves a risk. Energy generation profit cannot start unless and until the process is put into actual operation, and since for every system that proves profitable others do not, the profitability of the successful system should be sufficient to absorb the losses incurred by the unsuccessful ventures. What potential level of profit should seem probable before capital is invested in a new process or economy? A partial answer is provided in the next paragraph.

Money put into a money market account or invested in bonds may draw a 3% or 4% return without risk, effort, or worry. It follows that the investment of capital in a unique energy process or economy that would yield as little as 3–4% return would be unsound since this would represent the assumption of business risk *with no profit at all*. How far in excess and above the potential rate of return should it be to justify the investment of capital depends upon the nature of both the industry and the industrial process. When a new hydrogen energy process is being considered, a potential profit of 30% or more may be necessary to make a plant design attractive. On the other hand, starting a plant to produce energy by a process with a long record of successful performance, e.g., ethanol, for which detailed economic data are available is another matter and may be attractive at a much lower profitability potential.

21.4 The Traditional Economic Evaluation Process

Before the cost of an energy program can be evaluated, the factors contributing to the costs must be identified. There are two major contributing factors to the overall cost of a project: *capital costs* and *operating costs*. In addition to traditional capital and operating costs of engineering projects, there is a wide range of potential *hidden costs* associated with concerns related to liability, permitting, monitoring, insurance, image and relationship costs, and so forth, that are addressed under the general topic area of accounting.

Once the total cost of the various processes and options has been estimated, the engineer and applied scientist must then determine whether or not a specific project will be profitable. This involves converting all costs to an annualized basis, a method that is also discussed later in this chapter. If more than one project proposal is under study, this method provides a basis for comparing alternate proposals and for choosing the best option on a total cost basis. Since project optimization is another subject of concern, a brief description of the perturbation process is presented in a later chapter.

An economic evaluation of process alternatives is carried out using standard measures of profitability. Each company and organization has its economic criteria for selecting projects for implementation. In performing an economic evaluation, various costs and savings must be considered. The economic analysis described in this chapter represents a preliminary, rather than detailed, cost analysis. Detailed cost estimates are beyond the scope of this text, and although such procedures are capable of producing accuracies in the neighborhood of ±5%, they generally require many months of engineering work and project evaluation.

For smaller facilities with only a few and perhaps simple energy processes, the entire assessment procedure will tend to be informal. In this situation, several obvious options, such as the installation of flow controls and good operating practices, may be implemented with little or no economic evaluation. In these instances, no complicated analyses are necessary to demonstrate the advantages of adopting the selected options. A proper perspective must be maintained between the magnitude of savings that a potential option may offer, and the amount of manpower required to carry out the technical and economic feasibility analyses.

Besides the traditional cost factors involved in energy projects, some other important costs and benefits need to be quantified if a meaningful economic analysis of a proposed project is going to be performed. Table 21.1 summarizes various traditional cost accounting methods that have evolved to begin to attempt to capture the intangible costs and the potentially significant benefits that can accrue over time.

The Total Systems Approach (TSA) referenced in Table 21.1 attempts to quantify not only the economic aspects but also the social costs associated with the production of a product or service from the cradle to grave – often referred to as *life-cycle analysis*. TSA also attempts to quantify less tangible benefits such as the reduced risk of proven technologies. More emphasis will certainly be placed on the TSA approach in future project assessments. For example, at present, a utility considering the option of converting from a coal-fired boiler to hydrogen energy is usually not too concerned with the environmental effects and implications associated with such activities as mining, transporting, and storing energy prior to its usage as an energy feedstock. As industry will be required to conduct more complete carbon emission analyses, more complete life-cycle considerations will be incorporated into future energy assessments and business decisions.

Table 21.1 Developments in economic analysis approaches.

Period	Economic Analysis Approach
Prior to 1945	Capital costs only
1945–1960	Capital and some operating costs
1960–1970	Capital and operating costs
1970–1975	Capital, operating, and some environmental control costs
1975–1980	Capital, operating, and environmental control costs
1980–1985	Capital, operating, and more sophisticated environmental control casting
1985–1990	Capital, operating and environmental controls, and some life-cycle analysis (TSA)
1990–1995	Capital, operating and environmental control costs, and life-cycle analysis (TSA)
1995–2000	Widespread acceptance of Total Systems Approach (TSA)
2000–2025	Widespread acceptance of Total Systems Approach (TSA)
>2025	Levelized Full System Cost of Electricity (LFSCOE)

21.5 Capital and Operating Costs

The usual technique for determining capital costs (i.e., total capital costs, which include equipment design, purchase, and installation) for a facility is based on the factored method of establishing direct and indirect installation costs as a function of the known equipment costs. This is a modified Lang method, whereby cost factors are applied to known equipment costs (Neveril 1978; Vatavuk and Neveril 1980).

The first step is to obtain from vendors (or, if less accuracy is acceptable, from one of the estimation techniques previously discussed) the purchase prices of the primary and auxiliary equipment. The total base price, designated by X, which should include instrumentation, control, taxes, freight costs, and so on, serves as the basis for estimating the direct and indirect installation costs. The installation costs are obtained by multiplying X by the cost factors, which are available in the literature (Neveril 1978; Vatavuk and Neveril 1980; Ulrich and Vasudevan 2004). For more refined estimates, the cost factors can be adjusted to model the proposed system by using adjustment factors that consider the complexity and sensitivity of the system.

The second step is to estimate the direct costs by summing all the cost factors involved in the direct installation, which include piping, insulation, foundation and supports, and so on. The sum of these factors is designated as the direct installation cost factor (*DCF*). The direct installation costs are then the product of the *DCF* and X.

The third step consists of estimating the indirect installation costs. This procedure is the same as for the direct installation costs; that is, all the cost factors for the indirect installation (engineering and supervision, start-up, construction fees, etc.) are added; the sum is designated by the indirect installation cost factor (*ICF*). The indirect installation costs are then the product of *ICF* and X.

Once the direct and indirect installation costs have been calculated, the *TCC* may be evaluated as:

$$TCC = X + (DCF)(X) + (ICF)(X) \tag{21.18}$$

This is then converted to *ACC* with the use of the *CRF*, which is described in Section 21.2.7. The *ACC* is the product of the *CRF* and *TCC* (Equation 21.13) and represents the total installed equipment cost distributed over the lifetime of the facility.

Operating costs can vary from site to site since these costs, in part, reflect local staffing practices, labor, and any utility costs. Operating costs, like capital costs, may be separated into two categories: direct and indirect costs. Direct costs are those that cover material and labor and are directly involved in operating the facility. These include labor, materials, maintenance and maintenance supplies, replacement parts, waste disposal fees, utilities, and laboratory costs. Indirect costs are those operating costs associated with, but not directly involved in, operating the facility. Costs such as overhead (e.g., building and land leasing and office supplies), administrative fees, local property taxes, and insurance fees fall into this indirect cost category.

The major direct operating costs are usually those associated with labor and materials. Materials costs for systems involve the cost of any chemicals needed for the operation, replacement parts, and so on. Labor costs differ greatly, depending primarily on the degree of controls and/or instrumentation. Typically, there are three working shifts per day, with one supervisor per shift. On the other hand, at some energy plants, the systems may be manned by a single operator for only one-third or one-half of each shift. At some sites, only an operator, supervisor, and site manager are necessary to run a facility. Salary costs vary from state to state and depend significantly on the location of the facility.

Annual maintenance costs can be estimated as a percentage of the capital equipment costs. The annual costs of replacement parts can be computed by dividing the costs of the individual part by its expected lifetime. Such life expectancies can be found in the literature (Vatavuk and Neveril 1980; Ulrich and Vasudevan 2004). Applicable laboratory costs depend on the number of samples tested and the extent of the tests. These costs can be estimated as 10–20% of the operating labor costs.

Indirect operating costs consist of overhead, local property tax, insurance, and administration, less any credits. Overhead comprises payroll, fringe benefits, social security, unemployment insurance, and other compensation that is indirectly paid to plant personnel. This cost can be estimated as 50–80% of the operating labor, supervision, and maintenance costs (Ulrich and Vasudevan 2004). Local property taxes and insurance can be estimated as 1–2% of the *TCC*, and administration costs can be estimated as 2% of the *TCC*. The total operating cost is the sum of the direct operating costs and the indirect operating costs less any credits that may be recovered (e.g., the value of recovered steam). Unlike capital costs, operating costs are always calculated on an annual basis.

21.6 Project and Process Evaluation

Alternate facilities or options (e.g., two different processes for generating the same amount of power, or a baghouse versus an electrostatic precipitator for air pollution control) may be compared using the *CRF* (Section 21.2.7). *Note*: A small flaw in this procedure is the assumption that the operating costs remain constant throughout the lifetime of the facility. However, since the analysis is geared toward comparing different alternatives, the changes with time should be somewhat uniform among the various alternatives, resulting in little loss of accuracy.

Investment and operating costs can be accounted for in other ways, such as using a present worth analysis (Section 21.2.3). However, this capital recovery method is preferred because of its simplicity and versatility. This is especially true when comparing somewhat similar systems having different depreciable lives. In such decisions, there are usually other considerations besides economic ones, but if all other factors are equal, the proposal with the lowest *TAC* should be the most viable one.

One difficulty in this analysis is estimating the revenue generated from a facility because both technology and costs can change from year to year. Also affecting the revenue generated is the amount of power to be processed by the facility. Usually, the more power produced, the greater the revenues. If an applicable estimate can be made of the revenue that will be generated from the facility, a rate of return can be calculated.

21.7 Hydrogen Energy Economy Considerations

For a hydrogen economy to become a reality, hydrogen energy will have to become economically competitive with conventional energy fuels, e.g., fossil fuels. But, as discussed in previous chapters, the cost of producing, storing, transporting, converting, and distributing hydrogen at the levels of production envisioned for the future is currently difficult to assess. Although it is realized that cost estimates are difficult to make and are occasionally unreliable and misleading, the following estimates and factors need to be taken into consideration when evaluating hydrogen energy options:

1) The location of the process
2) The proximity to cheap fuel and power

3) The size of the plant
4) The purity of the hydrogen required
5) The pressure and temperature at which it is employed
6) The type of process by which it is made
7) The safety and control measures used
8) Regulatory issues
9) Extent of product recovery
10) Production efficiency

The primary uses of hydrogen today are as a chemical feedstock in ammonia, food, and drug production, as well as in petrochemical and refinery processing. Oil refinery hydrogen demand is a major component of the overall hydrogen production. Hydrogen is used to remove sulfur and to hydrotreat and hydrocrack heavy crude oil constituents into more valuable, lighter products. It is also used in crystal growth, glass manufacturing, chemical tracing, metal fabrication, polysilicon and semiconductor manufacturing, metal production, and thermal processing. Hydrogen from fossil energy could be an economically competitive carbon-neutral alternative to traditional fuels used in the electricity, industrial, and transportation sectors.

Others have listed the main hydrogen utilization sectors as:

1) Industrial
2) Transportation
3) Power generation
4) Energy storage/grid balancing

In terms of processing, steam methane reforming (SMR) is a mature production process that builds upon the existing natural gas pipeline delivery infrastructure. Another well-developed, but more expensive approach for hydrogen production is splitting water. Methods used include electrolysis, photo-electrochemical cells, or solar thermochemical systems. Industrial technologies for hydrogen production include catalytic steam reforming (800–1,000°C) and partial oxidation (600–900°C) of hydrocarbons (e.g., natural gas) or renewable fuels (e.g., bioethanol); coal or coal blends with biomass and waste plastics gasification; water electrolysis; thermochemical water splitting at around 900°C; and biological production. Today the technical community has concluded that the most cost-effective process option is SMR with pressure swing adoption (PSA) hydrogen purification (U.S. DOE 2023).

Technical challenges for its use have generally focused on two areas: the storage of hydrogen and the transportation of hydrogen. Due to its small molecular size, hydrogen gas is difficult to contain, which makes its storage or transportation more challenging. In addition, at room temperature and standard atmospheric pressure, hydrogen has a low energy density per volume. In addition, storing hydrogen requires the input of external energy to cool or compress the gas.

All projections of the economics associated with hydrogen use are, of course, highly sensitive to assumptions about the costs of the numerous components of a hydrogen energy system that will decline over time as the technology develops and matures. Past experience suggests that it is hard to be sure about what might reasonably be expected in the way of any cost analysis for hydrogen technologies. As noted earlier, government incentives and regulations, including carbon penalties to encourage investment in hydrogen infrastructure, will also play a key role in determining the future of the hydrogen energy economy.

21.8 Concluding Remarks

It is fair to say that cost is the principal barrier to investment in hydrogen. No private firm will invest in a commercial hydrogen venture unless it believes that it will be able to compete against existing fuels and turn a profit. Hydrogen is still far from being competitive today in most applications, but that could change with technological breakthroughs and government incentives or mandates. If that is the case, the opportunities for profitable development of a hydrogen energy economy would expand over time. Initially, the investment may be limited to a few remote locations, where the costs of fuel distribution and electricity infrastructure are relatively high, where public concern about environmental sustainability is especially strong, and where the government provides large incentives for hydrogen adoption.

The sheer scale of investment in a hydrogen project, together with inherent technical and financial risks, could also discourage private companies. Government intervention in the form of financial incentives and regulatory measures that tilt the playing field in favor of hydrogen will almost certainly be necessary to get around this problem, especially where the economic case for hydrogen is marginal. A strong, long-term commitment by the government to the development of hydrogen infrastructure will be essential in giving fuel suppliers, equipment manufacturers, and consumers confidence that they will be able to make a reasonable return on the investments required to switch fuels (U.S. DOE 2023).

Because hydrogen is a long way from becoming competitive, the focus of government action today in the field of hydrogen is on research and development (R&D). Research into the use of hydrogen for energy purposes goes back many decades, but the scale of public (and private) funding for hydrogen and fuel cell R&D and demonstration activities has increased enormously in the last few years. Significant technological advances as well as a growing urgency on the part of policymakers to seek out sustainable energy solutions that address environmental and energy-security concerns may make it more likely that hydrogen fuel will become a viable energy solution in the not-too-distant future. Many governments now expect the transition to the hydrogen economy to begin before the middle of this century and are looking to speed up the process, often through collaborative international and joint private-public sector programs.

It is extremely hard to predict how soon the transition to the hydrogen energy economy might begin and how long it might take, as it depends critically on the technological breakthroughs in several different areas. It is impossible to know when these might occur and the extent to which they will lower costs and enhance the competitiveness of hydrogen vis-a-vis conventional forms of energy. As a result, all long-term projections of hydrogen use are probably of limited value.

Major technological and cost breakthroughs are needed before the hydrogen energy economy can become a reality. Replacing fossil fuels with hydrogen in providing energy services could bring major environmental benefits, on condition that hydrogen is used in nonpolluting fuel cells and that any carbon dioxide and noxious gases emitted in the hydrogen production process are reduced or eliminated. At the point of use, burning hydrogen in the presence of oxygen in a fuel cell produces no harmful emissions, just electrical energy, and water. The production of hydrogen can be emission-free if the energy used in the production process is derived from nuclear power or renewables, or if based on fossil fuels if the carbon dioxide emitted is captured at the point of production and stored permanently (U.S. DOE 2023).

Road transportation is an important and growing source of both air pollutants and climate-destabilizing greenhouse gases. There is clear evidence of the harmful impact on human health of exposure to pollutants emitted by cars and trucks. As a result, local air quality has become a major policy issue in almost all countries. Air pollution in many major cities and towns in the developing world has reached unprecedented proportions. Most rich, industrialized countries have made substantial progress in reducing pollution caused by cars and trucks through improvements in fuel economy, fuel quality, and the installation of emission-control equipment in vehicles. But rising road traffic has offset at least part of the improvements in emissions performance.

Because of increasing pollution from road traffic, road vehicles have been the focus of efforts to develop fuel cells. Replacing internal combustion engines fueled by gasoline or diesel with hydrogen-powered fuel cells would, in principle, eliminate pollution from road vehicles. Fuel cells can also be used to provide electrical energy services in industrial processes and buildings, replacing the direct use of petroleum products, natural gas, and coal.

Hydrogen could also contribute to reducing or eliminating emissions of carbon dioxide and other greenhouse gases. For this to happen, the process of manufacturing hydrogen would have to be carbon-free or at least less carbonintensive than current energy systems based on fossil fuels. This could be achieved in one of three ways: through electrolysis using electricity derived solely from nuclear power or renewable energy sources; through steam reforming of fossil fuels combined with new carbon capture and storage technologies; or through thermochemical or biological techniques based on renewable biomass.

Despite the potential local and global environmental benefits of switching to hydrogen, there are several uncertainties about other environmental consequences of a large-scale shift toward a hydrogen economy. These concern mainly the potential effects of significant amounts of hydrogen being released into the atmosphere. The widespread use of hydrogen would make such releases inevitable, but the effects are very uncertain because scientists still have a limited understanding of the hydrogen cycle. Any buildup of hydrogen concentrations in the atmosphere could have several effects, the most serious of which would be increased water vapor concentrations in the upper atmosphere and, indirectly, the destruction of the ozone layer. Increased hydrogen releases could also lower the oxidizing capacity of the atmosphere, and so increase the lifetime of air pollutants and greenhouse gases such as methane, hydrochlorofluorocarbons (HCFCs), and hydrofluorocarbons (HFCs). More research is needed to obtain a better understanding of hydrogen sources and sinks.

Safety is also a critical issue. Contrary to popular opinion, hydrogen is less flammable than light oil products, such as gasoline, and most other fossil fuels. But the need to transport and store it under high pressure or at very low temperatures brings other hazards. There is plenty of evidence that, with proper handling and controls, hydrogen can be as safe as the fuels in use today. Indeed, hydrogen has a long history of safe use in industry. But, for it to become widely accepted in other applications, it will become increasingly important to develop harmonized safety protocols and standards for the use of hydrogen and hydrogen-based fuels and feedstocks in such diverse applications as domestic appliances and in the transportation sector.

Finally, economic analysis will continue to incorporate costs and benefits to both society and the environment. This trend is continuing to expand with the inclusion of both sustainability effects and green chemistry/engineering (Dupont, Ganesan, and Theodore 2017). The widespread adoption of design specifications for operations taking these considerations into account would ensure that economic evaluations are prepared to be more inclusive of these emerging concepts. This will hopefully result in economic assessments that are comprehensive, easily understood, and readily put to use.

References

Dupont, R.R., Ganesan, K., and Theodore, L. (2017). *Pollution Prevention: Sustainability, Industrial Ecology, and Green Engineering*, 2nd Edition. Boca Raton, FL: CRC Press/Taylor & Francis Group.

Neveril, R.B. (1978). *Capital and Operating Costs of Selected Air Pollution Control Systems*. EPA 450/5-80-002. Research Triangle Park, NC: U.S. Environmental Protection Agency, Office of Air Quality Planning and Standards.

Theodore, L. (2014). *Chemical Engineering: The Essential Reference*. New York: McGraw-Hill.

Theodore, L. and Neuser, K. (1996). *Engineering Economics and Finance. A Theodore Tutorial*. East Williston, NY: Theodore Tutorials. Originally published by the U.S. EPA/APTI, RTP, NC.

U.S. Corps of Engineers (2018). *Civil Works Construction Cost Index System (CWCCIS)*. EM 1110-2-1304. Washington, DC: Department of the Army. www.publications.usace.army.mil/Portals/76/Publications/EngineerManuals/EM_1110-2-1304_31_Mar_2018.pdf?ver=PIjVa1Vj-m6k-7hSzM9NDA%3d%3d.

U.S. Department of Energy (2023). *U.S. National Clean Hydrogen Strategy and Roadmap*. Hydrogen Program. Washington, DC: U.S. Department of Energy. https://www.hydrogen.energy.gov/clean-hydrogen-strategy-roadmap.html / https://www.hydrogen.energy.gov/pdfs/us-national-clean-hydrogen-strategy-roadmap.pdf (accessed 25 August 2023).

Ulrich, G.D. (1984). *A Guide to Chemical Engineering Process Design and Economics*. Hoboken, NJ: John Wiley & Sons.

Ulrich, G.D. and Vasudevan, P.T. (2004). *Chemical Engineering Process Design and Economics: A Practical Guide*, 2nd Edition. Boca Raton, FL: CRC Press/Taylor & Francis Group.

Vatavuk, W.M. and Neveril, R.B. (1980). Estimating costs of air-pollution control systems-2. Factors for estimating capital and operating costs. *Chemical Engineering* 157–162.

Vogel, G.A. and Martin, E.J. (1983). Hazardous waste incineration, part 3-estimating capital costs of facility components. *Chemical Engineering* 87–90.

22

Optimization Considerations

22.1 Introduction

Optimization? Here are some terms for *optimal* that can be found in a dictionary: most favorable or desirable; best, best possible option; most cheerful view of matters or expect the best outcome; bright side; to make the most of; develop or realize to the utmost extent; obtain the most efficient or optimum use of; the best or most favorable degree, condition, or amount; etc. Added to this are some of the classic sayings of yesteryear that have appeared in the literature:

- *The optimist proclaims that we live in the best of all possible worlds; and the pessimist fears this is true* (Cabell 1926).
- *Optimism is a mania for maintaining that all is well when things are going badly* (Voltaire 1759).
- *The optimist sees the doughnut, but the pessimist sees the hole* (Wilson 1915).
- *An optimist is a guy that has never had much experience* (Marquis 1927).

The term optimization has come to mean different things for different applications, i.e., it could involve a simple two-step calculation or one that requires the use of detailed numerical methods. To take this a step further, the authors were undecided as to how to present optimization in this chapter. After much deliberation and meditation, it was decided to present both qualitative and quantitative material. This decision was primarily influenced by the desire that this be material not only on the introduction to optimization but also that addresses the topic quantitatively.

The optimization process has been described by Aris (1964) as "getting the best you can out of a given situation." Problems amenable to solutions by mathematical optimization techniques generally have one or more independent variables whose values must be chosen to yield one or more viable solutions and use a measure of "goodness" to distinguish among the many viable solutions generated by different choices of these variables. Mathematical optimization techniques are also used for guiding the problem solver to that choice of variables that maximize a "goodness" measure (e.g., profit) or that minimizes a "badness" measure (e.g., cost).

As noted in its most elementary form, one may say that optimization is concerned with the determination of the "best" solution to a given problem. This process is required in the solution of many general problems in engineering and applied science in the maximization (or minimization) of a given function(s), in the selection of a control variable to facilitate the realization of a desired condition, in the scheduling of a series of operations or events to control completion dates of a given project, in the development of optimal layouts of organizational units within a given design space, etc.

Hydrogen Energy: Principles and Applications, First Edition. Vincent J. DelGatto, Louis Theodore, R. Ryan Dupont, and Matthew C. Ogwu.

One of the more important areas for the application of mathematical optimization techniques is engineering design. The process of optimization in energy production and waste generation is required in the solution of many problems and often involves the maximization or minimization of a mathematical function. Once a particular subject or process scheme has been selected for study, it is common practice to optimize the process from a capital cost and operation and maintenance (O&M) standpoint.

Investing in all aspects of energy production, services and infrastructure (development, operation, and maintenance) can generate significant social and economic benefits. The authors believe that optimally appropriating capital toward this activity will prove highly cost-effective.

Traditional mathematical assessments of energy resource management are often inadequate for addressing the optimization of energy utilization and demands. Optimization studies are necessary to make informed investment and management decisions, facilitate cross-sector decision-making, and address compromises and trade-offs among the multitudes of stakeholder groups involved in energy resource planning. Increased financial investment and support for energy "development," e.g., hydrogen energy, is essential. When this support is efficiently managed and optimized, these investments can generate social, economic, financial, and other benefits that greatly outweigh the costs of this process. In addition to adequately financing the development, operation, and maintenance of infrastructure, increased funding is also needed to develop institutional capacity and ensure well-functioning governance structures. Greater financial support is equally essential to enhance the knowledge base and provide reliable and objective information about the state of all energy resources, not only hydrogen energy. The use of optimization technicians can help managers, engineers, and administrators inform decision-makers about policy/process options and their associated potential impacts. Finally, one's ability to understand optimization and any associated risks is the key to solving some of the more complex energy-related problems in the future.

There are many optimization procedures available, most of them too detailed for meaningful application in a text of this nature. These sophisticated optimization techniques, some of which are routinely used in the design of conventional chemical and petrochemical plants, invariably involve computer calculations. Although the use of these techniques in the majority of energy management applications was not warranted in the past, more and more real-world problems are requiring the use of such optimization techniques. In almost every situation, optimum conditions can be reduced to a consideration of costs or profits. Thus, an optimum design could be based on conditions giving the least cost per unit of time, or the maximum profit per unit of production. When one design variable is changed, it is often found that some costs increase and others decrease. Under these conditions, the total cost may go through a minimum at one value of· a design variable. This value would be considered optimum.

Optimum design can take many forms. One example is fluid flow, where piping and pipe fittings can amount to a significant portion of the total investment costs for a plant (AWWA 2017; Dupont, Ganesan, and Theodore 2017). It is necessary, therefore, to choose pipe sizes that yield a minimum total cost for pumping and fixed charges. For any given set of flow conditions, the use of an increased pipe diameter will increase fixed costs for the piping system, while decreasing pump or blower operating costs. Therefore, an optimum economic pipe diameter must exist. The value of this optimum diameter can be determined by combining the principles of fluid mechanics with cost considerations. The optimum economic pipe diameter is found at the point at which the sum of pump or blower operating costs and fixed costs is a minimum.

A second example is in heat exchanger design, where the objective is to obtain a specified duty with water as the cooling medium. The temperature of the cooling water affects the log mean temperature-difference (LMTD) driving force for heat transfer. The cooling water can either be circulated at a high rate with a small change in water temperature, or at a low rate with a large change in water temperature (Happel 1955; Flynn, Akashige, and Theodore 2019). An increased water flow rate allows a reduction in the heat transfer area, which will result in a decrease in capital investment (and fixed costs). On the other hand, water costs increase if more water is used. An economic balance between conditions of high-water flow rate/low transfer surface area and low water flow rate/high transfer surface area is based on the fact that an optimum flow rate of cooling water exists. It occurs at the point of minimum total cost for cooling water and equipment fixed costs. Furthermore, increased fluid velocities result in larger heat transfer coefficients, and consequently, less heat transfer area and heat exchanger costs for a given rate of heat transfer. On the other hand, the increased fluid velocities cause an increase in pressure drop and greater pumping costs. The optimum economic design occurs at the conditions where the total cost is a minimum for a given duty. The basic problem, therefore, is to minimize the sum of the system capital costs and variable annual costs for the operation of the heat exchanger.

A final example involves mass transfer, where the design of a distillation unit is based on a required separation for a given feed, at a known composition, temperature, and flow rate. The engineer must determine the size of the distillation column and the reflux ratio necessary to meet the specifications. As the reflux ratio is increased, the number of theoretical stages required for a given separation decreases. An increase in reflux ratio also results in lower fixed costs for the distillation column. However, optimization is necessary because an increased reflux results in increased costs for the reboiler heat supply and condenser coolant.

Difficulty in predicting the future is one of the problems that an optimization proposal involving capital investment must address. A statistical approach is being employed by some engineers responsible for process plant design in an attempt to evaluate the risks that occur. In the past, uncertainties confronting investors were resolved by intuition or "'professional judgment." In estimating a projected return, the design engineer may already have used a life expectancy and a minimum acceptable return rate based on such judgments. Strict quantitative interpretation of the results may not be justified in most instances because basic assumptions of statistical theory are violated. The bottom line is that optimization produces a substantially clearer picture of the effect of errors on profitability than is possible by a nonanalytical approach.

Section numbers and their accompanying titles for this chapter concerned with optimization are listed below:

22.2 History of Optimization

This brief overview of the history of optimization traces the evolution of mathematical ideas and concepts beginning in prehistory. Mathematics is nearly as old as humanity itself; evidence of a sense of geometry and interest in geometric patterns has been found in the designs of prehistoric pottery, textiles, and cave paintings. Primitive counting systems were almost certainly based on using the fingers of both hands, as evidenced by the predominance of the number 10 as the base for many number systems employed today.

The subject of mathematics encompasses the study of relationships among quantities, magnitudes, and properties of logical operations by which unknown quantities and magnitudes may be deduced. At one point in history, mathematics was regarded as a science involving only geometry and algebra. Toward the middle of the 19th century, mathematics came to be regarded as the science of relations that draws necessary conclusions. This latter view encompasses mathematical or symbolic logic – the science of using symbols to provide a logical deduction and reference based on definitions, axioms, postulates, and rules for transforming primitive elements into more complex relations and theorems. Enter optimization.

Interestingly, the mathematics of the late 19th and 20th centuries is characterized by an interest in unifying elements across numerous fields of mathematical endeavor, especially logic. For example, group theory has proven a particularly effective unifier. The unified set theory, intuitive geometry, modular arithmetic and measurement, and the development of number systems contributed to increased accuracy, precision, and error study. It also included studies of linear algebra, modern algebra, vectors, matrices, logic, truth tables, the nature of proofs, Venn and Euler diagrams, relations, functions, probability and statistics, linear programming, and computer programming.

As to the origin of optimization, it depends on who responds, as there are many aspects of optimization of interest to the practitioner. Some claim it was Thomas Edison who developed a long-lasting, high-quality light bulb in the 1870s. His success was primarily the result of an extensive trial-and-error search for the optimum filament material. Some now refer to this trial-and-error process as the *Edisonian* approach.

The ability to perform complicated calculations efficiently has not only reduced the cost and time required to perform each iteration but also provided a better understanding of how this method works. The success of this approach still relies heavily on the limitations of the user's intuition and experience because one cannot often predict or even comprehend the effects of changing numerous variables in a complex system. Despite this barrier, advanced computational methods have expanded and improved the classical methods of optimization.

With the promise of reducing design time and cost while improving product quality, design optimization has tremendous potential. Starting with a suboptimal design, a numerical optimization algorithm can be used to iteratively adjust a set of preselected design parameters to achieve a set of design targets. This new class of optimization technology enables broader, more comprehensive, and faster searches for innovative designs than was possible using previous generations of tools. Moreover, it requires no expertise in optimization theory, so it is easier to use for nonexperts and experts alike. By leveraging an engineer's potential to discover new design concepts, this new class of optimization technology overcomes the limits of human intuition and extends the designer's professional capability to achieve breakthrough designs and accelerated innovation.

As noted previously, the current level of computational horsepower allows optimization to easily select the best element from a set of available alternatives. In the simplest case, an optimization problem consists of maximizing (or minimizing) an objective function by systematically changing input values from an allowed set of variables. Mathematically speaking, optimization allows one to find the best available value(s) of an objective function, given a defined domain (or a set of constraints). This also includes the optimization of an array of objective functions with an array of domains. This process is discussed later in this chapter.

The increased capability of computational power has also had a tremendous impact on engineering design and data processing. The ability of computers to handle large quantities of data and perform mathematical operations permits the analysis of many more applications and engineering variables than could have ever been imagined only 20 years ago. Scientific calculations previously estimated in lifetimes of computational time are currently generated in seconds.

22.3 Scope of Optimization

Optimization is mathematical, and typically involves the maximization or minimization of a function that represents the "performance" of a "system." This is carried out by quantifying the values of the input variables that produce an optimal value for the function.

Perhaps the most important tool employed in optimization by engineers and scientists is *linear programming*. Linear programming consists of a linear, multivariable function, which is optimized (maximized or minimized), and subject to a particular number of constraints. The constraints are normally expressed in linear form. Integer linear programming refers to optimization problems in which at least some of the variables assume integer values. The reader should note that the terms *linear* programming and *nonlinear* programming are similar from an application perspective.

One of the major responsibilities in optimization is to construct an accurate objective function. For example, in an environmental problem, the objective function might be the profit associated with the operation of a landfill that treats two categories of waste: "inert" and "non-inert" waste. Essentially, the optimization problem is to maximize profit (from handling the two waste streams) by employing an operating schedule subject to the landfill's capacity and environmental regulatory requirements. Rephrased, the objective function should represent a mathematical description of the combination of two wastes that will produce maximum profit subject to the constraints connected with the landfill's processing capabilities and the regulatory requirements imposed on each type of waste.

The objective function might be simple to identify and quantify, or may be very elusive as a result of the presence of conflicting or dimensionally incompatible objectives. One might be asked to optimize the profit for the aforementioned landfill, which not only minimizes air and water pollution but is also *aesthetically pleasing*. Thus, it may not always be possible to quantify an objective function using strictly mathematical terms.

Optimization problems can sometimes be divided into parts where it is possible to attain the optimum for the original problem by optimizing the constituent parts. This process is very powerful because it allows one to solve a series of smaller, easier problems rather than one large one. One of the best-known techniques to attack these problems is *dynamic programming*. This approach is characterized by a process that is performed in stages. One such example is the manufacturing process. Rather than solving the problem as a whole, dynamic programming optimizes one stage at a time to produce an optimal set of decisions for the entire process. Although dynamic programming has applications in some systems or processes, it is not reviewed in this chapter. If the assumption of linearity cannot be made, a collection of techniques called *nonlinear programming* is used. This is addressed later in this chapter.

Finally, most engineering optimization applications involve economics: maximizing profit, minimizing cost, or both. Environmental and energy applications can involve the purification of water, minimizing toxic emissions, maximizing energy use efficiency and conservation, minimizing health and safety risks, etc. The reader should not lose sight, however, that most real-world industrial applications involving optimization often require simple solutions.

The following is a comment from a retired process engineer after 30 years in the industry (R. Altomare, personal communication), "Generally, the consumer goods industry where I worked does not rely on sophisticated models. Optimization was conducted to determine how to use minimum resources to minimize costs, subject to constraints. In the determination of consumer preference, we typically designed experiments to examine the broad range of attributes that led to maximum 'product liking' or decision to purchase. Occasionally, we modeled processes and used regression models. In that case, we could search or compute the desired maximum. The last set of techniques used is what the business world calls *Operations Research*. For example, linear programming would be used in the production of America's favorite barbecue item – the hotdog. The objective function was to produce a combination of meats that maximized flavor, met or exceeded U.S. Food and Drug Administration (FDA) requirements for protein, fat, and water content, while minimizing overall production costs."

22.4 General Analytical Formulation of the Optimum

When more than one independent variable is involved in determining the objective function (e.g., profitability), the objective function y that depends on a system of $x_1, x_2, x_3, \ldots, x_n$ independent variables are written as follows,

$$y(\text{profit}) = f(x_1, x_2, x_3, \ldots, x_n) \tag{22.1}$$

In other words, one needs to specify $x_1, x_2, x_3, \ldots, x_n$ so that y, profit, will be a maximum. In general, this will involve some type of trial-and-error calculation.

It is possible in some cases to set up an explicit function that will relate the objective function y to the independent variables. Differentiation will then result in a series of partial derivatives,

$$\frac{\partial y}{\partial x_1}, \frac{\partial y}{\partial x_2}, \text{ etc.} \tag{22.2}$$

Setting these partial derivatives equal to zero results in the same number of simultaneous equations as the number of variables involved. These equations, often subject to "limitations" and "constraints," can be solved for the corresponding optimum values of these variables. The maximum occurs at a point where the first derivatives equal zero, accounting for restrictions on the second derivative, continuity of the function throughout the range of independent variables, and absence of optimum conditions at limiting values of one or more of the variables. These conditions can, on occasion, restrict the applicability of this approach (Happel 1955). For example, in the case of a two-variable function where $y = f(x_1, x_2)$, points $x_1 = a$ and $x_2 = b$ exist for a maximum value of $y = f(x_1, x_2)$ if the following conditions hold:

$$\frac{\partial y}{\partial x_1} = \frac{\partial y}{\partial x_2} = 0 \tag{22.3}$$

with,

$$\frac{\partial^2 y}{\partial x_1^2} < 0; \frac{\partial^2 y}{\partial x_2^2} < 0 \tag{22.4}$$

and,

$$\left(\frac{\partial^2 y}{\partial x_1^2}\right)\left(\frac{\partial^2 y}{\partial x_2^2}\right) > \left(\frac{\partial^2 y}{\partial x_1 \partial x_2}\right)^2 \quad (22.5)$$

In contrast, if $\partial^2 y/\partial x_1^2$ and $\partial^2 y/\partial x_2^2$ are positive, and Equation 22.5 holds, there will be a minimum. However, if,

$$\left(\frac{\partial^2 y}{\partial x_1^2}\right)\left(\frac{\partial^2 y}{\partial x_2^2}\right) < \left(\frac{\partial^2 y}{\partial x_1 \partial x_2}\right)^2 \quad (22.6)$$

the point will be a "saddle" point. The case where

$$\left(\frac{\partial^2 y}{\partial x_1^2}\right)\left(\frac{\partial^2 y}{\partial x_2^2}\right) = \left(\frac{\partial^2 y}{\partial x_1 \partial x_2}\right)^2 \quad (22.7)$$

is open and the value may be a maximum, a minimum, or neither.

Now consider an objective function specified as:

$$f(x_1, x_2) = \left(x_1^2 - 2x_1 + x_2^2 + x_1 x_2 + 1\right) \quad (22.8)$$

To determine if the function has a maximum or a minimum, proceed as follows:

$$\frac{\partial f}{\partial x_1} = 2x_1 - 2 + x_2 = 0 \quad (22.9)$$

$$\frac{\partial f}{\partial x_2} = 2x_2 + x_1 = 0 \quad (22.10)$$

When Equations 22.9 and 22.10 are solved, the following equations result:

$$x_1 = \frac{4}{3} \quad (22.11)$$

$$x_2 = \frac{1}{3} \quad (22.12)$$

Since

$$\frac{\partial^2 f}{\partial x_1^2} = 2; \frac{\partial^2 f}{\partial x_2^2} = 2 \quad (22.13)$$

the above solutions for x_1 and x_2 represent a *minimum* with

$$y = f(x_1, x_2) = 5 \quad (22.14)$$

Consider now a traditional mathematician's exercise to determine the maximum or minimum of a function dependent on three variables:

$$P = 8x + 6y + 8z + xy + xz + yz - 2x^2 - 3y^2 - 4z^2 \quad (22.15)$$

Here

$$\left.\frac{\partial P}{\partial x}\right|_{y,z} = 8 + y + z - 4x = 0 \tag{22.16}$$

$$\left.\frac{\partial P}{\partial y}\right|_{x,z} = 6 + x + z - 6y = 0 \tag{22.17}$$

$$\left.\frac{\partial P}{\partial z}\right|_{x,y} = 8 + y + x - 8z = 0 \tag{22.18}$$

The solution to the above three linear simultaneous equations can be shown to be:

$$x = -4.62;\ y = 0.94;\ z = 0.54 \tag{22.19}$$

for which

$$P = -84.5 \tag{22.20}$$

To determine whether the above P is maximum or minimum, calculate the three second-order partial derivatives as:

$$\frac{\partial^2 P}{\partial x^2} = -4;\ \frac{\partial^2 P}{\partial y^2} = -6;\ \frac{\partial^2 P}{\partial z^2} = -8 \tag{22.21}$$

Since the three derivatives are negative, the above value of $P = -84.5$ is a maximum, i.e., any other value of P for a different x, y, and z would produce a value of P less than (or more negative) than −84.5. The "proof" is left as an exercise for the reader. However, if one were to select values of x, y, and z straddling the supposed minimum set of values, P should be below −84.5. For example, the value of P for $x = -5$, $y = 1$, and $z = -0.5$ is −95.0.

The following general comments relate the preceding expressions to both economics and engineering practice.

1) For a process with a fixed output that involves several pieces of equipment (e.g., an absorber, a heat exchanger, a preheater, a stripper, and a cooler), several design variables affect the cost of several of the operations (Theodore 2007; Ricci and Theodore 2010; Theodore 2010; Flynn, Akashige, and Theodore 2019). Assume that there are two major independent variables x_1 and x_2. For a given value of x_2, the total costs are tabulated for various values of x_1, and the minimum total cost, C, is noted. This procedure is repeated for other values of x_2 and the combination of values of x_1 and x_2 is selected, which gives the minimum of the minima. If the values of x_1 and x_2 can be varied by differential increments, the same result can be obtained by first setting the partial derivatives $\partial C/\partial x_1$ and $\partial C/\partial x_2$ equal to zero, and solving simultaneously for the optimum values of x_1 and x_2 (as discussed previously). Similar procedures could be used where there are more than two design variables. A careful study of this process might show that certain design variables affect the costs of several but not all parts of the process. This allows the problem to be simplified.
2) In the important problem of estimating the optimum production of a plant, similar methods may be followed. In this case, maximum annual profit/minimum annual loss are the criteria. However, the output can be varied, whereas in the cases considered before, the output was fixed (Aris 1964). Hence, the minimum sum of variable costs could be used as the objective

function. For example, in the case of operating at a loss, the plant should not be shut down until the out-of-pocket income falls below the out-of-pocket expenses, excluding fixed costs. If continuity of supply is stipulated in a contract, the plant may be operated at a loss on an out-of-pocket basis.

3) In the case of a decision regarding a proposed expenditure that is not essential to the success of a process, the decision is often based on whether or not the *payback* time is attractive. The payback time is defined as that period in which the actual savings because of the proposed change would equal the original investment. The novice sometimes credits more toward savings than actually would be realized. For example, the cost of 1,000 lb/h of steam at a water treatment facility might be \$0.35/h. If the steam lines were insulated, the required steam load on the plant might be reduced by 1,000 lb/h. At first inspection, the insulation is credited with a saving of \$0.35/h. However, a more in-depth analysis of the steam costs shows that the \$0.35/1,000 lb is broken down into \$0.10 for fuel and \$0.25 for fixed costs. Because the fixed costs would not change by insulating the steam lines, the true savings is only the value of fuel saved.

22.5 Mathematical Concepts in Linear Programming

Linear programming is a technique for optimization that has evolved into a useful procedure for the technical community. As noted earlier, the title results from the assumption inherent in the method that linear relationships describe the system under consideration. It deals with the determination of an optimum solution to a problem expressed in linear relationships where there are a large number of possible solutions. In general, the methods employed are trial-and-error, where the procedure systematically follows a mathematical treatment that minimizes the labor involved and ensures that a correct result has been obtained.

In order to apply linear programming to a problem, the relationships among variables must be expressed as a set of linear equations or inequalities. The variables must be otherwise independent and must exceed the number of equations. Mathematically expressed,

$$\sum_i a_{ij} = b_j; \quad \left(i = 1,2,\ldots m;\ j = 1,2,\ldots n;\ n > m\right) \tag{22.22}$$

where the sum of the coefficient a_{ij} times the value of each variable x_j equals the requirement b_j in the ith equation. The number of variables must exceed the number of equations so that an infinite number of solutions is possible. The best one is selected. The relationship in Equation 22.22 is often expressed as an inequality. "Dummy" additional variables can be introduced to convert these inequalities into appropriate equations. Another requirement is that no quantity can be negative. Thus, mathematically,

$$x_j \geq 0; \ b_j \geq 0 \tag{22.23}$$

This requirement will ensure that only useful answers are obtained. The solution of a problem by linear programming also requires, for example, a profit or objective function to be maximized (or minimized). This requirement is stated as:

$$P = \sum_i c_j x_i = \text{maximum}\left(\text{or minimum}\right) \tag{22.24}$$

where c_j is the profit (or cost) per unit x_i used. The solution is carried out by an iterative process, which at each stage of the calculation assigns either zero or positive values to all x_i variables. As the calculation proceeds, the values selected, while meeting the constraints imposed, will tend to increase the summation to be maximized at each stage. If a solution exists, a maximum (profit) case will be realized.

22.6 Applied Concepts in Linear Programming

Consider now an example that involves minimizing an objective function, i.e., E (expenditures)

$$E = a_1x_1 + a_2x_2 \tag{22.25}$$

subject to two constraints

$$B = b_1x_1 + b_2x_2 \tag{22.26}$$

$$C = c_1x_1 + c_2x_2 \tag{22.27}$$

In addition,

$$x_1 \geq 0;\ x_2 \geq 0 \tag{22.28}$$

In terms of a solution, the above system of equations can first be examined graphically. The problem can be viewed in Figures 22.1 and 22.2. First note that only the positive quadrant needs to be considered because Equation 22.28 must hold. Equations 22.26 and 22.27 can be rewritten in the following form:

$$x_2 = \frac{B}{b_2} - \left(\frac{b_1}{b_2}\right)x_1 \tag{22.29}$$

$$x_2 = \frac{C}{c_2} - \left(\frac{c_1}{c_2}\right)x_1 \tag{22.30}$$

Each appears in Figure 22.1a and b. It is apparent that the only values of x_1 and x_2 that will satisfy Equations 22.29 and 22.30 lie above the solid lines in Figure 22.1c, i.e., the hatched area. This hatched area is termed the constraint set. Now a series of straight lines or contours corresponding to the expenditure Equation 22.25 with a different value of E used for each line are placed in Figure 22.2a. Superimposing Figure 22.2a onto Figure 22.1c produces Figure 22.2b. Of all these lines, the smallest E for which x_1 and x_2 remain in the specified region in Figure 22.1c will be the minimum expenditure. This will yield a proportion of x_1 and x_2 which minimizes the cost. In other words, one desires the lowest-value contour to have some point in common with the constraint set. For the particular values of a_1 and a_2 chosen for Equation 21.25, the minimum point is seen to correspond to E where $E_3 = E^*$. Further, if the constants c_1 and c_2 in Equation 22.25 are changed, it is possible that the minimum point would occur elsewhere.

Which point is the minimum is determined by constraints a_1 and a_2 of the problem. It should be pointed out that if the expenditure equation was nonlinear, the contour lines would be curved, and if the constraint equations were nonlinear, the constraint lines would also not be straight. It should be intuitively obvious that locating the minimum point is more complicated in this case. The situation is essentially reversed for a maximization problem, i.e., for a profit P. This is demonstrated in Figure 22.3.

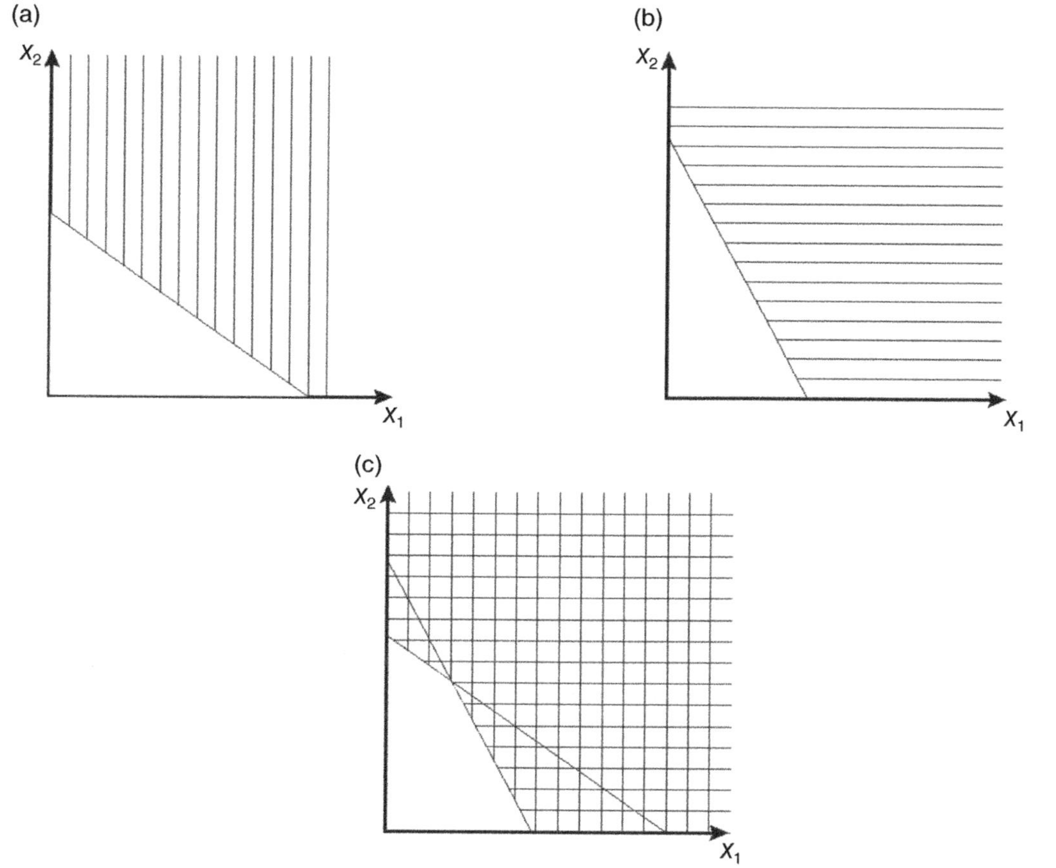

Figure 22.1 Two-variable minimization problem. (a) Equation 22.26; (b) Equation 22.27; (c) Equation 22.26 and 22.27.

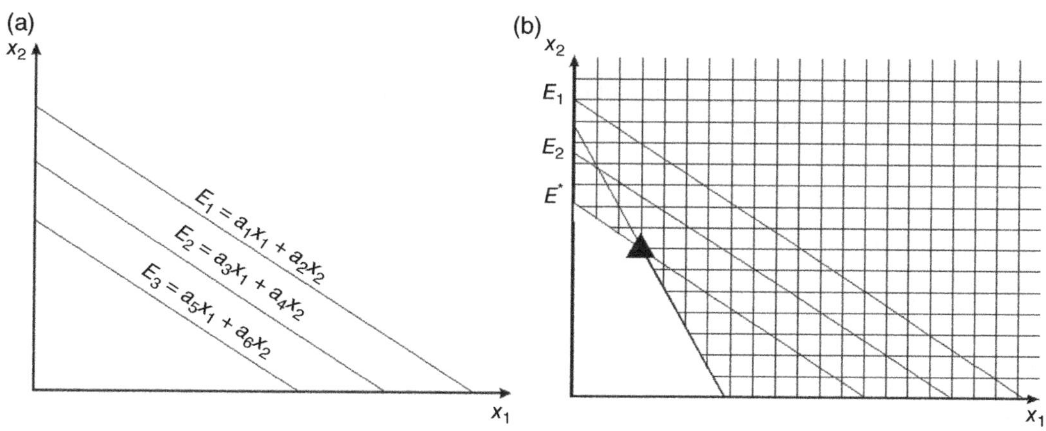

Figure 22.2 Solution to a two-variable minimization problem. (a) Expenditure equations; (b) determining minimum Expenditure meeting specified constraints.

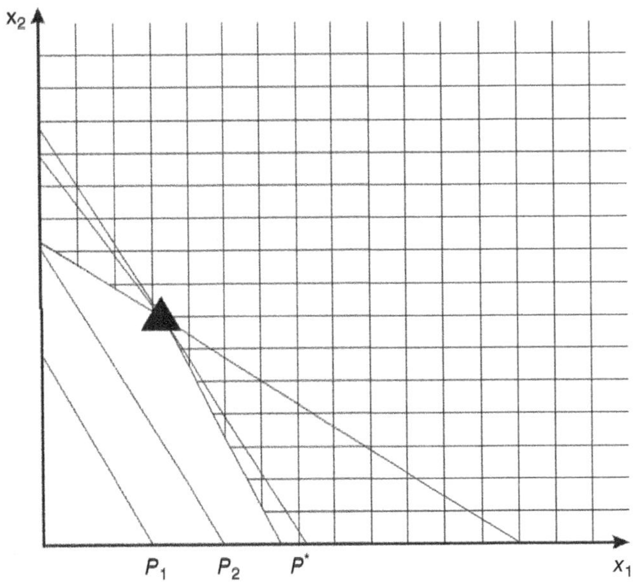

Figure 22.3 Solution to a two-variable maximization problem.

The above methodology can be extended to more than two variables and more than two constraints. The following would apply for three variables and four constraints:

$$P = a_1x_1 + a_2x_2 + a_3x_3 \tag{22.31}$$

$$B = b_1x_1 + b_2x_2 + b_3x_3 \tag{22.32}$$

$$C = c_1x_1 + c_2x_2 + c_3x_3 \tag{22.33}$$

$$D = d_1x_1 + d_2x_2 + d_3x_3 \tag{22.34}$$

$$E = e_1x_1 + e_2x_2 + e_3x_3 \tag{22.35}$$

$$x_1 \geq 0;\ x_2 \geq 0;\ x_3 \geq 0 \tag{22.36}$$

Naturally, this can be rewritten for n objective functions and m constraints.

Suppose you have been requested to obtain the maximum value of P subject to the following conditions. The objective function is:

$$P = 3x_1 + 5x_2 \tag{22.37}$$

Subject to the following constraints:

$$x_2 = -0.5\,x_1 + 3 \tag{22.38}$$

$$x_2 < -3x_1 + 6 \tag{22.39}$$

$$x_1 \leq 3 \tag{22.40}$$

$$x_2 \leq 4 \tag{22.41}$$

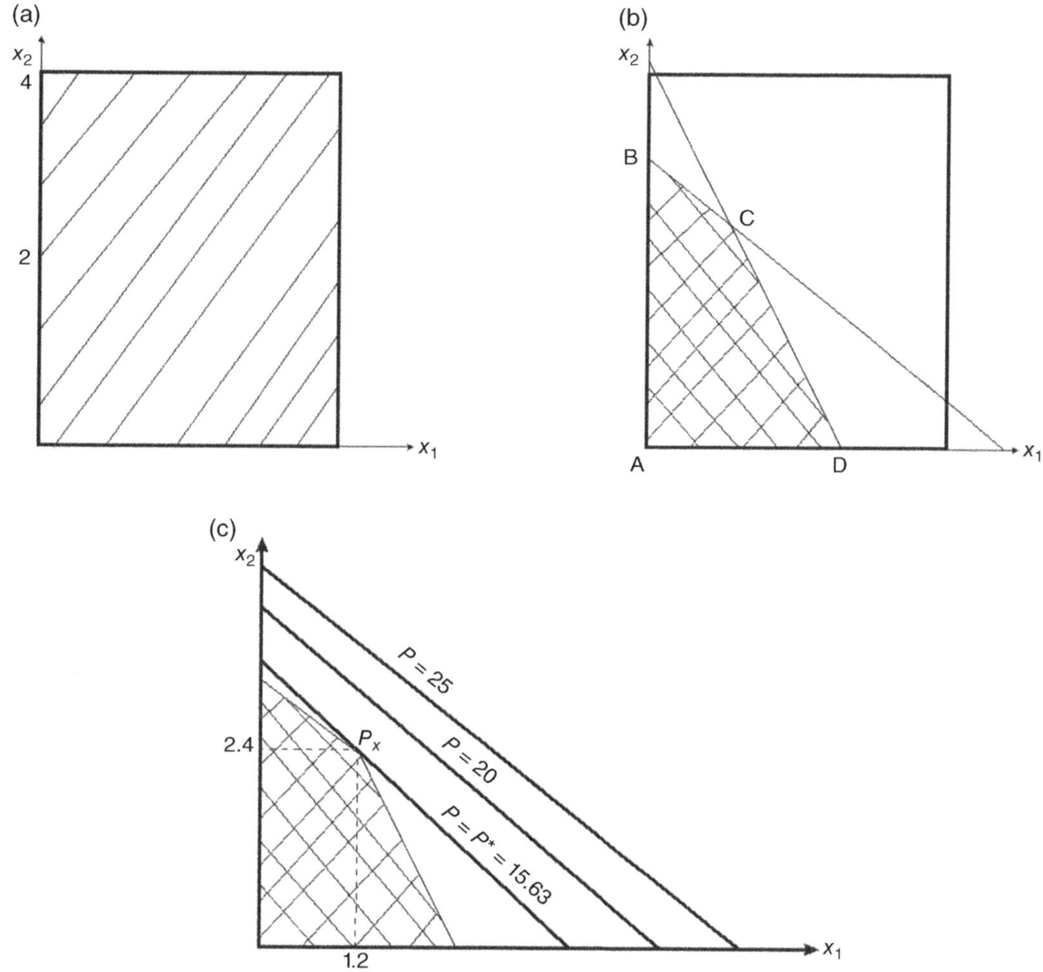

Figure 22.4 Graphical solution to maximization problem. (a) Conditions specified by Equations 22.40 and 22.41; (b) conditions specified by Equations 22.38 and 22.39; (c) determination of the maximum profit meeting specified constraints.

Once again, the solution is first provided in graphical form. Refer to Figure 22.4. Conditions in Equations 22.40 and 22.41 require that the solutions, i.e., the values x_1 and x_2 that will maximize P lie within the rectangle provided in Figure·22.4a.

Conditions in Equations 22.38 and 22.39 require that the solution be within the area ABCD in Figure 22.4b. When the objective function equation is superimposed on Figure 22.4b and c results. The maximum P is located at $x_1 = 1.2$ and $x_2 = 2.4$, for which $P = 15.6$. Excel provides the same solution.

22.7 Optimization of Existing Systems

The objective of this section is to provide a general discussion of the importance of proper O&M to overall plant performance. A plant, including all equipment, must be recognized as a highly specialized and complex facility producing an acceptable product, e.g., power from fuel supply. It is the plant management's responsibility to produce this product at the lowest unit cost and at the highest quality possible. The key to fulfilling this responsibility is a sound O&M management program.

The initial cost objective when constructing a power generation system is to achieve a technically reliable system capable of meeting energy production goals and achieving all of this at the minimum annualized cost. Once the system has been constructed, this goal cannot be realized without a conscientious and determined O&M effort. It is obvious, therefore, that once the optimum or near optimum level of performance is achieved, it must be maintained day in and day out. Unfortunately, such achievements rarely happen on their own in actual practice. High levels of performance result from the application of not only sound engineering practice but also robust and intentional O&M procedures, both of which are addressed in this section.

Effective operation must be accompanied by a routine maintenance program. The program should provide for scheduled inspections of all equipment components as well as for necessary monitoring of operating parameters to ensure correct operation and optimum performance of all of the process equipment. Monitoring of the fuel streams, power generation, etc., will provide valuable information about the performance of the process.

Routine maintenance of equipment is also important because equipment failure can be expensive in terms of lost production, degradation of resources, and potential health effects on employees. Each component in a process must operate properly to ensure steady, efficient output and desired results from the system.

Establishing an equipment preventive maintenance program need not be elaborate or complicated. The actual work in routine inspection and servicing may, in large part, be performed by personnel operating the equipment. Of course, extensive repairs or rebuilding should be accomplished by skilled and trained maintenance personnel.

An adequate scheduled maintenance program will repay itself many times over by reducing production time losses and costly major repairs. Effective maintenance is easily accomplished if the following basic rules are followed:

1) At initial installation, provide easily accessible access for cleaning and removal of any part of the system. Provision for convenient maintenance begins on the drawing board. The designer should allow for the eventual removal of parts that may wear out. Special material should be designated for those parts normally exposed to the greatest wear or possible abuse.
2) Establish an information file for the history of all equipment. The file should include actual operating data, such as flow volumes, operating temperatures, inlet and outlet velocities, pressure differentials, and any other data affecting operation.
3) Maintain conditions as designated on the nameplate.
4) Maintain copies of arrangement or construction drawings, including parts lists and a spare parts reference drawing. These contain commercial descriptions of such items as gaskets, bolts, nuts, and washers.

At least one copy of the information file should always be available to maintenance personnel, and a master copy should always be kept in the plant engineering department.

In a large, complex, and continuously operating process unit, these maintenance programs have evolved almost into a science, which cannot be discussed fully in this section. Two general areas of maintenance are recognized: emergency and preventive. Emergency maintenance follows an unexpected failure of a plant component, which often disrupts plant operation, sometimes causes a major shutdown, and almost always results in an abnormally high repair cost. Preventive maintenance is scheduled so as not to disrupt operations, and it does not involve overtime use of maintenance personnel and the high costs of emergency procurement of parts or tools. The purpose of preventive maintenance is to minimize emergency maintenance, and a judicious program of preventive maintenance is vital to the continued satisfactory performance of a system.

An effective maintenance program is based on an inventory of every component of the plant, categorized with respect to its potential for failure and its availability for service. Some components wear more rapidly than major process vessels or utility piping. Preventive maintenance programs must assign a frequency of attention to all these items; although the components least likely to fail will be examined less frequently, they must not be ignored entirely. Forced outages to repair defective equipment are costly. For this reason, specifications demand high component reliability and design redundancy, with the exception that sufficient performance reliability will be attained so that major maintenance can be performed during regularly scheduled outages. Any action on a piece of equipment that is planned and executed to prolong availability by detecting or preventing impending failures is preventive maintenance. In addition to a potential reduction in downtime cost and improved reliability, a major advantage of preventive maintenance is that since it is planned, it can be scheduled. As noted above, key elements in a successful preventive maintenance program are promoting the concept of recognizing trouble before it happens, following the preventive maintenance schedule without exception, and recording data at appropriate times.

O&M problems have plagued users since the installation of the first device in a process, and over time there are now numerous reasons for an O&M program (Theodore and Buonicore 1992). The most important are:

1) Continuously meeting present quality requirements.
2) Maintaining productivity (reducing production interruptions) of the process.
3) Prolonging equipment life.
4) Prolonging process life.
5) Reducing operating costs.
6) Promoting better public relations and avoiding community alienation by reducing accidental releases.
7) Promoting better relations with regulatory officials.
8) Improving worker health and safety conditions.

The following is a list of the basic features that should be included in any O&M program, and should apply to all plants, including power plants, regardless of type or size (Theodore and Buonicore 1992):

1) Equipment record system
2) Planning and scheduling
3) Storeroom and inventory system
4) Maintenance personnel
5) Cost and budgets for maintenance operations
6) Special tools and equipment
7) Lubrication
8) Major equipment information
9) Warranty and contract provisions

To troubleshoot equipment performance problems adequately, a multiphase program should be implemented. As a guideline, the following procedures are recommended (Theodore and Buonicore 1992):

- Phase 1 – Problem identification
- Phase 2 – Implementation
- Phase 3 – Testing and sampling
- Phase 4 – Preventive maintenance

A well-conceived maintenance program replete with appropriate spare parts and trained personnel, coupled with a systematic approach to monitoring and control of process operations, will pay significant dividends to overall process performance and cost management.

Additional information on the role of operation, maintenance, and inspection (O&M&I) in optimization is available in the work of Theodore and Buonicore (1992). The Electric Power Research Institute (EPRI) provides details on operation and maintenance standards and practices specifically for electric utilities in their Equipment Maintenance Optimization Manual Applications document (EPRI 2001).

References

American Water Works Association (2017). *Water Treatment Plant Operation*, Volume 1, 7th Edition. Sacramento, CA: Sacramento State Office of Water Programs.

Aris, R. (1964). *Discrete Dynamic Programming: An Introduction to the Optimization of Staged Processes*. Waltham, MA: Blaisdell Publishing Co.

Cabell, J.B. (1926). *The Silver Stallion*. New York: Robert McBride & Company.

Dupont, R.R., Ganesan, K., and Theodore, L. (2017). *Pollution Prevention: Sustainability, Industrial Ecology, and Green Engineering*, 2nd Edition. Boca Raton, FL: CRC Press/Taylor & Francis Group.

Electric Power Research Institute (2001). *Equipment Maintenance Optimization Manual Applications*. Report 1004590. Palo Alto, CA: EPRI. https://www.epri.com/research/products/1004590.

Flynn, A.M., Akashige, T., and Theodore, L. (2019). *Kern's Process Heat Transfer*, 2nd Edition. Hoboken, NJ: John Wiley & Sons.

Happel, J. (1955). *Chemical Process Economics*. Hoboken, NJ: John Wiley & Sons.

Marquis, D. (1927). *Archy and Mehitabel*. New York: Doubleday & Company, Inc.

Ricci, F. and Theodore, L. (2010). *Mass Transfer Operations for the Practical Engineer*. Hoboken, NJ: John Wiley & Sons.

Theodore, L. (2007). *Fluid Flow for the Practicing Chemical Engineer*. Hoboken, NJ: John Wiley & Sons.

Theodore, L. (2010). *Heat Transfer Applications for the Practicing Engineer*. Hoboken, NJ: John Wiley & Sons.

Theodore, L. and Buonicore, T. (1992). *Air Pollution Control Equipment: Selection, Design, Operation, and Maintenance. A Theodore Tutorial*. East Williston, NY: Theodore Tutorials.

Voltaire (1959) [1759]. Bair, L. ed. *Candide* (trans. L. Bair; with an appreciation by A. Maurois; illustrations by S. Beckett). New York: Bantam Dell.

Wilson, M. (2016) [c. 1915]. *Optimist and Pessimist*. Ratcliff, S. ed. *Oxford Essential Quotations*. Oxford, England: Oxford University Press.

23

Illustrative Examples

23.1 Introduction

This last chapter is devoted to presenting illustrative examples directly or indirectly related to hydrogen energy and the topics covered in the preceding chapters. Section numbers and their accompanying titles for this chapter concerned with illustrative examples related to hydrogen energy are listed below.

- 23.1 Introduction
- 23.2 Energy Principles
- 23.3 Thermodynamics
- 23.4 Energy Systems
- 23.5 Environmental Issues
- 23.6 Ethics
- 23.7 Economics
- 23.8 SDS Information
- 23.9 Optimization
- References

23.2 Energy Principles

Three illustrative examples complement the presentation on energy principles. One is concerned with the ideal gas law, one with the conservation law for mass, and the last with stoichiometry.

23.2.1 The Ideal Gas Law

The following information is provided for an ideal gas: pressure = 1.0 atm, temperature = 60°F, molecular weight of the gas = 2.0 lb/lbmol. Determine the density of the gas in lb/ft^3.

Solution. Rewrite the ideal gas law, $PV = n\ RT$ in terms of the density, $\rho = m/V$. It should be recognized that $n = m/\text{MW}$, where m = mass of gas, lb; and MW = molecular weight, lb/lbmol. Substituting yields:

$$PV = \left(m/\text{MW}\right)RT \tag{23.1}$$

Hydrogen Energy: Principles and Applications, First Edition. Vincent J. DelGatto, Louis Theodore, R. Ryan Dupont, and Matthew C. Ogwu.

Rewriting this equation yields:

$$P(\mathrm{MW})/RT = (m/V) = \rho \tag{23.2}$$

Substituting the given values into Equation 23.2, using a value for R = 0.73 atm-ft^3/lbmol-°R for consistent units (Table 5.1) gives:

$$\begin{aligned} P(\mathrm{MW})/RT &= (1\ \mathrm{atm})\,(2.0\ \mathrm{lb/lbmol})/[(0.7302\,\mathrm{atm\text{-}ft^3/lbmol\text{-}^\circ R})(60+460^\circ\mathrm{R})] \\ &= (2\ \mathrm{atm\text{-}lb/lbmol})/(379.6\,\mathrm{atm\text{-}ft^3/lbmol}) = 0.0053\ \mathrm{lb/ft^3} \end{aligned}$$

Since the molecular weight of the given gas is 2.0 lb/lbmol, the calculated density can be assumed to apply to hydrogen. The effect of pressure, temperature, and molecular weight on density can also be obtained directly from the ideal gas law. Increasing the pressure and molecular weight increases the density, while increasing the temperature decreases the gas density.

23.2.2 Mass Conservation Law

An external gas stream is fed into a hydrogen energy process at a rate of 10,000 lb/h in the presence of 20,000 lb/h of air. Due to environmental constraints, 1,250 lb/h of a conditioning agent is added to assist in the treatment of the gas stream. Determine the rate of product gases exiting the unit in lb/h.

Solution. First, write the conservation law for mass for the control device:

$$(\mathrm{mass}_{in}) - (\mathrm{mass}_{out}) + (\text{mass generated}) = (\text{mass accumulated}) \tag{23.3}$$

Apply this equation subject to the given values in the problem statement. Remember that mass is not generated in the process and steady-state conditions apply; therefore, (mass_{in}) = (mass_{out}).

$$\mathrm{mass}_{in} = 10{,}000\ \mathrm{lb/hr} + 20{,}000\ \mathrm{lb/hr} + 1250\ \mathrm{lb/hr} + 31{,}250\ \mathrm{lb/hr}$$

Since (mass_{in}) = (mass_{out}), the product gas exit flow rate is then, mass_{out} = 31,250 lb/h.

23.2.3 Stoichiometry

The reaction equation for the combustion of hydrogen is as follows:

$$H_2 + O_2 \rightarrow H_2O \tag{23.4}$$

Balance this equation and determine the ratio of reactants to products. Using the balanced equation, determine the air required for stoichiometric combustion of 1.0 scf of hydrogen and calculate both the stoichiometric air requirements and flue gas produced from the complete combustion of this hydrogen.

Solution. The balanced stoichiometric equation is:

$$H_2 + \tfrac{1}{2}O_2 \rightarrow H_2O \tag{23.5}$$

The ratio of hydrogen and oxygen to water is 1.5–1.0. Although the number of moles on either side of the equation does not balance, the masses of reactants and products (in accordance with the conservation law for mass) must balance.

The scf of O_2 required for the complete combustion of 1 scf of hydrogen can now be determined. From the balanced equation, there are 0.5 mol of O_2 required per mole of hydrogen. Since the moles of an ideal gas are directly proportional to their volume, the scf of O_2 required is 0.5 (1 scf of hydrogen) = 0.5 scf of O_2.

From a mass balance, 2 lb of hydrogen reacted with 16 lb of O_2 to form 18 lb of H_2O, with an initial (reactant) and final (product) total mass of 18 lb. The amount of N_2 in a quantity of air that contains 0.5 scf of O_2 is given by the following knowing that the nitrogen to oxygen volume (or mole) ratio in air is 79/21:

$$\text{scf of } N_2 = (79/21)(\text{scf of } O_2) = (79/21)(0.5) = 1.88 \text{ scf}$$

The stoichiometric amount of air is then:

$$\text{scf of air} = \text{scf of } N_2 + \text{scf of } O_2 = 1.88 + .05 = 2.38 \text{ scf}$$

In order to calculate the amount of flue gas produced, one must include the nitrogen from the air in addition to the product of combustion, that is, water vapor. The equation here is written as:

$$H_2 + [1.88N_2] + 0.5O_2 = H_2O + [1.88N_2] \tag{23.6}$$

So that

$$\text{scf of flue gas} = \text{scf of } N_2 + \text{scf of } H_2O = 1.88 + 1.0 = 2.88 \text{ scf}$$

23.3 Thermodynamics

Three illustration examples follow, which are related to thermodynamics. One is concerned with partial pressure, one with the gross heating value of a fuel, and the last with a material and energy balance for a process.

23.3.1 Partial Pressure

A storage tank contains a gaseous mixture comprised of 70% H_2, 5% CO_2, 5% H_2O, 10% N_2, and 10% O_2, by volume. What is the partial pressure of each component if the total pressure is 2 atm? What are their pure component volumes if the total volume is 10 ft^3? What are their concentrations in ppm_v (parts per million by volume)?

Dalton's law states that the partial pressure, P_a, of an ideal gas is given by (Theodore, Ricci, and VanVliet 2009):

$$P_a = Y_a P \tag{23.7}$$

where Y_a = mole fraction of component a; and P = total pressure. Thus,

$$P_{H_2} = 0.70(2\,\text{atm}) = 1.40 \text{ atm}$$

$$P_{CO_2} = 0.05(2\,\text{atm}) = 0.10 \text{ atm}$$

$$P_{H_2O} = 0.05(2\,\text{atm}) = 0.10 \text{ atm}$$

$$P_{N_2} = 0.10(2\,\text{atm}) = 0.20 \text{ atm}$$

$$P_{O_2} = 0.10(2\,\text{atm}) = 0.20 \text{ atm}$$

Avogadro's law states that the pure component volume, V_a, of an ideal gas, is given by:

$$V_a = Y_a V \tag{23.8}$$

where V is the total volume.

Thus,

$$V_{H_2} = 0.70\left(10\ \text{ft}^3\right) = 7.0\ \text{ft}^3$$

$$V_{CO_2} = 0.05\left(10\ \text{ft}^3\right) = 0.50\ \text{ft}^3$$

$$V_{H_2O} = 0.05\left(10\ \text{ft}^3\right) = 0.50\ \text{ft}^3$$

$$V_{N_2} = 0.10\left(10\ \text{ft}^3\right) = 1.00\ \text{ft}^3$$

$$V_{O_2} = 0.10\left(10\text{ft}^3\right) = 1.00\text{ft}^3$$

By definition, the concentration in parts per million by volume is given by:

$$\text{ppm}_v = Y_a 10^6 \tag{23.9}$$

Thus,

$$\text{ppm}_v\ H_2 = 0.70\left(10^6\right) = 7.0\times10^5\ \text{ppm}_v$$

$$\text{ppm}_v\ CO_2 = 0.05\left(10^6\right) = 5.0\times10^4\ \text{ppm}_v$$

$$\text{ppm}_v\ H_2O = 0.05\left(10^6\right) = 5.0\times10^4\ \text{ppm}_v$$

$$\text{ppm}_v\ N_2 = 0.10\left(10^6\right) = 1.0\times10^5\ \text{ppm}_v$$

$$\text{ppm}_v\ O_2 = 0.10\left(10^6\right) = 1.0\times10^5\ \text{ppm}_v$$

Unless otherwise stated, ppm for gases is by mole or volume and is usually designated as ppm_v. When applying the term to liquids or solids, the basis is almost always by mass; the notation may appear as ppm, mg/L, ppm_w, or ppm_m.

23.3.2 Gross Heating Value of a Fuel

Given the following composition and combustion properties of a gas mixture at a hydrogen plant in Table 23.1, determine the gross heating value, HHV, of the mixture in units of Btu/scf.

Solution. The gross heating value (HHV) represents the enthalpy change or heat released when a gas is stoichiometrically combusted at 60°F, with the final (flue) products at 60°F and any water present in the liquid state. Stoichiometric combustion requires that no oxygen be present in the flue gas following the combustion process. The gross heating value of the gas mixture, $HHV_{mixture}$, may now be calculated using the equation:

$$HHV_{mixture} = \sum_{i=1}^{n} X_i\ HHV_i \tag{23.10}$$

Table 23.1 Values of R in various units.

Species	Mole Fraction	HHV, Btu/scf
N_2	0.0515	0
CH_4	0.8111	1,013
C_2H_6	0.0967	1,792
C_3H_8	0.0351	2,590
C_4H_{10}	0.0056	3,370

where X_i is the mole fraction of ith component, and HHV_i is the gross heating value of the ith component, Btu/scf. Thus,

$$HHV_{mixture} = (0.0515)\,(0\text{ Btu/scf}) + (0.8111)\,(1{,}013\text{ Btu/scf}) + (0.0967)\,(1{,}792\text{ Btu/scf}) + (0.0351)\,(2{,}590\text{ Btu/scf}) + (0.0056)\,(3{,}370\text{ Btu/scf}) = 1{,}105\text{ Btu/scf}$$

The net heating value (LHV) is similar to HHV except the water is in the vapor state. The net heating value is also known as the lower heating value, and the gross heating value is also known as the higher heating value.

23.3.3 Material and Energy Balance Calculations

Draw a flow diagram representing the process as shown in Figure 23.1. Express the volumetric flow lost by evaporation from the process in terms of that entering the process as:

$$q_{lost} = 0.5\, q_{in} \tag{23.11}$$

Express the process outlet temperature and the maximum river temperature in terms of the upstream temperature as:

$$T_{out} = T_{up} + 60°\text{F} \tag{23.12}$$

$$T_{mix} = T_{up} + 10°\text{F} \tag{23.13}$$

Using the conservation law for mass, express the process outlet flow in terms of the process inlet flow. Also expresses the flow bypassing the process in terms of the upstream and process inlet flows.

$$q_{out} = 0.5\, q_{in} \tag{23.14}$$

$$q_{byp} = q_{up} - q_{in} \tag{23.15}$$

$$q_{mix} = q_{up} - 0.5\, q_{in} \tag{23.16}$$

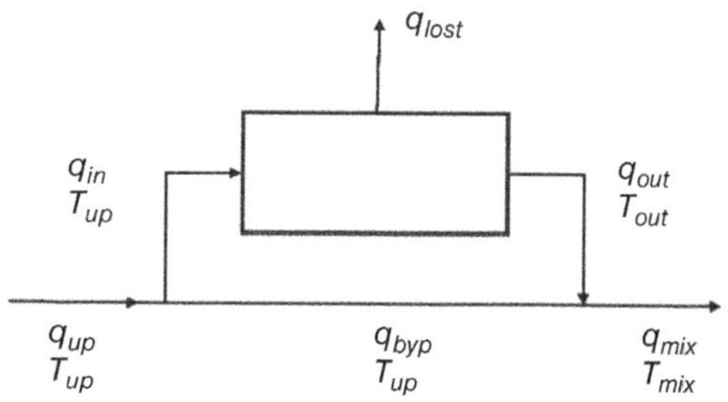

Figure 23.1 Flow diagram for Illustration Example 23.3.3.

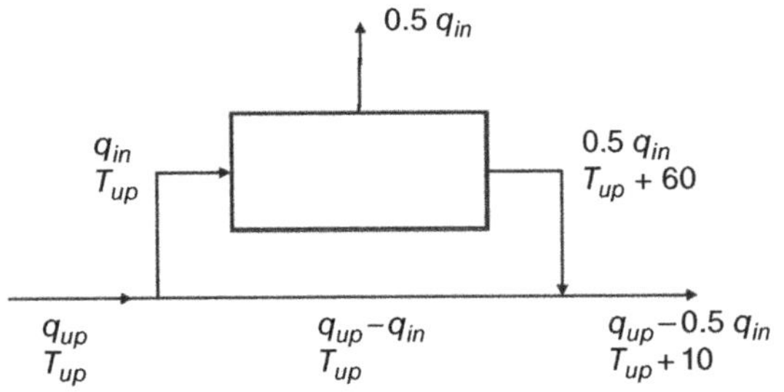

Figure 23.2 Flow diagram after applying mass balances.

Solution. The flow diagram with the expressions developed above is shown in Figure 23.2.

Noting that the enthalpy of any stream can be represented by $qC_p\rho(T - T_{ref})$, an energy balance around the downstream mixing point leads to:

$$\begin{aligned}&(q_{up} - q_{in})C_p\rho(T_{up} - T_{ref}) + 0.5q_{in}C_p\rho(T_{up} + 60 - T_{ref})\\&= (q_{up} - 0.5q_{in})C_p\rho(T_{up} + 10 - T_{ref})\end{aligned} \tag{23.17}$$

Note that T_{ref} is arbitrary and indirectly defines a basis for the enthalpy calculations. Setting $T_{ref} = 0$ and assuming that density and heat capacity are constant yields:

$$(q_{up} - q_{in})(T_{up}) + 0.5q_{in}(T_{up} + 60) = (q_{up} - 0.5q_{in})(T_{up} + 10) \tag{23.18}$$

The equation may now be solved for the inlet volumetric flow to the process in terms of the upstream flow by carrying out the multiplication and canceling terms to yield a final result as follows:

$$q_{up}T_{up} - q_{in}T_{up} + 0.5q_{in}T_{up} + 30q_{in} = q_{up}T_{up} + 10q_{up} - 0.5q_{in}T_{up} - 5q_{in}$$

$$35q_{in} = 10q_{up}$$

$$q_{in} = 10/35q_{up} = 0.286q_{up}$$

Therefore, 28.6% of the original flow is available for cooling. Note that the problem can also be solved by setting $T_{ref} = T_{up}$. Since for this condition, $T_{up} - T_{ref} = 0$, the solution to the problem is greatly simplified.

23.4 Energy Systems

Three illustrative examples concerned with energy systems are included in this section. One is concerned with energy efficiency calculations, one with the energy–mass relationships, and the last with energy storage.

23.4.1 Energy Conversion Efficiency

In 1900, it took about 20,000 Btu fuel input to produce 1 kWh of electricity. Estimate this efficiency of conversion and compare it with a typical value for today's power industry.

Solution. The solution to this problem is based, in part, on unit conversions. From standard conversion tables, 1 kWh is equivalent to 3,412 Btu. Since only 1 kWh was being produced in 1900 from 20,000 Btu input, the energy efficiency was:

$$\text{Energy efficiency} = (3{,}412\ \text{Btu/kWh})/(20{,}000\ \text{Btu/kWh}) = 0.1706 = 17.1\%$$

Today's energy conversion efficiency has improved over this value from a century ago (thankfully!), but not as much as many would like. Typical values are in the 35% efficiency range, or approximately 100% better than before. However, as one can see, there is significant room for improvement.

23.4.2 Energy–Mass Relationships

One pound of U^{235} releases 3.6×10^{10} Btu of energy when it undergoes nuclear fission. One type of coal gives off 12,000 Btu/lb on burning. Calculate the mass-to-energy conversion rate in each case.

Solution. Einstein's equation relates mass and energy and can be used to solve this problem. With consistent units, one also needs to know that 1 Btu = 1.055×10^{10} erg = 1.055×10^{10} g-cm^2/s^2. With the given input values, the following results can be determined:

$$\Delta E = \Delta m c^2 \tag{23.19}$$

Rearranging yields:

$$\Delta m = \Delta E / c^2 \tag{23.20}$$

Substituting values for U^{235} yields:

$$\Delta m = (3.6 \times 10^{10}\ \text{Btu/454 g})(1.055 \times 10^{10}\ \text{g-cm}^2/\text{s}^2/\text{Btu})/(3 \times 10^{10}\ \text{cm/s})^2 = 0.00093 = 0.09\%$$

Therefore, the loss in mass is less than 0.1%, even with this tremendous release of energy.

Substituting values for coal yields:

$$\Delta m = (12{,}000\ \text{Btu/454 g})(1.055 \times 10^{10}\ \text{g-cm}^2/\text{s}^2/\text{Btu})/(3 \times 10^{10}\ \text{cm/s})^2 = 3.1 \times 10^{-10}$$

This extremely small loss of mass is essentially undetectable.

23.4.3 Energy Storage

A lake is located at the top of a mountain. The potential energy of the water traveling downhill can be used to spin turbines and generate electricity. This is the operating mode in the daytime during peak electrical demand. At night, when demand is reduced, the water is pumped back up the mountain to provide storage for the next day's power production. Determine how much power (watts) could be generated by water from the lake located at an elevation of 3,000 ft above the power plant. The flow rate of water through the conduit feeding the power plant is up to 50,000 gpm. The turbine efficiency is 30%. Neglect friction losses.

Solution. First, determine the mass flow rate of water that could be provided to the turbines:

$$(50{,}000\ \text{gal/min})\ (8.34\ \text{lb/gal}) = 417{,}000\ \text{lb/min}$$

The potential energy associated with this mass flow coming from 3,000 ft above the power plant is:

$$\left(417{,}000\ \text{lb/min}\right)\left(3{,}000\ \text{ft}\right) = 1{,}251 \times 10^{6}\ \text{ft-lb/min} = 1{,}607{,}620\ \text{Btu/min} = 28.3\ \text{MW}$$

Assuming that the potential energy is entirely converted to energy input to the turbine, accounting for energy conversion efficiency, the actual power output could be:

$$P = \left(0.30\right)\left(28.2\right) = 8.47\,\text{MW}$$

This is enough power for a small town. If green energy, wind or hydrogen, is used to pump at night, no pollution is generated because no fossil fuel is required. The initial construction expense for storage purposes would be quite high, but the long-term cost of producing electricity would be very low.

23.5 Environmental Issues

Three illustrative examples are included in this section concerned with environmental issues. One is concerned with hydrogen catalyst recovery, one addresses explosion overpressure, and the third demonstrates Weibull distribution calculations.

23.5.1 Catalyst Recovery

A process emits 50,000 acfm of gas-containing dust from a hydrogen plant (it may be considered catalyst particles) at a loading of 2.0 gr/ft^3. A particulate control device is employed for particle capture, and catalyst captured from the unit is worth \$0.03/lb. Experimental data have shown that the collection efficiency, E, is related to the system pressure drop, ΔP, by the relationship shown in Equation 23.21.

$$E = \frac{\Delta P}{\Delta P + 15.0} \tag{23.21}$$

where ΔP = pressure drop across the collector, lb/ft^2. If the overall energy use efficiency of the particle control device is 55% and electric power costs \$0.18/kWh, at what collection efficiency is the cost of power equal to the value of the recovered catalyst? What is the pressure drop in inches of water at this collection efficiency?

Solution. The total value of the recovered catalyst, RC, \$/min; may be expressed in terms of the collection efficiency E, unitless; the volumetric flow rate q, ft^3/min; the inlet dust loading ω, gr/ft^3; and the unit value of the catalyst, CV, \$/lb; using the following expression:

$$RC = q\ \omega\ CV\ E \tag{23.22}$$

Substituting the given values results in the following:

$$RC = (50{,}000\ \text{ft}^3/\text{min})\ (2.0\ \text{gr/ft}^3)\ (1\ \text{lb}/7{,}000\ \text{gr})\ (\$0.03/\text{lb})\ E = (\$0.43/\text{min})\ E$$

The recovered value can be expressed in terms of pressure drop by substituting E in Equation 23.22 with Equation 23.21. This results in the following:

$$RC = \frac{\left(\$0.43/\text{min}\right)\Delta P}{\Delta P + 15.0} \tag{23.23}$$

The total cost of power, CP, can be calculated based on the volumetric flow rate q, ft^3/min; the pressure drop across the control device, ΔP; the unit cost of electricity, CE, \$/kWh; and the energy conversion efficiency of the control device, E_f, unitless. This results in the following equation:

$$CP = q\Delta P\,CE/E_f \tag{23.24}$$

Substitution with given values yields the following:

$$\begin{aligned}CP &= \left(50{,}000\ \text{ft}^3/\text{min}\right)\left(\Delta P,\ \text{lb/ft}^2\right)\left(\$0.18/\text{kWh}\right)/\left(0.55\right)\\ &= \left(\$16{,}364\ \text{ft-lb/min}\right)/\left(\text{kWh}\right)\left(\Delta P\right) = \left(\$272.73\ \text{ft-lb/min}\right)/\left(\text{kWh}\right)\left(\Delta P\right)\\ &= \left(\$0.35\ \text{Btu/min}\right)/\left(\text{kWh}\right)\left(\Delta P\right) = \left(\$0.006\ \text{kW/kWh}\right)\left(\Delta P\right)\\ &= \left(\$0.006/\text{h}\right)\left(\Delta P\right) = \left(\$0.0001/\text{min}\right)\left(\Delta P\right)\end{aligned}$$

The pressure drop at which the total cost of power is equal to the total value of the recovered catalyst is found by equating RC to CP. This results in the following:

$$\frac{\left(\$0.43\Big/\text{min}\right)\Delta P}{\Delta P + 15.0} = \left(\frac{\$0.0001}{\text{min}}\right)\Delta P$$

Solving this equation for ΔP results in the following:

$$\left(\$0.43/\text{min}\right)/\left(\$0.0001/\text{min}\right) = \Delta P + 15.0$$

$$\Delta P = 430 - 15 = 415\ \text{lb/ft}^2 = 2.88\ \text{psi} = 79.8\ \text{in}\ H_2O$$

The collection efficiency corresponding to this ΔP is determined from Equation 23.21 as:

$$E = \left(415\right)/\left(415+15\right) = 0.965 = 96.5\%$$

The reader should note that operating below this efficiency (or the corresponding pressure drop) will produce a profit; operating above this value leads to a loss. For example, at a $\Delta P = 200\ \text{lb/ft}^2$, the total value of recovered catalyst, RC, is \$0.40/min, while the total cost of power, CP, is only \$0.20/min. At this operating pressure drop, the efficiency of particle capture is 93%, not as high as the breakeven point, but is still profitable.

23.5.2 Explosion Overpressure

Explosions are frequently calibrated in terms of the equivalent mass of TNT. One lb of TNT releases 1,983 Btu (500 Kcal) on detonation; therefore, any explosion that releases 1,983 Btu is said to be an "1 lb of TNT-equivalent" explosion.

The overpressure at any distance from an explosion can be estimated using the following method. The overpressure from 1 kg of TNT, as a function of distance, is given by (Theodore and Dupont 2012):

$$P = \frac{1{,}737}{D^2} + \frac{1.875}{D} - 0.01566 \tag{23.25}$$

where P is the total overpressure, psi; and D is the distance from the point of explosion, ft, for a range of D from 3 to 400 ft. The distance for equal overpressures scales as the 1/3 power of the energy of the explosion, as indicated in Equation 23.26 (Theodore and Dupont 2012):

$$\left(\frac{D_1}{D_2}\right) = \left(\frac{E_1}{E_2}\right)^{\frac{1}{3}} \tag{23.26}$$

Consider the following hypothesis. The hydrogen vapor from a plant at Hydrogen Unlimited is suddenly discharged, mixes with air, and explodes.

1) Estimate the maximum possible energy release (E_{max}) and the TNT equivalency.
2) Estimate the radius from the center of the explosion at which overpressures of 1 psi (sufficient to injure many people and cause damage to buildings), and 5 psi (sufficient to break wooden telephone poles) would occur.
3) Based on these results, discuss how the hazard risks from the discharge can be minimized.

The reactor has a volume of 800 ft^3. Assume the gas at the time of release consists of hydrogen at 25 psi and 350°F. The average heat of combustion in air of the hydrogen in the tank is 50,000 Btu/lb.

Solution. At the given input conditions, the mass of hydrogen involved in the explosion can be calculated using the ideal gas law, assuming the entire content of the tank explodes. The number of moles of hydrogen is calculated as follows:

$$\begin{aligned}\text{lbmol H}_2 &= n = PV/RT = (25\text{ psi})(800\text{ ft}^3)/[(10.73\text{ psi-ft}^3/\text{lbmol-°R})(809.7\text{ °R})]\\ &= 2.30\text{ lbmol; or } 4.6\text{ lb H}_2\end{aligned}$$

The maximum possible energy release (E_{max}) is then calculated as:

$$E_{max} = (4.6\text{ lb H}_2)(50{,}000\text{ Btu/lb}) = 230{,}000\text{ Btu}$$

The corresponding TNT-equivalent is calculated as follows:

$$\text{TNT-equivalent} = (230{,}000\text{ Btu})/(1{,}983\text{ Btu/lb TNT}) = 116\text{ lb} = 52.62\text{ kg TNT-equivalent}$$

The radius from the center of the explosion at which overpressures of 1 psi occur for a 1 kg TNT explosion can be determined by solving Equation 23.25 in the form shown as Equation 23.26:

$$1 = \frac{1{,}737}{D^2} + \frac{1.875}{D} - 0.01566 \tag{23.27}$$

Substitution and iteration into Equation 23.27 yield an overpressure distance of 42.29 ft for a 1 kg TNT explosion. Based on the TNT-equivalent of the hydrogen explosion, Equation 23.26 can be used to determine its 1 psi overpressure distance. Substituting appropriate values into Equation 23.26 leads to the following:

$$\left(\frac{42.29\text{ ft}}{D_2}\right) = \left(\frac{1\text{ kg}}{52.62\text{ kg}}\right)^{\frac{1}{3}}$$

$$D_2 = 42.29/(0.267) = 158.5\text{ ft}$$

For the 5 psi overpressure distance, these calculations can be repeated as follows:

$$5 = \frac{1{,}737}{D^2} + \frac{1.875}{D} - 0.01566 \tag{23.28}$$

Substitution and iteration into Equation 23.28 yield an overpressure distance of 18.10 ft for a 1 kg TNT explosion. Based on the TNT-equivalent of the hydrogen explosion, Equation 23.26 can be used to determine its 5 psi overpressure distance. Substituting appropriate values into Equation 23.26 leads to the following:

$$\left(\frac{18.8\text{ ft}}{D_2}\right)=\left(\frac{1\text{ kg}}{52.62\text{ kg}}\right)^{\frac{1}{3}}$$

$$D_2 = 18.8/(0.267) = 70.4\text{ ft}$$

It has been suggested that in most vapor explosions, only 10–25% of the maximum possible combustion energy is released. This results from the vapor/air mixture being ignited when the mixture is still "rich," that is, before the amount of air for complete combustion has had a chance to mix with the explosive gas. For hydrogen, the range of explosive limits is so large that this would have little impact on the outcome of the event. Thus, efforts should be directed to minimizing the chance of a release and/or its subsequent contact with an ignition source, and/or surrounding the tank with barriers to minimize the effects of an explosion if one does occur.

23.5.3 Weibull Distribution Calculation

The life in hours of a valve in a hydrogen plant is a random variable having a Weibull distribution (Shaefer and Theodore 2007) with $\alpha = 0.025$ and $\beta = 0.50$. What is the probability that the valve will last more than 4,000 h?

Solution. Frequently, the failure rate of equipment exhibits three stages: a break-in stage with a declining failure rate, a useful life stage characterized by a fairly constant failure rate, and a wear-out period characterized by an increasing failure rate. An example of this type of failure rate is the death rate of humans. A failure rate curve exhibiting these three phases is called a "bathtub curve." The Weibull distribution provides a mathematical model of all three stages of the bathtub curve. The probability distribution function (PDF) for the Weibull distribution is given by:

$$f(t) = \alpha\beta t^{(\beta-1)}\exp\left(-\int_0^t \alpha\beta t^{(\beta-1)}dt\right) = \alpha\beta t^{\beta-1}\exp\left(-\alpha t^{\beta}\right);\ t>0,\ \alpha>0,\ \beta>0 \tag{23.29}$$

where α and β are constants and t = time. Let t denote the life of the valve in hours. The PDF of t is obtained by applying the above formula and substituting in $\alpha = 0.025$ and $\beta = 0.50$ to yield:

$$f(t) = (0.025)(0.50)t^{(0.5-1)}\exp\left(-0.025\,t^{0.50}\right) = 0.0125\,t^{(-0.5)}\exp\left(-0.025\,t^{0.50}\right)$$

The probability that the valve will last more than 4,000 hours is given by:

$$\begin{aligned}P(t>4{,}000) &= \int_{4{,}000}^{\infty} f(t)dt = \int_{4{,}000}^{\infty} 0.0125t^{(-0.5)}\exp\left(-0.025t^{0.50}\right)\\ &= 0.0125\left(4000^{(-0.5)}\right)\exp\left(-0.025\left(4{,}000^{0.50}\right)\right)\\ &= 0.0125\,(0.0158)\exp\left(\left(-0.025(63.25)\right)\right)\\ &= 0.0002\exp(-1.58) = 0.0002(0.206) = 0.0000412; = 0.004\%\end{aligned}$$

As indicated by this result, the probability of the valve lasting more than 4,000 hours is extremely low, and inspection and maintenance of this valve should be a safety priority.

23.6 Ethics

Three illustrative examples are included in this section on ethics principles. The first relates to common, domestic ethics issues, while the last two are case studies that have been adapted from Wilcox and Theodore (1998).

23.6.1 Domestic Ethical Issues

Provide some normal, everyday, domestic applications involving the general topic of ethics.

Solution. Ethical decisions crop up in daily lives frequently and take the form of deciding between lying and telling the truth, following the rules or cheating, being honest and fair to others, or manipulating them for your benefit, etc. Some issues are clear, i.e., it is not ethical to cheat on an exam, plagiarize, or take credit for another's work. Other decisions are a bit less clear, i.e., telling a friend less than the truth about how his haircut looks so his feelings are not hurt. Perhaps the Hippocratic oath is relevant when considering ethical behavior in normal life, that is, "*primum non nocere*" or "first, do no harm."

23.6.2 Production Increase Demands

Rufus works for a utility and is responsible for managing the production of hydrogen. His boss, Vito, informs him of a need to increase the output of the hydrogen because orders have increased. Rufus continues to socialize with his coworkers as usual. He thinks that speeding production is just a matter of cutting some procedures short and decides to eliminate an extra step in the process to increase hydrogen production. The company later received customer complaints regarding the quality of the hydrogen; it was contaminated with catalyst particles that were supposed to be removed prior to shipment. Vito meets with Rufus to discuss the issue.

"Rufus, we've been getting customer complaints regarding the quality of the hydrogen," says Vito. "They have found the hydrogen contaminated with fine particles. What's going on?"

Rufus is left confused about how to get out of this predicament. What can he say to Vito without risking his job and losing his production team? Should he blame his team? Or should he tell Vito the truth?

Questions for Discussion. The following questions should be used to discuss the details of the case and to determine an ethical remedy for Rufus' problem:

1) What are the facts in this case?
2) How should Rufus respond to Vito's question?
3) How could Rufus have avoided the problem?

23.6.3 ISO 14000 Consulting Dilemma

Julio has a PhD in Mechanical Engineering, a Professional Engineer (P.E.) license, and a job as an Associate Professor at a local university. Julio recently became a certified ISO 14000 auditor. Serving as a consultant for a hydrogen plant, during an audit Julio observed a significant violation of regulation related to hydrogen safety at the plant. (Note that ISO 14000 is a voluntary standard related to environmental management systems, i.e., it is not a regulation and ISO 14000 audits are confidential.)

What does Julio do? Does he forget about the significant violation since he is not an inspector, and it does not appear to violate the ISO 14000 standard he is there to audit? Does he insist on his observation of this violation being part of the audit report? Does he call it to the attention of the plant management? Does he report it to a regulating body?

Solution. The question of professional ethics presented here can be generalized as follows: does a professional have an ethical duty to report any violation of which he/she becomes aware? Most licensing organizations have codes of ethics to which their professionals must subscribe. These codes generally call upon the professional to serve their client/employer with loyalty and to protect the welfare of the public. The codes, however, are sufficiently vague as to not provide a clear-cut answer to Julio's situation. In the absence of such a clear-cut code, there is no one right answer and one's answer may vary from the one presented here.

Julio must first investigate further to make certain of his facts. If the violation is confirmed and has never been reported, he must insist that it be part of the findings presented to management at the exit meeting and in the report to the ISO Registrar, who makes the final decisions on awarding certification. Although ISO 14000 does not require compliance, knowingly failing to comply does affect the evaluation of the facility's commitment to good environmental management system practices and continuous improvement that the standard requires.

Having done all that, does Julio go the last step and report it elsewhere? One aspect of the problem that Julio must evaluate is whether the violation poses an imminent threat to public health or the environment. Such a report would violate the confidentiality of the audit and might even precipitate a lawsuit. Julio is also concerned that he may never be hired again as an auditor. He senses a conflict between two sets of ethics: the ethics of a professional bound to confidentiality. Keeping silent would be easier, but ethics is not easy. It would be easy to answer, "Report it!" but the real question is what would YOU (the reader) do?

23.7 Economics

Three illustrative examples are included in this section concerned with economics. One is concerned with optimum pipe diameter calculations, one with optimum profit, and the third is associated with economic decision-making based on tax credit availability.

23.7.1 Optimum Pipe Diameter Considerations

When plants are designed and developed, designers try to minimize the cost of operations and maximize efficiency. One way this is commonly done in plants is by strategically selecting the piping and pipe fittings to maximize flow and minimize cost. For any given set of flow conditions, as pipe diameter increases, capital cost correspondingly increases. However, with increasing pipe diameter, comes decreasing pumping or fan charges due to lower frictional resistance. Conversely, an increase in pressure drop will generally result from a decreased pipe size. Therefore, an optimum economic pipe diameter must exist. This optimum pipe diameter is found at the point when the sum of pumping or fan costs plus fixed costs is at a minimum. Outline how to determine the optimum economic pipe diameter for a given flow system.

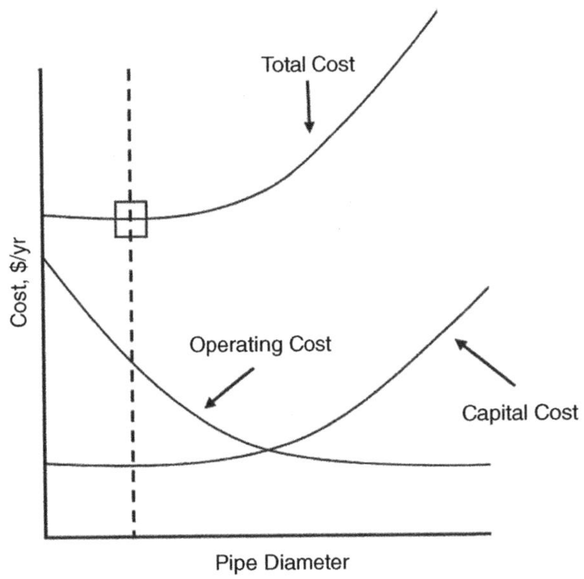

Figure 23.3 Optimal pipe diameter selection for Illustrative Example 23.7.1.

Solution. The usual calculation procedure is as follows:

1) Select a pipe diameter.
2) Obtain the annual operating cost.
3) Obtain the capital equipment cost.
4) Convert the capital cost to an annual basis.
5) Sum the two annual costs in Steps 2 and 4.
6) Return to Step 1, selecting a different pipe diameter.

The only variable that will appear in the resulting total cost expressions is the pipe diameter. The economically optimum pipe diameter can be generated analytically by taking the derivative of the total annual cost expressed in terms of pipe diameter, setting the result equal to zero, and solving for the diameter. The derivative operation can be replaced by a trial-and-error procedure that involves calculating the total cost for various diameters and simply selecting the minimum. These calculations can be used to generate a graph of the form provided below in Figure 23.3.

23.7.2 Optimum Hydrogen Plant Profit

Plans are underway to construct and operate a small plant to generate hydrogen. The company is still undecided as to whether to employ Processes A, B, or C. The A process plant is less expensive to purchase and operate than either a comparable B or C system. However, the projected income from the B and C processes is higher.

Based on the economic and financial data provided in Table 23.2, select the process that will yield the highest annual profit. Calculations should be based on an interest rate of 5% and a process lifetime of 20 years for the three hydrogen production systems.

Table 23.2 Costs associated with hydrogen production Processes A, B, and C for Illustrative Example 23.7.2.

Costs/Credits	Process A	Process B	Process C
Capital, $	26,250	29,750	34,350
Installation, $	15,750	17,000	17,550
Operation, $/yr	4,000	5,500	5,900
Maintenance, $/yr	6,500	7,750	8,450
Income, $/yr	20,000	25,000	25,100

Table 23.3 Annual and total costs associated with hydrogen production Processes A, B, and C for Illustrative Example 23.7.2.

Costs/Credits	Process A	Process B	Process C
Capital, $/yr	2,100	2,382	2,748
Installation, $	1,260	1,360	1,404
Operation, $/yr	4,000	5,500	5,900
Maintenance, $/yr	6,500	7,750	8,450
Income, $/yr	20,000	25,000	25,100
Total Profit, $/yr	6,140	8,008	6,598

Solution. To determine an annual cost for the fixed costs in this problem, the capital recovery factor (CRF) needs to be calculated. The *CRF* is given as:

$$CRF = \frac{i}{\left(1-(1+i)^{-n}\right)} \tag{23.30}$$

where i = interest rate, decimal; and n = number of payment periods of a loan. Substituting for the given input values yields the following:

$$CRF = \frac{0.05}{\left(1-(1+0.05)^{-20}\right)} = \frac{0.05}{(1-0.377)} = \frac{0.05}{0.623} = 0.080$$

Using this *CRF*, the annualized costs for each process can be determined for their capital and installation costs. These values along with a total annual cost for each process are summarized in Table 23.3. As indicated there, Process B is the option that should be selected for this plant because it provides the best profit, \$8,008/yr based on all fixed and operating costs plus expected income.

23.7.3 Plant Selection Based on Tax Credit Availability

An engineer has compiled the data for two different project options (Table 23.4) – one that includes a new hydrogen production technology for which there are government tax credits provided and one that does not. From an economic point of view, which project should the engineer select? The lifetime of the equipment is 10 years, and the interest rate is 10%.

Solution. For the project with the new hydrogen production process, calculate the total capital cost, TCC_A, as follows:

$$TCC_A = \$1,294,000 + \$786,000 = \$2,080,000$$

In a similar manner, the total capital cost for the conventional production process, TCC_B is calculated as follows:

$$TCC_B = \$1,081,000 + \$659,000 = \$1,740,000$$

To determine an annual cost for the fixed costs in this problem, the CRF needs to be calculated using Equation 23.30 as follows:

$$CRF = \frac{0.10}{\left(1-(1+0.10)^{-10}\right)} = \frac{0.10}{(1-0.386)} = \frac{0.10}{0.614} = 0.163$$

Calculate the annualized capital cost, ACC_A, of the new hydrogen generation process as:

$$ACC_A = (TCC_A)(CRF) = (\$2,080,000)\,(0.163) = \$339,040/\text{yr}$$

Calculate the annualized capital cost, ACC_B, for the conventional hydrogen production process as follows, noting that the *CRF* for this option is identical to that calculated for the option with the new hydrogen production process:

$$ACC_B = (\$1,740,000)\,(0.163) = \$283,620/\text{yr}$$

Table 23.4 Costs associated with hydrogen production Processes A and B for Illustrative Example 23.7.3.

Costs/Credits	Process A with Tax Credits	Process B without Tax Credits
Equipment, $	1,294,000	1,081,000
Installation, $	786,000	659,000
Operation Labor, $/yr	39,900	8,500
Maintenance, $/yr	43,000	17,000
Utilities, $/yr	958,000	821,000
Overhead, $/yr	51,300	13,900
Taxes, Insurance, and Administration, $/yr	86,200	72,600
Tax Credits, $/yr	380,000	0

Table 23.5 Annual and total costs associated with hydrogen production Processes A and B for Illustrative Example 23.7.3.

Costs/Credits	Process A with Tax Credits	Process B without Tax Credits
Annual Capital Costs, \$/yr	339,040	283,620
Operation Labor, \$/yr	39,900	8,500
Maintenance, \$/yr	43,000	17,000
Utilities, \$/yr	958,000	821,000
Overhead, \$/yr	51,300	13,900
Taxes, Insurance, and Administration, \$/yr	86,200	72,600
Tax Credits, \$/yr	−380,000	0
Total Annual Costs, \$/yr	1,137,440	1,216,620

The better choice for the project can then be chosen based on the summary of costs presented in Table 23.5. Based on these numbers, the project using the new hydrogen production process is the better choice on a total annual cost basis by \$79,180/yr, saving approximately 6.5% a year over the total annual cost of Process B.

23.8 SDS Information

Three illustrative examples are included in this section concerned with safety data sheets. One provides a layman's definition of an SDS, one explores the limitations of an SDS, and the last evaluates physical and chemical characteristics included in SDS forms. Reviewing this material, the reader should understand the importance of each subsection in helping to identify a chemical hazard.

23.8.1 Layman's Definition of an SDS

Qualitatively describe in "layman's language" the health and safety information that is provided on an SDS.

Solution. An SDS serves as a reference source for information on hazardous substances. The SDS identifies the substance, the producer or seller of the substance, and the location of the producer or seller. The SDS explains why the substance is hazardous and how a person can be exposed to the substance; identifies conditions that increase the hazard; explains safe handling procedures; identifies proper protective clothing or devices to be used when working with the substance; explains the steps that should be taken if a person is exposed to the substance; and explains the steps that should be taken if there is a spill or emergency.

23.8.2 Limitation of SDSs

Discuss some of the limitations of SDSs and their use.

Solution. The information contained in an SDS is highly variable and is dependent on the supplier's knowledge and expertise. Work is in progress to improve the quality of information in

SDSs and (as noted earlier) to standardize the SDS format. The American National Standards Institute (ANSI) has developed a recommended SDS format. Understanding the terminology used in SDSs can be a problem for some employees. Additional training and the use of supplemental information can help employers and employees in dealing with this problem.

23.8.3 Physical and Chemical Characteristics Contained in SDSs

Describe the physical and chemical characteristics included in an SDS. Explain why this information should be included in an SDS.

Solution. The physical and chemical characteristics of a chemical that are included in an SDS are the boiling point, melting point, vapor pressure, vapor density, evaporation rate, solubility in water, specific gravity, and a description of the normal appearance and odor of a chemical. The boiling and melting points are included in an SDS to make the user aware that protective measures may need to be taken to prevent a potentially dangerous change in state. The vapor pressure, vapor density, and evaporation rate are included to make the user aware that protective measures may need to be taken against flammable or toxic gases and vapors. The solubility in water and specific gravity inform the user of the state of the chemical when added to water, i.e., whether the chemical may dissolve in water, sink, or float. A description of the normal appearance and odor of a chemical in an SDS helps the user recognize a potentially dangerous situation.

23.9 Optimization

Three illustrative examples are included in this section concerned with optimization. One is concerned with a profit model, one with hydrogen plant operations, and the third with the optimization of utility conversion to hydrogen to meet air pollution goals at the least cost.

23.9.1 Profit Model Optimization

Consider the following application. Farber consultants have provided Hydrogen and Theodore (HAT) Associates with the following model to determine a company's profits (P) gained from the installation of two different flow control devices. Assume the number of valves installed are x_1 and x_2. The model is given by:

$$P = 2{,}000x_1 + 2{,}750x_2 \tag{23.31}$$

$$2x_1 + 3x_2 \leq 30 \tag{23.32}$$

$$x_1 + 10x_2 \leq 25 \tag{23.33}$$

with $x_1 \geq 0$ and $x_2 \geq 0$. Qualitatively explain the above profit model, and how to determine the number of the two devices that will maximize profits.

Solution. Equation 23.31 is an objective function and describes the profit that will be produced from the installation of the control valves based on the profit generated from each. Equations 23.32 and 23.33 are part of the constraints on the limitation of the number that can be installed because of space and other issues, such as supply chain issues. The final two constraints require that the number of each valve be zero or a positive number.

The calculations are carried out to maximize the value of the objective function within the constraints established for the problem. These calculations are left as an exercise for the reader. (Hint x_1 =15 valves $x_2 = 0$ valves to maximize the profit from their installation of $30,000).

23.9.2 Hydrogen Plant Operation

The cost associated with the operation of a hydrogen plant, *CO*, is linearly related to both the operating temperature *T* and operating pressure *P*. That relationship (in consistent absolute units) is given by:

$$CO = 0.18\,T + 0.05\,P;\ \text{consistent units, \$/hr} \tag{23.34}$$

However, for ignition to occur, the following constraints must be satisfied:

$$a_1 T + b_1 P \geq I \tag{23.35}$$

along with the energy component of the feed (gaseous fuel) constrained by:

$$a_2 T + b_2 P \geq E \tag{23.36}$$

If the coefficients for these constraint equations are: $a_1 = 0.2$, $b_1 = 0.05$, $a_2 = 0.4$, $b_2 = 0.02$, and $I = 600$, $E = 450$, $T \geq 0$, and $P \geq 0$, calculate the operating temperature and pressure that minimizes the cost of operation of the hydrogen plant and satisfies the constraints associated with the ignition and energy constraints.

Solution. This is a typical linear programming problem. For this example, the objective function is Equation 23.34, subject to the constraints as follows:

$$0.2\,T + 0.05\,P \geq 600 \tag{23.37}$$

$$0.4\,T + 0.02\,P \geq 450 \tag{23.38}$$

Microsoft Excel can be used to solve this problem with the inputs shown in Figure 23.4, along with the optimal solution that minimizes the objective function subject to the stated constraints.

	A	B	C	D	E	F	G	H	I
1									
2	**Illustrative Example 23.9.2**								
3									
4		Cost							
5	*CO* =	$ 540	/hr						
6									
7	Variables	*T*	*P*						
8	Objective Coefficients	0.18	0.05						
9	Values	3,000	0		<--Optimal values appear here after solving				
10									
11	Coefficients	*T*	*P*		Total		Limit		
12	Constraint Equation 1	0.2	0.05		600	=>	600		
13	Constraint Equation 2	0.4	0.02		1200	=>	450		
14	Constraint Equation 3	1			3000	=>	0		
15	Constraint Equation 4		1		0	=>	0		
16									

Figure 23.4 Excel solution for optimization problem Illustrative Example 23.9.2.

As indicated in Figure 23.4, the optimal solution indicates an operating temperature, $T = 3{,}000$, and an operating pressure, $P = 0$, to yield a minimum cost of operation of \$540/h.

23.9.3 Optimization of Utility Conversion to Hydrogen

A statewide utility has proposed to convert some of its 50 coal-fired boilers to oil-fired to meet new environmental regulations. DuPont Associates was hired to generate a model (objective function) that would minimize the cost of the conversion to oil-fired or hydrogen-fired boilers. The project also requested that an attempt be made to salvage as many of the old boilers already firing coal as possible while still meeting new environmental regulations. The cost associated with each fuel varies with its composition, heating value, required boiler modifications, etc. The engineers at DuPont Associates proposed the following total cost, *TC*, model in units of \$10,000 for making the conversion:

$$TC = 3x_1 + 5x_2 - x_3 \tag{23.39}$$

where x_1 = number of boilers burning oil 1, x_2 = number of boilers burning hydrogen, and x_3 = number of remaining coal-fueled boilers. Equation 23.39 indicates that salvaging existing coal-fired boilers results in a \$10,000 credit each.

The constraints due to emission rules for NO_x (nitrous oxide), SO_x (sulfur dioxide), and RO_x (particulates) are:

$$1.1x_1 + 0.5x_2 + 1.5x_3 \leq 100; \text{ due to } NO_x \tag{23.40}$$

$$1.2x_1 + 3x_3 \leq 70; \text{ due to } SO_x \tag{23.41}$$

$$x_1 + 0.1x_2 + x_3^2 \leq 1{,}000; \text{ due to } RO_x \tag{23.42}$$

$$x_1 + x_2 + x_3 = 50 \tag{23.43}$$

Solution. Excel provides the following solution:

$$x_1 = 0;\ x_2 = 27;\ x_3 = 23$$

	A	B	C	D	E	F	G	H	I
1									
2	**Illustrative Example 23.9.3**								
3									
4		Cost							
5	Min *TC* =	$ 110							
6									
7	Variables	x1	x2	x3					
8	Objective Coefficients	3	5	-1					
9	Values	0	27	23	<--Optimal values appear here after solving				
10									
11	Coefficients	x1	x2	x3	Total		Limit		
12	Constraint Equation 1	1.1	0.5	1.5	48	<=	100		
13	Constraint Equation 2	1.2		3	70	<=	70		
14	Constraint Equation 3	1	0.1	1	547	<=	1000		
15	Constraint Equation 4	1	1	1	50	=	50		

Figure 23.5 Excel solution for optimization problem Illustrative Example 23.9.3.

With these values, the objective function is a minimum at:

$$TC = \$1,100,000$$

and the solution requires the purchase of 27 new boilers that will burn hydrogen and the continued use of 23 of the old coal-fueled boilers, all at a cost of $1,100,000 that represents the minimum expenditure meeting all of the upgrade and pollution control constraints (Figure 23.5).

References

Shaefer, S. and Theodore. L. *Probability and Statistics in Environmental Subjects,* CRC Press/ Taylor and Francis Group, Boca Raton, FL, 2007.

Theodore, L. and Dupont, R.R. (2012). *Environmental Health and Hazard Risk Assessment: Principles and Calculations*. Boca Raton, FL: CRC Press/Taylor & Francis Group.

Theodore, L., Ricci, F., and VanVliet, T. (2009). *Thermodynamics for the Practicing Engineer*. Hoboken, NJ: John Wiley & Sons.

Wilcox, J. and Theodore, L. (1998). *Environmental and Engineering Ethics*. Hoboken, NJ: John Wiley & Sons.

Index

Hydrogen Energy: Principles and Applications, First Edition. Vincent J. DelGatto, Louis Theodore, R. Ryan Dupont, and Matthew C. Ogwu.

d

e

f

g

h

i

u

v

w

y

Printed and bound by CPI Group (UK) Ltd, Croydon, CR0 4YY
17/09/2026
14964313-0004